Tribology and Characterization of Surface Coatings

Scrivener Publishing
100 Cummings Center, Suite 541J
Beverly, MA 01915-6106

Publishers at Scrivener
Martin Scrivener (martin@scrivenerpublishing.com)
Phillip Carmical (pcarmical@scrivenerpublishing.com)

Tribology and Characterization of Surface Coatings

Edited by
Sarfraj Ahmed
National Institute of Technology, Raipur, India
and
Vinayak S. Dakre
Department of Mechanical Engineering, G H Raisoni University, Amravati, India

Scrivener Publishing

WILEY

This edition first published 2022 by John Wiley & Sons, Inc., 111 River Street, Hoboken, NJ 07030, USA and Scrivener Publishing LLC, 100 Cummings Center, Suite 541J, Beverly, MA 01915, USA

For more information about Scrivener publications please visit www.scrivenerpublishing.com.

Wiley Global Headquarters
111 River Street, Hoboken, NJ 07030, USA

For details of our global editorial offices, customer services, and more information about Wiley products visit us at www.wiley.com.

Library of Congress Cataloging-in-Publication Data

ISBN 978-1-119-81878-6

Cover image: Pixabay.Com
Cover design by Russell Richardson

Set in size of 11pt and Minion Pro by Manila Typesetting Company, Makati, Philippines

Printed in the USA

Contents

Preface **xiii**

1 Overview of Coating Deposition Techniques **1**
Neeraj Mehta
1.1 Introduction 2
1.2 Thin-Film Characteristics 4
1.3 Techniques Based on Physical Deposition Processes 5
1.3.1 Evaporation 6
1.3.1.1 Mean-Free Path: A Significant Parameter for High-Quality Evaporation 8
1.3.2 Sputtering 9
1.3.2.1 Characteristics of Sputtering 12
1.3.2.2 Types of Magnetron Sputtering Technique 14
1.3.3 Other Techniques Based on Physical Deposition Processes 15
1.4 Techniques Based on Chemical Deposition Processes 18
1.4.1 Chemical Vapor Deposition (CVD) 18
1.4.1.1 Atmospheric Pressure Chemical Vapor Deposition (APCVD) 19
1.4.1.2 Low-Pressure Chemical Vapor Deposition (LPCVD) 20
1.4.1.3 Metal Organic Chemical Vapor Deposition (MOCVD) 21
1.4.1.4 Plasma Enhanced Chemical Vapor Deposition (PECVD) 23
1.4.1.5 Laser Chemical Vapor Deposition (LCVD) 23
1.4.2 Atomic Layer Deposition (ALD) 25
1.4.2.1 ALD Versus CVD Deposition Technique 27
1.4.3 Other Techniques Based on Chemical Deposition Processes 27

1.5 Present Status and Future Scope 29
1.6 Conclusion 30
References 30

2 Tribological Investigation of Developed Nanocomposite Mos_2-Tio_2-Zro_2 Coating Material 33
Avinash V. Borgaonkar, Deepak Rai and Ismail Syed
2.1 Introduction 34
2.2 Materials and Methods 37
2.3 Experimental Details 40
2.4 Results and Discussions 41
2.5 Conclusions 45
References 45

3 Methods of Microstructural Characterization of Coatings 49
A. M. Kamalan Kirubaharan, D. Dinesh Kumar, K. Gobi Saravanan and Subhenjit Hazra
3.1 Introduction 49
3.2 X-Ray Diffraction 52
3.2.1 Coating Texture 54
3.2.2 Residual Stress Measurement 55
3.3 Atomic Force Microscopy 56
3.4 Scanning Electron Microscopy 57
3.5 Energy Dispersive X-Ray Spectroscopy 59
3.6 Transmission Electron Microscopy 60
3.7 X-Ray Photoelectron Spectroscopy 62
3.8 Electron Probe Microanalysis 63
3.9 Secondary Ion Mass Spectroscopy 64
3.10 Raman Spectroscopy 65
3.11 Conclusion 66
References 67

4 Wear Mechanisms and Methods of Wear Testing 71
Ahmed Abdelbary
4.1 Introduction 72
4.2 Mechanisms of Wear 72
4.2.1 Adhesion 72
4.2.2 Abrasion 74
4.2.3 Surface Fatigue 78
4.2.4 Corrosion 79
4.2.5 Erosion 80
4.2.6 Fretting 82

4.2.7 Impact 83
4.2.8 Thermal 84
4.3 Methods of Wear Testing 84
4.3.1 Sliding Wear Test 86
4.3.2 Rolling Wear Test 88
4.3.3 Scratch Wear Test 89
4.3.4 Abrasion Wear Test 90
4.3.5 Erosion Wear Test 92
4.3.6 Impact Wear Test 93
References 94

5 Mechanical Behavior of Thermal Spray Coatings 99
Jayant Gopal Thakare
5.1 Introduction 99
5.2 Elastic Modulus and Fracture Toughness 101
5.2.1 Effect of Microstructure on Fracture Toughness 108
5.3 Evaluation of Residual Stress 109
5.4 Conclusion 111
References 111

6 Tribological Properties of Carbon-Based Coatings 115
D. Dinesh Kumar, P. Kuppusami, Gobi Saravanan Kaliaraj, Siva Sankar Sana and A.M. Kamalan Kirubaharan
6.1 Introduction 116
6.1.1 Classification of Carbon-Based Coatings 116
6.1.1.1 DLC Coatings 117
6.1.1.2 Diamond Coatings 119
6.2 Deposition of Carbon-Based Coatings 120
6.2.1 Physical Vapor Deposition Techniques 121
6.2.2 Chemical Vapor Deposition Techniques 122
6.3 Tribo-Mechanical Characteristics of Carbon-Based Coatings 123
6.3.1 H-Free a-C or DLC Coatings 124
6.3.2 Hydrogenated a-C (H-DLC) Coatings 125
6.3.3 ta-C Coatings 126
6.3.4 Diamond Coatings: MCD, NCD, and UNCD 127
6.3.5 Metal-Doped Carbon-Based Coatings 129
6.4 Factors Influencing Tribological Properties of Carbon-Based Coatings 130
6.5 Conclusions 132
References 133

7 Performance Evaluation of Journal Bearing Under Thin-Film Lubrication **139**
Sandeep Soni
7.1 Introduction 140
7.2 Governing Equations 141
7.2.1 Hydrodynamic Lubrication Modeling 141
7.2.2 Thin-Film Lubrication Modeling 143
7.3 Solution Strategy 144
7.4 Performance Characteristics 144
7.4.1 Load Carrying Capacity and Attitude Angle 145
7.4.2 Frictional Characteristics 145
7.4.3 Oil Flow Rate 146
7.4.4 Frictional Torque and Energy Loss 146
7.5 Results and Discussion 147
7.6 Conclusions 156
Nomenclature 157
Dimensionless Parameters 157
References 158

8 Bioactive Coatings for Biomedical Applications **159**
Gobi Saravanan Kaliaraj, D. Dinesh Kumar and A.M. Kamalan Kirubaharan
8.1 Introduction 160
8.2 Surface Coating 161
8.3 Polyether Ether Ketone 163
8.4 Nitride Composites Coating 164
8.5 Binary Transition Metal Nitride Coatings 166
8.6 Hydroxyapatite-Based Coating 169
8.7 Bioactive Glass Coatings 171
8.8 Composite Coatings 172
8.9 Conclusions 174
References 174

9 Study of Tribological Behavior and Wear Mechanism of Nanocomposite Coatings **181**
Lailesh Kumar, Pankaj Shrivastava, Deepankar Panda, Arka Ghosh and Syed Nasimul Alam
9.1 Introduction 182
9.1.1 Techniques Employed for Depositing Nanocoatings 187
9.1.1.1 Electrodeposition Coating 188

9.1.1.2 Plasma Detonation Technology 188
9.1.1.3 Vacuum Arc Deposition 189
9.1.1.4 Sol-Gel Technique 189
9.1.1.5 Cold Spray Method 189
9.1.1.6 Chemical Vapor Deposition 190
9.1.1.7 Physical Vapor Deposition 190
9.1.1.8 Thermal Spray Method 191
9.1.1.9 Solution Dispersion 191
9.1.1.10 Spray Coating 192
9.1.1.11 Dip Coating 192
9.2 Applications of the Different Nanocomposite Coating Techniques 193
9.2.1 The Future of Nanocoatings 194
9.3 Effect of Different Coating Techniques on the Mechanical Properties of the Materials 195
9.4 Corrosion Behavior in Nanocoatings 198
9.5 Wear Mechanism in Nanocoatings 200
References 205

10 Investigation of Automotive Disc Brake's Material Based on Tribological Parameters by Using Computational Structural Analysis 211
Arul Prakash R., Vijayanandh R., Ramesh G., Hariaran S., Janardhanan Y., Senthil Kumar M. and Jagadeeshwaran P.
10.1 Disc Brake 212
10.1.1 Introduction 212
10.1.2 Materials Used in Disc Brakes 212
10.2 Literature Survey—Problem Description and Solution Technique 214
10.2.1 Problem Descriptions 214
10.2.2 Methodology Used—Computational Structural Analysis 215
10.3 Structural Analysis 215
10.3.1 Conceptual Design 215
10.3.2 Pre-Processing 216
10.3.3 Mathematical Modeling Used in this Finite Element Analysis 218
10.3.4 Generation of Grid and Its Convergence Study 220
10.3.5 Numerical Results on Disc 223
10.3.6 Comparative Analysis of Disc 226

10.3.7 Numerical Simulation on Brake Pad 226
10.3.8 Comparative Analysis of Brake Pad 231
10.4 Theoretical Investigations 231
10.4.1 Investigation-1 232
10.4.2 Investigation-2 233
10.4.3 Investigation-3 234
10.5 Conclusion 235
References 235

11 Studies of Nanomechanical and Wear Testing Methods of Thermal Sprayed Coatings 239
Zulfequar Ali, Ashish Kumar Singh, Omprakash Thakare and Chitrakant Tiger
11.1 Introduction 240
11.2 Nanowear Testing Methods 240
11.2.1 Nanoscratch Testing 241
11.2.2 Nanofretting Wear Testing 242
11.3 Nanomechanical Testing Methods 242
11.3.1 Nanoindentation Technique 243
11.3.2 Nanoimpact Test 246
11.3.3 Micropillar Compression 246
11.4 An Overview of Past Research 246
11.5 Design of a New Coating 248
11.6 General Discussions 249
References 249

12 Overview of Biological Tissue Properties for Biotribology 253
L. DhiviyaLakshmi and V. Muthukumaran
12.1 Introduction 254
12.2 Tissue Interaction With Implant 255
12.2.1 Bone 257
12.2.2 Compact Bone 259
12.2.3 Trabecular Bone 260
12.2.4 Mechanical Property of Bone 260
12.3 Anisotropy 261
12.4 Heterogeneity 263
12.5 Aging 263
12.6 Hysteresis 264
12.7 Bone Remodeling 265
12.8 Joint 266
12.9 Biomechanics of Cartilage Degeneration 273

12.10 Biotribology of Total Hip Replacement (Metal-on-Metal Articulation) 274
12.11 Conclusion 278
References 278

13 Mechanical Behavior of Single and Multilayer Hardfacing on Ferrous Alloys 281
B. Venkatesh
13.1 Introduction 281
13.2 Experimental Details 282
13.2.1 Hardfacing Procedure 282
13.2.2 Characterization of the Stellite Hardfaced Surface 284
13.2.3 Erosion Investigation 284
13.2.4 Eroded Sample Characterization 285
13.3 Outcome 285
13.3.1 Hardfaced Layer Behavior 285
13.3.2 Rate of Erosion 287
13.3.3 Eroded Sample Inspection 290
13.4 Discussion 291
13.4.1 Literature Observations 291
13.4.2 Common Erosion Behavior 292
13.4.3 Deformation Depth 295
13.4.4 Efficiency of Erosion 297
13.4.4.1 Abrasive Wear 299
13.4.4.2 Sliding Wear 299
13.4.4.3 Erosive Wear 299
13.4.4.4 Microstructures for Fe-Cr Hardfaced Layers 303
13.5 Conclusions 303
References 303

Index 307

Preface

The book provides the updated information on friction and wear behavior of coatings used in different industrial applications. It covers the various coating deposition techniques, tribological response of nanocomposite coatings, multilayer hardfacing and wear testing methods for coatings at nanoscale. Use of the nanostructures may alter the tribological and mechanical properties of the materials. Thermal spraying is a most widely used technique in industries for the deposition of coatings and their tribological properties need to be determined. The mechanical and metallurgical engineer need to study the tribological behavior of engineering materials for their selection in particular applications. It contains a total of 13 chapters giving detailed information on the characterization and testing methods of the coating materials.

In order to develop advanced coating materials, researchers need to understand the wear and mechanical testing methods of the coatings at nanoscale. Combination of different elements yields composite coatings with enhanced mechanical, thermal and tribological properties. Various coating deposition techniques e.g. physical vapor deposition (PVD), sputtering and thermal spraying are described, as well as various characterization techniques with respect to the microstructure of coating materials. Different instruments like X-Ray diffraction, scanning electron microscope, and transmission electron microscope are used to study the elements and morphology are detailed as well. This book also includes the recent trends in Biotribology and the materials used in implants against the abrasive wear. The syllabus of this book is designed in such a way that it fulfills the purposes of research scholars, engineers and scientists.

Sarfraj Ahmed
Vinayak S. Dakre
Editors
November 2021

1

Overview of Coating Deposition Techniques

Neeraj Mehta

Department of Physics, Institute of Science, Banaras Hindu University, Varanasi, India

Abstract

The current age of science and technology is influenced tremendously by thin-film (TF) coatings due to their countless benefits in solid-state devices. The coating deposition techniques have become a pillar for state-of-the-art fabrication techniques in diversified areas, for instance, optical memories (CDs and DVDs), sensing elements, fiber lasers, LED displays, optical filters, medical implants, environmental equipment, and energy-storing devices. The critical issues for any established or proposed utilization of TFs in science and technology depend upon the various factors like the technique employed for coating, cost, time scale, and durability. There are various techniques for coating deposition through the physical or chemical pathways. The science of TF deposition has grown quickly in such an impressive way that now it has opened a new branch of technology named coatings and surface engineering.

There is a very huge number of coating techniques and they are extensively different from each other. However, no exceptional way exists to categorize such techniques. Thus, the same procedure may be considered into one or more categories based on this perspective. Every method possesses its specific benefits and drawbacks. The selection of a particular method of coating deposition depends upon the numerous factors and the selection criteria (e.g., the definite application and accessible resources).

This chapter enlightens about some innovative coating techniques and their corresponding principles. This chapter deals with various deposition techniques such as evaporation under vacuum (e.g., thermal/flash evaporation), the evaporation by a beam of electrons or a laser source, magnetron sputtering [either DC (direct current) or RF (radio frequency–based sputtering], and coating deposition

Email: dr_neeraj_mehta@yahoo.co.in

Sarfraj Ahmed and Vinayak S. Dakre (eds.) Tribology and Characterization of Surface Coatings, (1–32) © 2022 Scrivener Publishing LLC

via chemical routes. The aspects of various components, approaches, and apparatus that are essential to organize or upgrade the coating deposition for different kinds of materials have been also discussed.

***Keywords*:** Coating deposition, thin films, evaporation, sputtering, chemical vapor deposition, atomic layer deposition

1.1 Introduction

Thin film (TF) is a very thin layer of a substance or a coating that is deposited on a supporting material. The coating deposition is a skill of achieving a very TF of a substance having a thickness between arranging from hundred micrometers to a few nanometers. In the recent era, the thickness of TFs can be limited to that of few atomic layers. Thus, TFs are just a layer of materials, thickness ranging between ~10 and 500 nm. The surface of a suitable material is used to accomplish the deposition of such films. The substances that are used to be coated are known as "substrate". If there is no substrate used, then such thin film is a "foil". In ancient times, the art of preparing the gold leaf was already known. The conversion of gold into a few micron gold leaves by hammer was in vogue. This type of gold leaf was used for coating different kinds of accessories. The main purposes of the coating deposition techniques are (i) deposition of the conductive or insulating layers on a substrate, (ii) creation of the local conductive channels that can be used to interconnect devices, and (iii) to build up more complex structures in a single layer or multiple layers.

Various technologies for coating deposition or preparing TFs have been industrialized in past and recent years [1–3]. These technologies are directly useful for solid-state device fabrications because they require less amount of the sample for deposition over the use of the comparatively large amount of the bulk materials in other existing techniques. Thus, the processing costs are reduced significantly. The coating deposition offers various fascinating and beneficial features such as surface uniformity, simple fabrication, reduced manufacturing costs, and strange electrical, optical, and mechanical properties.

There are a vast number of deposition methods available; all methods have their specific advantages as well as disadvantages based on substrate and source substance, the thickness of the deposited film. There are two main categories in which we can classify the methods that are employed for the coating deposition. One category consists of the physical deposition processes (PDPs). In physical deposition processes, there is a physical movement of the substance on the substrate's surface during the deposition. This means that any kind of chemical reaction neither participates in the method nor responsible for the coating deposition. The chemical deposition processes belong to the second category. In this category, the chemical reactions of a sample with the substrate material in the presence of some precursors (that are generally in the liquid and/or gas phase) play a dominant role in the direct production of TFs. Thus, these approaches use the reaction of the substrate's surface with precursors in either gas or liquid form for the procurement of the desired coating deposition. The solid material is typically not the only product formed by the reaction. Byproducts may be comprised of solids, liquids, or gases.

The PDPs have certain limitations such as poor conformality, low throughput, limited directional variation, and reduced compositional control. All of these drawbacks can be removed using chemical deposition techniques that provide additional benefits such as ultrathin film growth. During the last decade, chemical deposition techniques have played an exceptionally important role over the physical deposition techniques in the design and manufacture of novel advanced devices. The flow chart of both kinds of processes is shown in Figure 1.1.

Figure 1.2 shows the block diagrams that display the schematics of typical physical and chemical deposition processes.

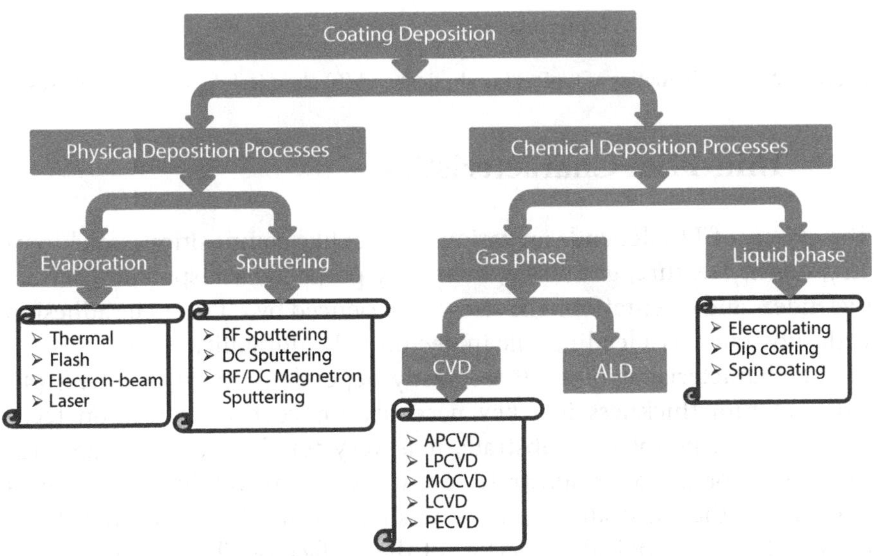

Figure 1.1 Flow chart showing some widely used methods of coating deposition.

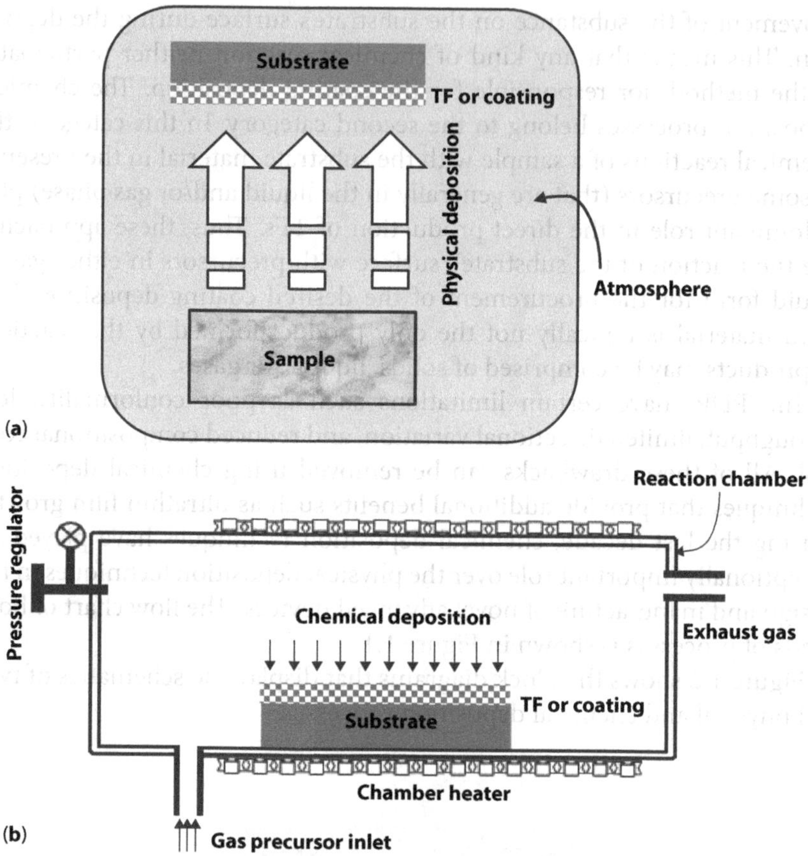

Figure 1.2 The schematics of typical (a) physical and (b) chemical deposition processes.

1.2 Thin-Film Characteristics

The quality of TFs depends on various factors like composition, stoichiometry, density texture, grain size, boundary property, transparency, refractive index, and orientation. The stress experienced by a TF and its adhesion with substrate has a leading role in deciding the durability of TF.

The characteristics of a TF are vastly dependent on its thickness [4]. Uniform film thickness is a key necessity for coating deposition techniques. For a nonplanar substrate, it is very tough to attain a uniform thickness. For a planar substrate too, there is a possibility of thickness gradient on the substrate. The thickness has a crucial effect on the structural and morphological properties of the surface of TF because its close interaction with the substrate can affect different characteristics of TF.

For most of the coating deposition techniques, the thickness depends on the temperature of deposition and the chemical affinity between substrate and film. The geometry of the evaporation source and its location concerning the substrate also has a significant role in deciding the uniformity of TFs [5]. Thickness can be measured by two kinds of methods: (i) the methods that are applied during deposition (known as "*in situ*" methods) and (ii) the methods that are employed for the determination of thickness after completion of a coating deposition (known as "*ex situ*" methods). The second significant property is the roughness of TFs. Surface roughness is the degree of unevenness of the film surface. The surface of the TF may diverge from the perfect plane on different scales constantly or arbitrarily. When such a scale is related to a particular coating deposition technology, then it is compulsory to take into consideration such surface irregularity or unevenness. This can frequently support controlling the quality of the deposition techniques and modify the deposition conditions so that the desirable film smoothness can be achieved. The roughness of a thin film can affect all kinds of its properties such as electrical, mechanical, and optical. Therefore, the knowledge about various qualitative (e.g., the categories of roughness and the mechanisms of its evolution) and quantitative (e.g., roughness measurement, depth profile, and surface geometry) aspects of roughness must be taken into consideration [6].

The chemical analysis of a TF after coating deposition is also necessary to know about its homogeneity and chemical composition. For this, various state-of-the-art techniques like atomic absorption spectral analysis, Auger electron spectrometry, X-ray fluorescence, and mass spectrometry are used. The homogenous distribution of the constituents (e.g., atoms and molecules) of a typical TF and their nucleation may bring out the regular surface. The surface tension forces present on the substrate surface due to impurities prevent the uniform distribution of TF and affect the homogeneity. To resolve these challenges, an appropriate chemical cleaning of the substrate surface is necessary. Sometimes, more nucleation may happen that causes the irregularities over the apparent surface and gives rise to surface roughness. The inhomogeneity of a TF influences noticeably its optical properties because of the scattering of free carriers.

1.3 Techniques Based on Physical Deposition Processes

In this section, we will discuss some significant and state-of-the-art techniques that employ the physical deposition process for the coating deposition. Generally, a physical deposition process is used to obtain the TFs of unadulterated substances (e.g., compounds, metallic or non-metallic alloys, amorphous materials, and ceramics) by converting them into vapor form. The thickness of such TFs is usually in the range of 1 to 10 μm. Thus, the physical vapor deposition (PVD) techniques, as its title indicates, involve the physical movement of constituents (e.g., atoms, ions, or molecules) of a substance for coating deposition on the substrate's surface. All the techniques based on PVD are commenced in a vacuum chamber that is equipped with the facility of controlling the atmospheric condition of its interior space. In the next sub-sections, we will discuss some important types of PVD methods.

1.3.1 Evaporation

In the 1850s, the first evaporation technique for coating deposition by evaporating the sample was developed by Faraday [7] which is known as thermal evaporation. This technique is a frequent way of PVD. Thermal evaporation is one of the straightforward methods of PVD. To evaporate a sample material, this method characteristically employs a source that can perform resistive heating in an inert atmosphere for the production of TFs. In a standard vacuum coating unit, the sample is heated as far as its vapor pressure is reached. The sample material, during evaporation, crosses the path between the boat (sample holder) and substrate holder inside the coating unit for deposition on the surface of the substrates.

We can produce TFs of both metallic and non-metallic substances (e.g., Al, Au, Ag, and In) by using thermal evaporation. This method of coating deposition is used mostly for industrial manufacturing of electric contacts by using Ag, Cu, or Al elements as coating materials. More sophisticated applications comprise the co-deposition of numerous segments that are attained by cautious control on the temperatures of separate vessels. The general construction of an apparatus using for thermal evaporation is shown in Figure 1.3a. This method works on the principle of resistive heating (i.e., Joule effect) to vaporize the constituents (e.g., atoms, clusters of atoms, or molecules) of the sample placed in a metallic crucible. This crucible is known as a boat. A boat containing the sample material is heated either directly by flowing an electric current across it or indirectly using a resistive element (e.g., Molybdenum and Carbon) as a heating filament.

The substrate is fixed at the top position with the help of a holder. A metallic vacuum chamber or a vacuum vessel made of glass is used to enclose all of these components. Thermal evaporation is found useful for the coating deposition of the metals as the contact layers in the integrated circuits (e.g., organic light-emitting diodes, TF transistors, and solar cells).

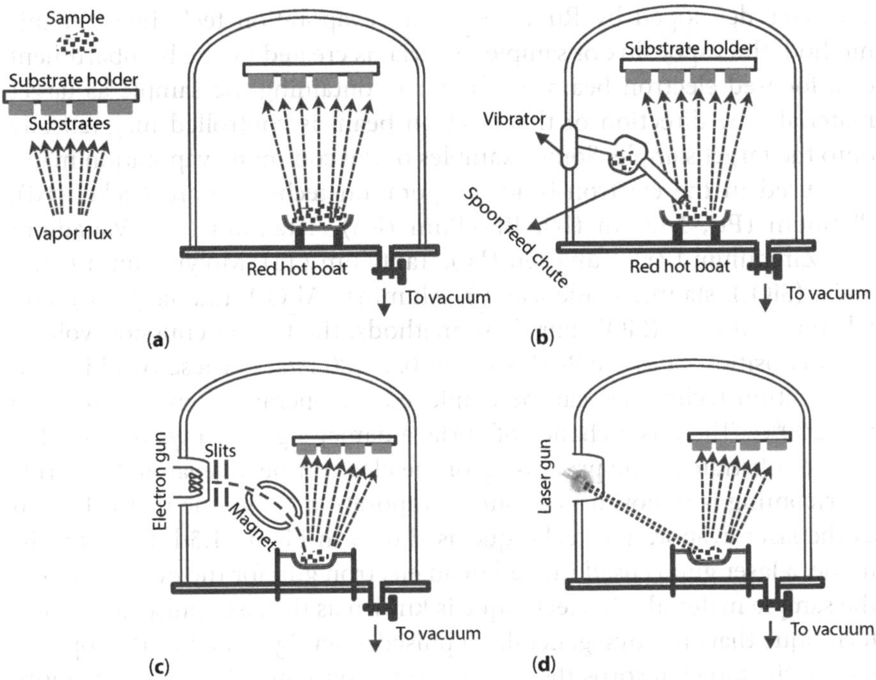

Figure 1.3 Different kinds of evaporations: (a) thermal, (b) flash, (c) electron-beam, and (d) laser evaporation.

Also, this type of evaporation might be utilized for the deposition of thick coatings of Indium to create the wafer bonding.

The second evaporation technique is known as flash evaporation as depicted in Figure 1.3b. In this technique, the substance is taken either in powdered form or in the form of a thin wire. When the powdered sample or sample wire is dispensed continuously onto a hot boat, it evaporates and deposits on the substrate.

Some examples of the common evaporation materials used in thermal and flash evaporation techniques as samples are Gold (Au), Silver (Ag), Aluminum (Al), Tin (Sn), Chromium (Cr), Antimony (Sb), Germanium (Ge), Indium (In), Magnesium (Mg), and metallic chalcogenides (e.g., CdS, PbS, CdSe, PbSe, CdTe, and PbTe). For these methods, the typical filament currents and deposition rates are 200–300 Amp and 1–20 Å/sec. These two kinds of evaporation techniques can be employed for operating temperatures up to 1,800°C.

Figure 1.3c shows the third evaporation technique that is known as electron-beam evaporation that was discovered accidentally by Pirani [8] and later developed by Ruhle [9] as an evaporation technique. In this method, the vapor flux of sample material is created by the bombardment of a focused electron-beam on the boat containing the sample as target material. The direction of this electron beam is controlled magnetically onto the target sample. Some examples of the common evaporation materials used in the electron-beam evaporation technique are Nickel (Ni), Platinum (Pt), Iridium (Ir), Rhodium (Rh), Titanium (Ti), Vanadium (V), Zirconium (Zr), Tungsten (W), Tantalum (Ta), Molybdenum (Mo), Silica (SiO_2), stannic oxide (SnO_2), Alumina (Al_2O_3), titania (TiO_2), and Silicon monoxide (SiO). For these methods, the typical emission voltage and deposition rates are 8–10 kV and 0.2–100 Å/sec. These two kinds of evaporation techniques can be employed for operating temperatures up to 3,200°C. There is a chance of surface damaging by x-rays or ions flux on the substrate in the processing of the electron-beam evaporation. This shortcoming is removed in a fourth evaporation technique that is known as the laser evaporation technique as shown in Figure 1.3d. In this technique, a laser gun is used instead of an electron gun for the evaporation of the sample material. This technique is known as the laser pulse deposition technique that employs, generally, a pulsed-laser light source. The top surface of the target absorbs the laser energy. Consequently, a small quantity of the sample is evaporated by a thermally energetic flash. The remarkable benefit of this approach is the deposition of the evaporated sample on the substrate's surface with an almost negligible probability of decomposition. Thus, the atomic ratios of the target material retain the same values before and after the evaporation (i.e., the atomic ratio of each constituent in the sample in TF form is the same as that was prior to deposition). This technique is highly recommendable when the coating deposition of material with complicated stoichiometry is mandatory.

The thickness of a TF deposited by evaporation can be measured unceasingly throughout the experiment by employing a vibrating crystal. The principle of the thickness measurement is based on the modification in the frequency of the vibrating crystal, which is caused by the material being evaporated onto the crystal. Thereby, the resonance frequency is decreased with increasing material being deposited.

1.3.1.1 Mean-Free Path: A Significant Parameter for High-Quality Evaporation

The mean free path λ plays a very significant role in achieving high-quality TFs through any kind of evaporation technique. Lower base pressures or high vacuum is needed to increase the value of λ considerably. This ensures that only a few gas molecules from the atmosphere have been retained in the chamber. Thus, the probability of their inclusion in the depositing films is decreased and reduces to a negligible value. When the vacuum in the chamber increase (i.e., the pressure decreases), the number of molecules bouncing around in the deposition environment goes down. Consequently, the mean free path (i.e., the mean distance traveled by a molecule before its collision with another molecule) increases.

The mean-free path λ for an atmospheric gas molecule having diameter σ at absolute temperature T and atmospheric pressure P is calculated by using the following relation:

$$\lambda = \frac{kT}{\pi\sqrt{2}\sigma^2 P} \quad (1.1); \text{ [k is Boltzmann's constant]}$$

The vacuum maintained in the evaporation chamber must be high enough so that the gap between the source (i.e., sample) and target (i.e., substrate) is smaller than the value of λ for atmospheric gas molecules.

1.3.2 Sputtering

In sputtering, the source or the sample is typically in the shape of a plate, called a target. A negative potential is applied to the sample plate so that it acts as a cathode. The substrates are placed on another electrode having positive potential, i.e., it acts as an anode. The electrons ejected from the negative electrode collide with and knock out electrons from the outermost shells of argon atoms. Thereby, ionization of Ar atoms takes place to produce Ar+ ions. The collection of Ar^+ ions and electrons constitute a plasma in the region between the two electrodes.

Initially, we generate a gaseous plasma. After that, the speeding ions move toward the sample from the generated plasma. The sample is dissipated by ions coming through energy transfer and ejected as neutral particles (e.g., separate atoms or clusters of atoms or molecules). After ejection, these neutral particles travel in a linear path until they deposit on the surface of the substrate. Thus, a TF of the sample material is coated on the substrate when it is positioned in the track of the ejected flux of the particles.

Figure 1.4 shows a simple sputtering apparatus. This figure shows that the main components of a sputtering apparatus are (i) vacuum chamber, (ii) power supply, and (iii) sputtering gas. The inert gases (e.g., argon, xenon, or krypton) are used to create the plasma. The argon gas is generally used as the sputtering gas because argon atoms do not respond to the

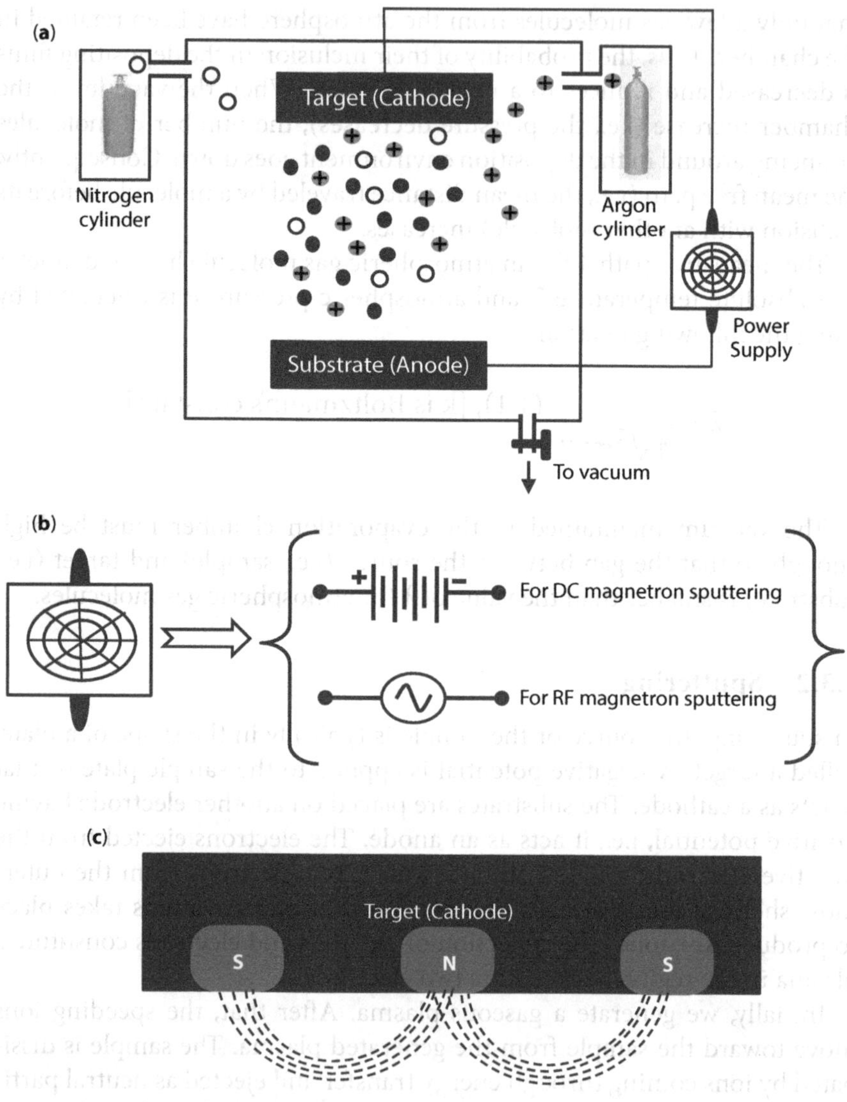

Figure 1.4 (a) A typical example of sputtering apparatus. (b) Power supply used in DC/RF magnetron sputtering. (c) Position of magnets in the target (cathode) of the magnetron type sputtering apparatus.

target specimen chemically or physically and interact with the gases used for processing. Further, Argon gas helps in enhancing the sputtering and deposition rates. A plasma is created by ionizing the sputtering gas.

When the bombarding of the target specimen by the sputtering gas occurs then this process sputters off the deposition of the specimen.

The fast-moving electrons deliver their energy to the neutral atoms through the collisions that cause their ionization. Thus, the role of plasma is to hit atoms out of the sample. These dynamic atoms are transported to the substrate and deposited on the surface of a substrate positioned close to the sample. The best step coverage is created by increasing the pressure. The surplus energy of the ions also supports the drifting of atoms toward the surface. Consequently, the increase in surface mobility takes place. Sometimes, a small quantity of responsive gas (O_2 or N_2) is also used for regulating the stoichiometry and structural properties of the TFs.

The benefits of sputtering are pointed below:

(i) It can be employed for large-area-targets which gives uniform thickness over the wafer.
(ii) We can control the thickness by deposition time and other parameters.
(iii) The sputtering of the substances having extremely high melting points is also possible.
(iv) The adhesive force between the sputtered TF and the substrate is high and so adhesion of good quality is achieved.
(v) Sputtering can be executed in both horizontal and vertical orientations.

For the fabrication of the TFs of the materials having multi-components, the sputtering provides an efficient control on the composition by using the compound targets. On other hand, it is a tough task to control the composition in the evaporation technique because of its dependency on the individual vapor pressure of each component. The sputtering technique provides better lateral thickness uniformity over the evaporation technique because of the superposition of multiple point sources. Furthermore, if we consider the same rate of sputtering and thermal evaporation, then the energy of atoms or molecules is 10 times greater for the former technique as compared to the later technique. This is another reason for achieving TFs of the best quality in the sputtering over evaporation.

There are some drawbacks to the sputtering technique. In this technique, the generated ions have sufficiently high kinetic energy (~0.5 to 1 keV) as compared to the energy (1 to 10 eV) of the atoms sputtered in the process. Thus, the capability of energy saving in sputtering is poor because the maximum part (>90%) of incoming energy is wasted in heating the target and generating the secondary electron. Another serious concern is about employing a well-organized cooling arrangement to prevent the damaging of the target materials from the overheating caused by the sputtering at a fast rate during the coating deposition.

1.3.2.1 Characteristics of Sputtering

In sputtering, there is an emission of the particles (e.g., neutral atoms, positively/negatively charged ions, electrons, and neutral/charged clusters of atoms) from the cathode. The neutral atoms have the major populations in the total number of particles. The energy of such atoms is greater as compared to the neutral atoms produced by thermal evaporation sources. It is believed that this surplus amount of energy is responsible for the good adhesion capability of TFs that are deposited by sputtering over thermal evaporation. The value of λ for the sputtered species is low in the case of sputtering as compared to evaporation because of the relatively low vacuum inside the sputtering chamber during the process. The sample experiences many gas-phase collisions between the gap of substrate and target during the process. Consequently, the depositing species lose a significant quantity of kinetic energy when they reach very near to the substrate for the deposition. This affects the structural configuration and the density of TFs formed in the sputtering technique.

When we divide the number of sputtered atoms by that of impinging ions, we get a new sputtering parameter: sputtering yield η. The significance of sputtering yield η is that we can control the deposition rate by varying the value of η. The mathematical expression for the determination of sputtering yield is as follows:

$$\eta = A \frac{mM}{(m+M)^2} \frac{K}{U} \tag{1.2}$$

Here, m and M are the respective masses of impinging ion and target atoms, while A is a proportionality constant that has a dependence on striking/incident angle. The symbols U and K denote the binding energy of the target and the kinetic energy of the incident ions, respectively. The value of η is independent of the temperature of the sputtering target. However, it increases at very high values of target temperature because of the occurrence of thermal evaporation.

In modern times, the proportion of electrons participating in the ionization events is enhanced by using magnets. This increases the number of the collision of electrons with the atoms of sputtering gas. Consequently, there is a remarkable rise in the path-length of the electrons and their ability to ionization. This advanced version of sputtering is known as magnetron sputtering. The benefits of using magnets are as follows:

(i) In magnetron sputtering, lower voltage is required to strike the plasma.
(ii) The heating of the substrate is reduced due to electron bombardment.
(iii) The deposition rate is increased and the control on it is improved.

Usually, if we want a TF having low adhesive force for the material of the substrate used, then thermal evaporation can be used without any hesitation. The quality of TFs is best in sputtering as compared to thermal evaporation if we think about the processing parameters (e.g., surface texture, thickness, and work function) because they are controlled effortlessly. Sputtering does not necessarily offer a better purity of TFs as compared to evaporation because it is mostly dependent on the deposition rate and atmosphere of the vacuum chamber. Thus, in some cases, sputtering is not recommended over thermal evaporation while it is a better option in other cases as compared to evaporation.

Table 1.1 depicts the comparison of evaporation deposition and sputtering techniques.

Table 1.1 Comparison between evaporation and sputtering deposition techniques.

Property	**Evaporation**	**Sputtering**
Deposition rate	Fast (10^3 atomic layer/sec)	Slow (1 atomic layer/sec)
Thickness control	Thickness control is possible but tough	Controlling of thickness is easy
Sample Materials	Limited	Almost unlimited
Purity	Better	Less good
Alloy composition	Little control or no control	Tightly controlled
Grain size and orientation	Larger size with fewer orientations	Smaller size with more orientations
Equipment cost	Medium	expansive
Adhesion	Often poor or medium	Very good

1.3.2.2 Types of Magnetron Sputtering Technique

The sputtering technique is generally separated into main classes: (i) DC magnetron sputtering, and (ii) RF magnetron sputtering. In both types of techniques, the magnets are fixed at appropriate positions in the negatively charged target material (i.e., sample cathode) to confine the electrons over the cathode (see Figure 1.4c). Consequently, the electrons experience a constraint during the bombardment on the surface of the substrate. This facilitates the coating deposition more rapidly by increasing its rate. A boundary is created by the magnetic field for the tunneling process so that the electrons can be confined in the vicinity of the cathode's surface. This strengthens the productivity of the ionization by increasing the rate of formation of the gas ions.

DC sputtering is the most elementary and low-cost class of sputtering for the coating deposition of the metallic samples and the conductive coating of the substances used for the target. There are two key benefits of employing a DC power supply: (i) its monitoring is straightforward and (ii) it is a cost-effective choice for the coating deposition of the metals. This class of sputtering is employed widely in semiconductor manufacturing for the creation of integrated circuits (i.e., microchips) on the molecular scale. It is also one of the most recommended choices for the gold coating by sputtering in the jewelry industry, circuit boards, watches, and medical implants. It is also used for coatings on other decorative textures, non-reflective coating deposition on optical components made of ceramics and glasses, and for metalized wrapping films. The fundamental configurations of the assemblies used for DC and RF sputtering are almost identical excluding the power supply employed in both as shown in Figure 1.4b.

The direct current is used to supply electric power in a DC magnetron sputtering coating system. The vacuum maintained in its chamber lies in the range of 0.001 to 0.1 Torr. Also, the DC supply is generally chosen for such target materials that are good conductors of electricity to increase the efficiency of the process and to reduce the expenses. This class of sputtering is uncomplicated in operating even when we process the huge numbers of big-sized substrates. For specific metallic targets having negligible impurity, the rate of coating deposition is quite high. In this method, when the fast-moving positive ions from the sputtering gas reach the target in DC sputtering, then the accumulation of the ejected atoms starts on the surface of the substrates for coating deposition.

The alternate current is used to supply electric power in an RF magnetron sputtering coating system. A radio frequency source of sufficiently high electric supply is used with a constant frequency between 10 and 15 MHz. The pk-pk (peak to peak) value of radio frequency voltage is of the order of 10^3 volts and the number density of electrons is 10^{15} to 10^{17} m^{-3}. This technique has an extensive range of applications, and it is appropriate for both kinds (i.e., either conductive or non-conductive nature) substances. However, this class of sputtering is frequently used when a dielectric material is chosen as a target. The rate of the coating deposition is slow in the RF class as compared to the DC class of sputtering. Also, the small-sized substrates are used in this class to reduce the processing cost. The cycles involved in the processing of the RF class have two major components. During the negative half-cycle, the target becomes a negative electrode (i.e., it plays the role of cathode). Consequently, the atoms of sputtering gas get polarized and they move toward the source to knock out its atoms. The ionized gas ions and source atoms retain their positions on the surface of the target because of the polarization.

During the positive half cycle, the target becomes an anode (i.e., it plays the role of a positive electrode). Consequently, the polarization goes upside down. This results in the removal of source atoms and gas ions from the target surface by accelerating them toward the surface of the substrate for the coating deposition.

1.3.3 Other Techniques Based on Physical Deposition Processes

The cathodic arc deposition technique is one of the older techniques based on physical deposition technique processes in which an electric arc is used to vaporize the substance from a target that acts as a cathode [10, 11]. This technique employs the triggering of an electric arc initiated by a sufficiently high voltage in most of the cases (see Figure 1.5a). The TF is formed when the condensation of the vaporized substance takes place on the surface of a substrate. The coating deposition of the composite and metallic substances can be achieved by employing this method. The coating deposition by creating a plasma in this method possesses a large density and adhesive force because the completely ionized plasma promotes these benefits through the generation of a large number of highly energetic ions. This approach has some shortcomings and glitches too that are related to the incorporation of apparent "macroparticles" due to the strain experienced by TFs under the influence of highly compressive internal stresses. This method is very useful for evolving stiff and durable coating depositions on cutting, trimming, and shaping materials of machine tools. Its additional

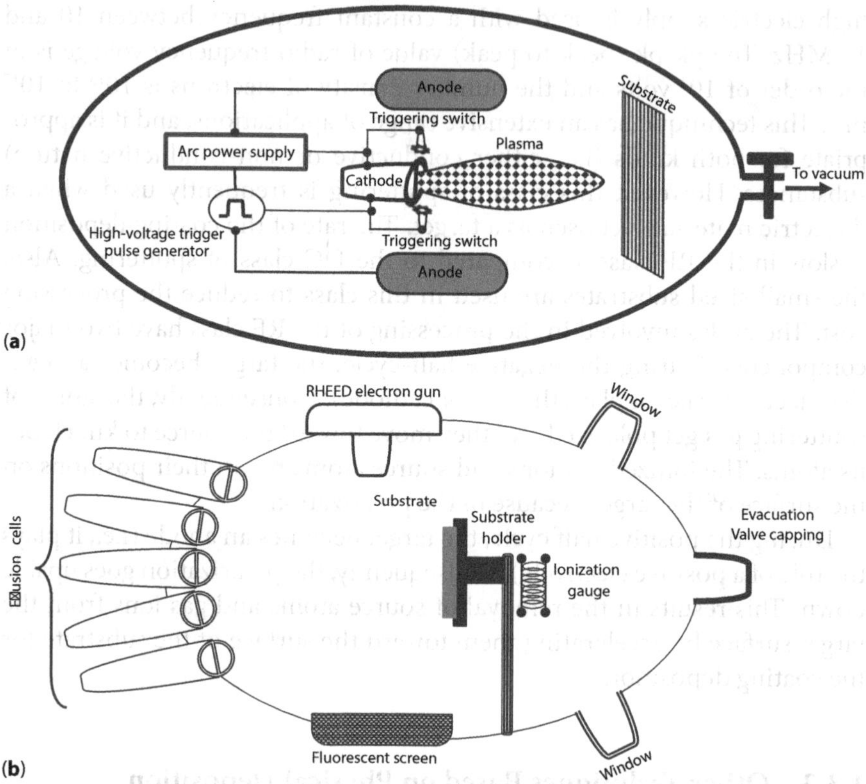

Figure 1.5 (a) Schematic arrangement of a typical cathodic arc deposition technique. (b) Schematic arrangement of a typical molecular beam epitaxy technique.

applications are the artistic decorative coatings, rust-unaffected coatings, and metallic protecting coatings.

Cathodic arc deposition technique has the following common features when we compare it with magnetron sputtering deposition technique:

(i) A target of the solid source material is used in both deposition techniques.
(ii) Both techniques are executed in the presence of diverse reactive and inert for coating deposition.
(iii) A vacuum chamber is used in both techniques.
(iv) Ionized species (plasma) are used in both techniques to eliminate material from the target.
(v) Both techniques have the capability of producing TFs for decorative and/or functional applications.

Table 1.2 Some comparative properties of the magnetron sputtering and cathodic arc deposition techniques.

Property	Cathodic arc deposition	Sputtering deposition
Target material	Conductive materials	Generally, all kinds of materials
Evaporation tool	Arc effect	Sputter effect
The electrical requirement for discharge	High current and low voltage	High voltage and low current
The approach adopted for material removal	Concentrated electrical discharge	Bombardment of gas ions

However, both techniques have distinctly different characteristics as mentioned in Table 1.2.

Another cutting-edge method is the molecular beam epitaxy (MBE) [12–14] that is used for preparing TFs of crystalline materials by using a turbo-molecular vacuum pump so that a very high inert atmosphere can be achieved during the coating deposition. It is employed for the deposition of epitaxial TF of diverse kinds of materials that are specially used in the fabrication of solid-state devices.

A flux of atoms or molecules from an effusion cell is incident on the well-sterilized hot substrates in the precisely controlled manner so that the deposition of TF of desired crystalline nature can be achieved.

A typical MBE chamber consists of numerous effusion cells having different elements (see Figure 1.5b). The evaporation of these elements in independent effusion cells occurs according to the requirement of the desired growth process. The selective opening of the shutter of these cells allows the ejection of desirable atomic beams toward the target that is a hot substrate. The amalgamation of the atomic beams is observed at the surface of the substrate. For the diffraction studies and growth monitoring in the MBE setup, a reflection high-energy electron diffraction (RHEED) apparatus is an essential component. Because of a gradual rate of deposition, we achieve an epitaxial growth of multiple atomic layers. We rotate the substrate's holder to ensure the homogenization of the grownup layers. In MBE, very thin coatings of sharp boundaries can be shaped since the controlling time of the shutters is very fast over the rate of the deposition.

1.4 Techniques Based on Chemical Deposition Processes

In this section, we will cover some noteworthy and state-of-the-art techniques that are used for the coating deposition and belong to this category. Generally, in the techniques based on chemical deposition processes, a liquid precursor experiences a chemical modification at the surface of a substance in a solid-phase that causes the departure of a solid layer. The frequent dilution of the precursor gases in the carrier gases occurs and then both are directed to enter in the reaction chamber at nearly ambient temperatures. The precursor gases go through some reaction or decomposition due to interaction with a heated substrate. Consequently, solid-phase deposition on substrates takes place. In the processing of CVD, the coating deposition of the substances from the gaseous state occurs. For such processing in CVD, the precursors must possess the volatile behavior but they should exist in a quite stable condition to ensure their transportation at the reactor. In the process of the CVD, the compound precursors normally allow the coating deposition of only one element while other constituents volatilize.

Thus, as the title reflects, the chemical vapor deposition is related to the chemical reactions that convert molecules of a gas (gaseous precursor) in the solid phase for the formation of TF by their coating deposition over the surface of a suitable substrate. The components of a vapor phase are frequently adulterated with a non-reacting inert carrier gas. A well solid TF is obtained after coating deposition due to the reactions in these components at the surface of a hot substrate. In the next sub-sections, we will discuss some important types of CVD methods.

1.4.1 Chemical Vapor Deposition (CVD)

CVD is a technique based on one or more chemical processes that are employed to produce TFs of solid substances having negligible impurity. For the industrial manufacturing of TF fibers and coating of homogeneous components, CVD is found very appropriate. By changing the experimental settings (e.g., the substance used as a substrate, its temperature, proportion of the constituents in the mixture of the reaction gas, and net pressure due to drift of gas molecules), we can grow the TFs of materials with different fascinating properties for multifunctional applications. It is significant to choose the correct precursors to acquire the desired coating. In the classical CVD system, the metal hydrides and halides were used as precursors. In recent times, a huge assembly of metal-organic compounds that includes metal alkoxides, metal alkyls, metal diketones, metal amidinates, and metal carbonyls is also used as precursors.

The following is the list of the elementary constituents of a typical CVD device:

(i) The precursors supply to the reactor chamber is achieved by a gas delivery system.
(ii) A reactor chamber for the occurrence of the coating deposition.
(iii) A substrate loading mechanism for inserting and eradicating wafers or substrates.
(iv) An energy source to deliver the heat to activate the reaction and decomposition of the precursors.
(v) A vacuum chamber for deletion of all unwanted vaporous species released due to various reaction/deposition.
(vi) A drainage arrangement for the elimination of volatile co-products from the reaction chamber.
(vii) An after-treatment system for the removal of such harmful compounds in some exhaust gases that are hazardous for the environment and cannot be released directly into the atmosphere.
(viii) Process control unit consisting of gauges and various control knobs for the monitoring of the benchmarks of the process like time-variation of temperature and pressure.
(ix) Warning and protection devices.

Several books are available in the literature regarding the science and technology of CVD [15] but the classic book written by Powell et al reveals that the knowledge of CVD of carbon on the electric filaments was known in the 1880s, and it was used in the manufacturing of luminous lamps to enhance their durability.

1.4.1.1 Atmospheric Pressure Chemical Vapor Deposition (APCVD)

This CVD operates at barometric (see Figure 1.6). This method is capable of producing the coating of a substance classically thick on different kinds of substrates or wafers. The thickness of TFs is of the order of numerous micrometers.

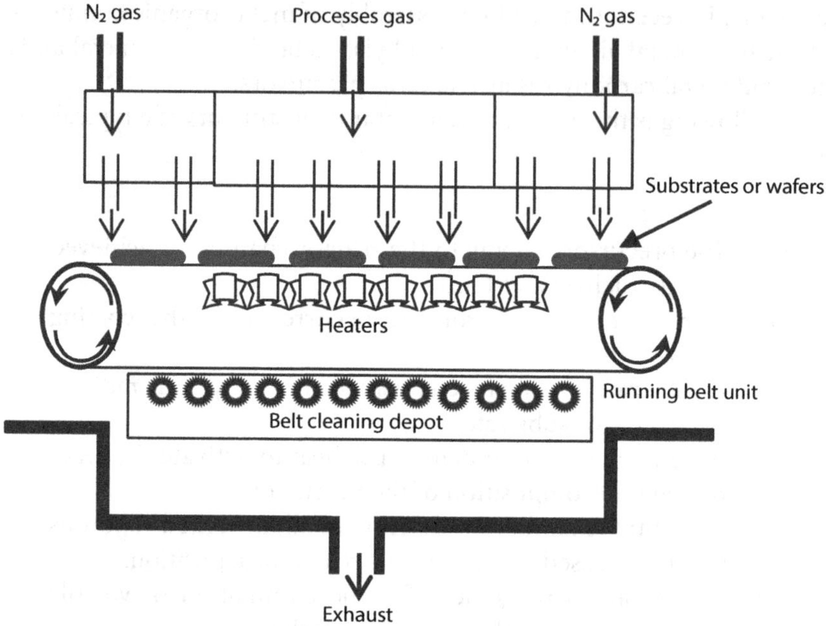

Figure 1.6 Schematic of APCVD.

It is also useful for the surface polishing process of several kinds of stuff (e.g., tools and turbine blades) to expand their durability and efficiency. Since there is no requirement of creating a vacuum in this CVD system, therefore this type of CVD system has comparatively inexpensive. However, such a CVD system is utilized infrequently because of high chances of oxidization because of the larger gas density and long durations of residency.

1.4.1.2 Low-Pressure Chemical Vapor Deposition (LPCVD)

In LPCVD, the pressure remains lower than the atmospheric pressure to about 100 Pa (see Figure 1.7). Consequently, the rate of diffusion was reduced by a factor of 1,000. Thus, the rate of mass transport is reduced and the reactants approach the surface of the substrate more nearly. Therefore, the coating deposited in this CVD displays outstanding consistency, homogeneity, high cleanliness, and excellent step coverage. It operates at sub-atmospheric pressures. The low pressure helps in decreasing the undesirable gas-phase reactions, and it enhances the consistency of the coating over the substrate. The lower pressure rises the diffusion of precursor

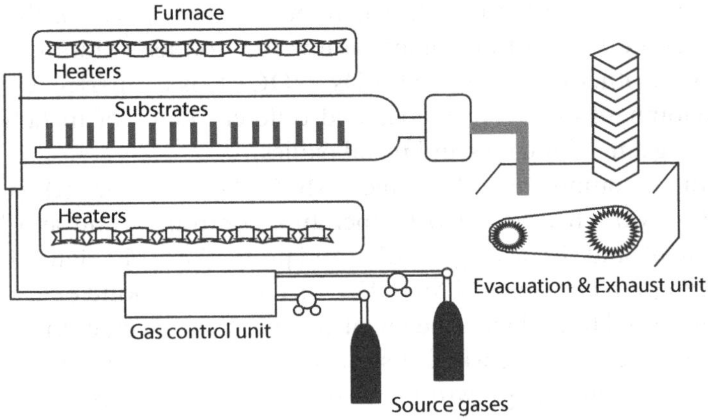

Figure 1.7 Schematic of LPCVD.

material via the gas. Consequently, the transport rate of the mass of the gaseous reactants is increased, and it becomes greater than the transport rate of surface-reaction. LPCVD is employed to deposit TFs of a wide range of materials (e.g., thick oxides to separate metallic interconnectors, SiH_4 for short interconnect lines, doped oxides for universal flattening, polysilicon for gate contacts in transistors, nitrides for encapsulation, and other dielectric materials).

The disadvantages of LPCVD are as follows:

(i) The deposition rate in LPCVD is less than that in APCVD
(ii) There is a necessity for higher temperatures for the CVD operation.
(iii) The surplus material that resides after the operation is detrimental.

1.4.1.3 Metal Organic Chemical Vapor Deposition (MOCVD)

The credit for developing the MOCVD system was to go to Manasevit [16, 17]. Manasevit and Simpson [18] designed an early version of the MOCVD unit and use metal-organics in this unit successfully for the preparation of TFs of various semiconducting and crystalline materials.

In this CVD, the metal-organic compounds play the role of molecular precursors for coating deposition. Thus, we can create an inclusive diversity of TFs of different materials for novel scientific and technological applications. The most attractive benefit of this kind of CVD is the facility of high vaporousness of precursors at low reaction temperatures as compared to the other conventional CVDs. MOCVD is the perfect choice for the development of TFs of metals and different kinds of metallic compounds (e.g., metal nitrides and metal oxides).

The main components of a typical MOCVD system are (i) gas delivery and growth chambers, (ii) temperature controlling unit and heating arrangement, and (iii) leakage finding and protection unit. Figure 1.8 illustrates a diagram of a classical MOCVD system. For this kind of CVD, a gas chamber is used to deal with the metal-organic substances, carrier gas, and mass flow regulators. The widely used carrier gases in this kind of CVD are H_2, N_2, and He for the growth processes. The carrier gases are extremely cleaned to remove the impurities (e.g., water vapor). The metal-organic substances are confined in a chamber furnished with a temperature controlling unit. Two different types of heating units are used to heat the substrate throughout the growth process. The first and most common kind

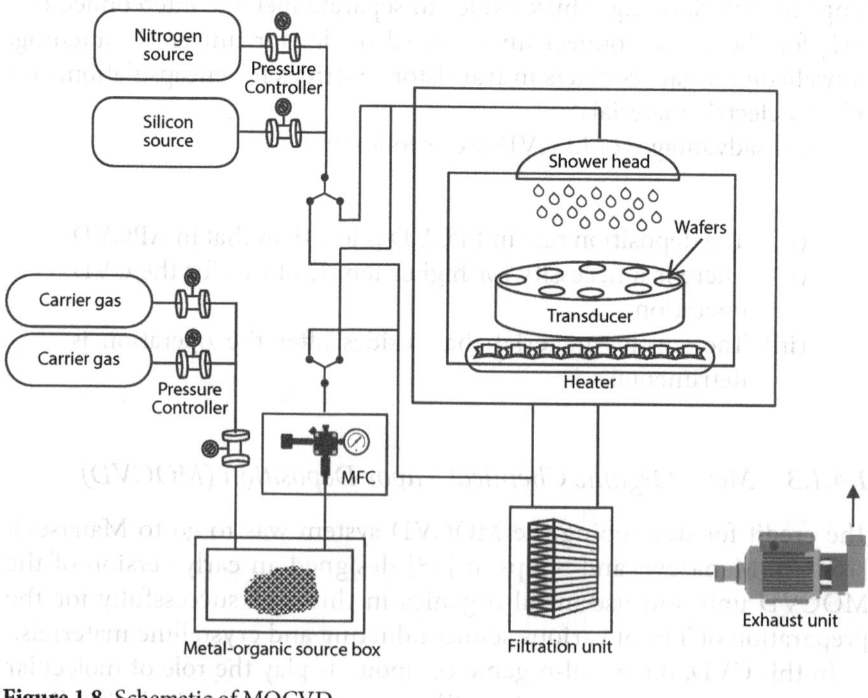

Figure 1.8 Schematic of MOCVD.

of heating unit is the radiofrequency heating unit that is widely used in large viable MOCVD systems. This unit employs the inductive coupling between a radio frequency coil and a transducer (i.e., substrates holder). The second kind of unit is the radiative heating unit that transforms the heat energy (created by the process of Joule heating of a current-carrying resistance element) into radiant energy. The transducer absorbs this radiant energy and retransforms it into heat energy. The protection unit employs the leakage finding detectors for the carrier gas, the purifier, and toxic gases.

The main shortcoming of MOCVD is its expansive design cost because it requires very expensive source material and highly precise pressure regulation for the purification of the system due to the high volatility of the precursors. The environmental disposal costs are also high for this kind of CVD.

1.4.1.4 Plasma Enhanced Chemical Vapor Deposition (PECVD)

Figure 1.9 shows the schematic diagram of PECVD. This kind of CVD is best for use in silicon industries for the coating deposition of the amorphous Si films and the wafers of silicon compounds (e.g., Silica, Silicon Nitride, and Silicon Oxynitride). The decomposition of such molecules can be done by using a plasma that shows a great resistance against the decomposition at an adequate value of the elevated temperature. It can also be used to decompose a thermally unstable molecule at a much lower temperature. Thus, PECVD has high reactivity and can process at lower temperatures and pressures. In plasma CVD, such materials (e.g., polymeric substances) might be employed as substrates that are unable to endure the high values of temperature (ranging between 373 and 773 K). The quality of deposited TFs is poorer than that in LPCVD.

1.4.1.5 Laser Chemical Vapor Deposition (LCVD)

This type of CVD employs a focused laser beam [19] for heating the wafers as shown in Figure 1.10. LCVD system has the facility to heat locally a portion of the wafer during the flow of the reactant gas [20]. Consequently, it induced the deposition of a coating by controlling locally the CVD reaction at the substrate's surface. This kind of CVD has several applications in the fabrication of the microscale solid patterns or 3D-structures on the wafers. In this type of CVD, two different mechanisms can be used for coating deposition [21–23]. These mechanisms are known as photolytic and pyrolytic

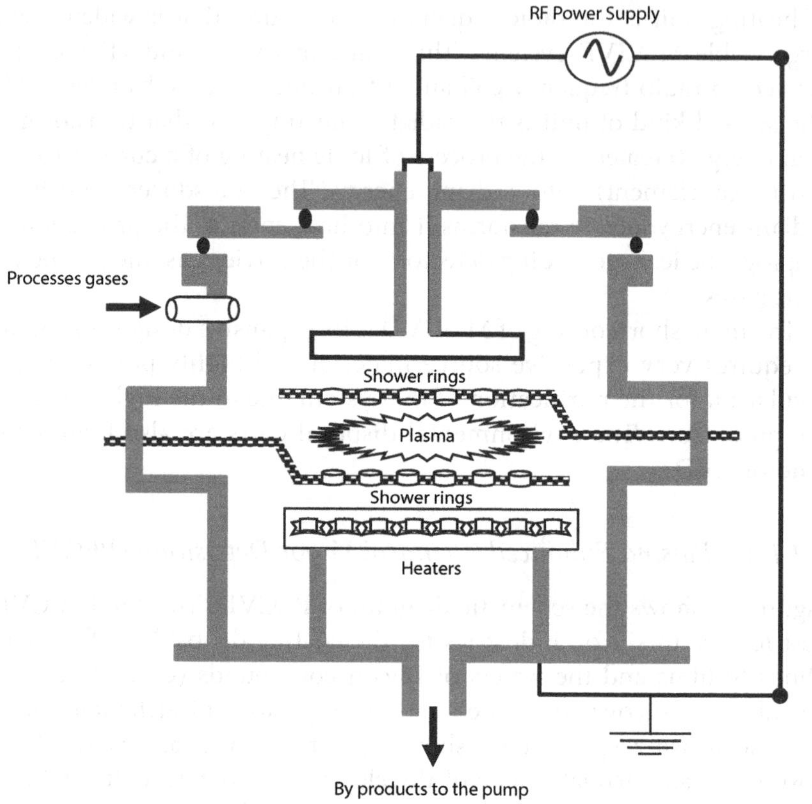

Figure 1.9 Schematic of PECVD.

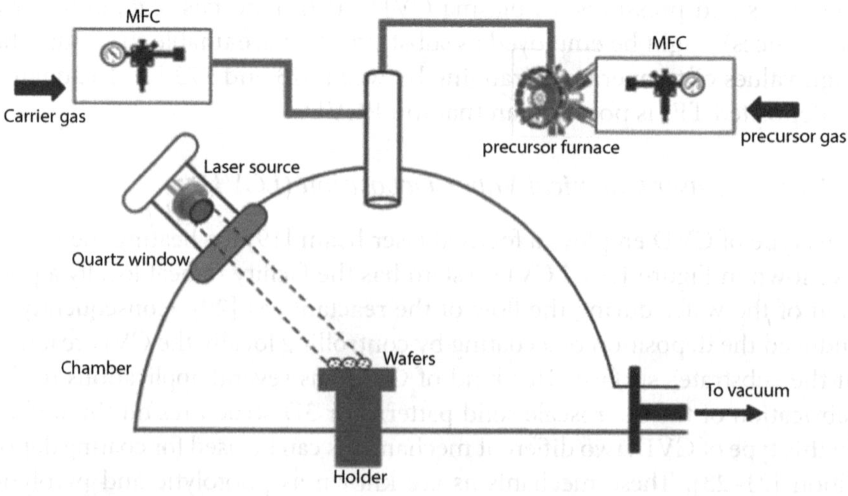

Figure 1.10 Schematic of LCVD.

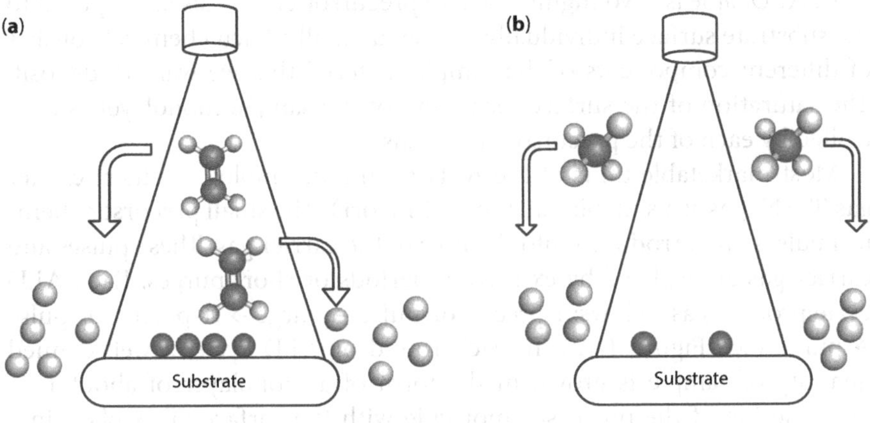

Figure 1.11 LCVD of carbon by using (a) photolytic mechanism and (b) pyrolytic mechanism.

mechanisms. In the photolytic mechanism, the laser-induced deposition of the reactant gas molecules happens directly. However, the rate of laser-induced deposition in the photolytic mechanism is slow. Thus, this mechanism is practicable when we want TF deposition without heating the wafers. This technique is used for the production and reparation of interconnecting lines or joints of integrated circuits [24]. The LCVD of carbon by the photolytic mechanism of ethylene gas precursor is shown in Figure 1.11a.

In the pyrolytic mechanism, the precursor gas is persuaded to deposit on the wafers indirectly by laser-induced heating of their surfaces. The rate of laser-induced deposition in this mechanism is comparatively higher than that in the photolytic mechanism. Thus, this technique is used for the freeform manufacturing of high shape-factor microstructures.

The LCVD of pyro-carbon by the pyrolytic mechanism of methane gas precursor is shown in Figure 1.11(b) through the dehydrogenation process.

1.4.2 Atomic Layer Deposition (ALD)

This technique is an exceptional type of CVD technique. In ALD, the successive and self-saturated chemical surface reactions are employed to obtain the materials in the expected TF form by flowing the precursors (i.e., gaseous reactants) into the reaction space. This method was developed by Suntola [25] in his pioneer work. They initially mentioned this technique as atomic layer epitaxy. A prominent characteristic of ALD is that the precursors are pulsed consecutively and detached by cleansing inert gas to keep away from gas-phase reactions.

In ALD, at least two highly reactive precursor chemicals are exposed to the substrate surface individually and sequentially. Each chemical consists of different components of the sample material that we want to deposit. The saturation of the surface formation of the sample monolayer is controlled by each of the precursor chemicals.

Most marketable units of the ALD technique employ the inert carrier gas like N_2 gas at a suitable vacuum (~1 mTorr). The small precursor chemical pulses are introduced into the flow of the carrier gas. These pulses and carrier gas are isolated by extraction periods or short purges. Each ALD cycle proceeds as follows: I-precursor-pulse → purge → II-precursor-pulse → purge (see Figure 1.12). In each round or ALD cycle, a well-defined quantity of sample is grown in the form of a monolayer of about 1 Å. The reaction of the precursor molecule with the surface takes place in a self-limiting mode for each alternate pulse. This confirms that the reaction halts after the use of all reactive sites on the surface of the substrate. A whole ALD cycle is designed by keeping in mind the behavior of the precursor-surface interaction. The ALD cycle can be executed several times to upsurge the layers of the TF, according to the necessity.

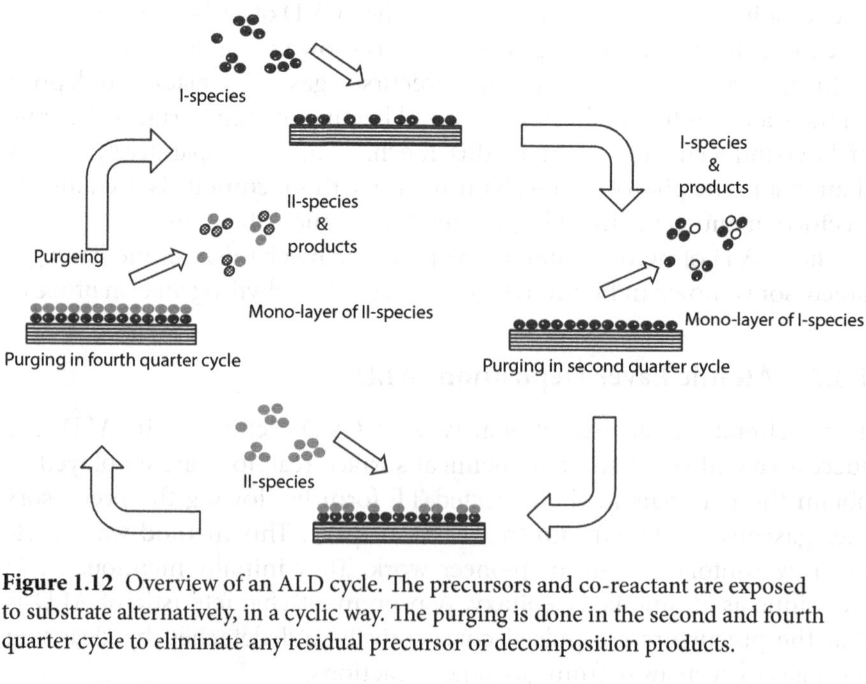

Figure 1.12 Overview of an ALD cycle. The precursors and co-reactant are exposed to substrate alternatively, in a cyclic way. The purging is done in the second and fourth quarter cycle to eliminate any residual precursor or decomposition products.

1.4.2.1 ALD Versus CVD Deposition Technique

The CVD technique is a more general term that is used for the coating deposition of specific materials on a particular substrate material from the gas phase by exploiting a chemical reaction. However, the ALD technique is a sub-group of the CVD technique, i.e., its working principle diverges from that of any kind of CVD technique because it employs a surface catalytic reaction with a gas-phase reactant that is referred a precursor. This precursor creates a deposition on the substrate under the influence of the interaction between a solid-state reactant and a cold or hot substrate. The surface reaction usually necessitates a dustless and impurity-free substrate surface because it proceeds across the catalytic locations on the surface. After the covering of the whole substrate's surface completely with the deposited material, the reaction halts because no additional dynamic catalytic sites are exposed to the gas-phase precursor. Therefore, the reaction is termed as self-terminate or self-limited. But the deposition might deliver the active locations for a catalytic reaction with other precursors, and so on. This consequence is put based on TF engineering with atomic layer accuracy and so the technique is named ALD.

1.4.3 Other Techniques Based on Chemical Deposition Processes

When the coating of material is deposited on an electrode under the influence of the electrolysis phenomenon, then the process obeys Faraday's laws. This is the working principle of electro-deposition or electroplating. In this technique, when an electric current passes in an electrolyte the chemical fluctuations take place [26, 27]. The extent of the chemical variation happening is proportional to the strength of electricity (i.e., the magnitude of electric current). The masses of diverse species are either deposited at electrodes or dissolved from the electrodes by the equal amount of current that varies in proportion to their equivalent weights. In electroplating, three electrodes are employed (see Figure 1.13a). The first electrode is the working electrode (WE) that is made of the sample of the corroding material. The second electrode is the reference electrode (RE) which is a voltage sensing electrode with a constant electrochemical potential. The third electrode is the counter electrode (CE) that is a current-carrying electrode and it completes the electric circuit. We cannot control the current flow through the RE, because all restrictions should be triggered by the desired process at the WE. That is why we use the CE.

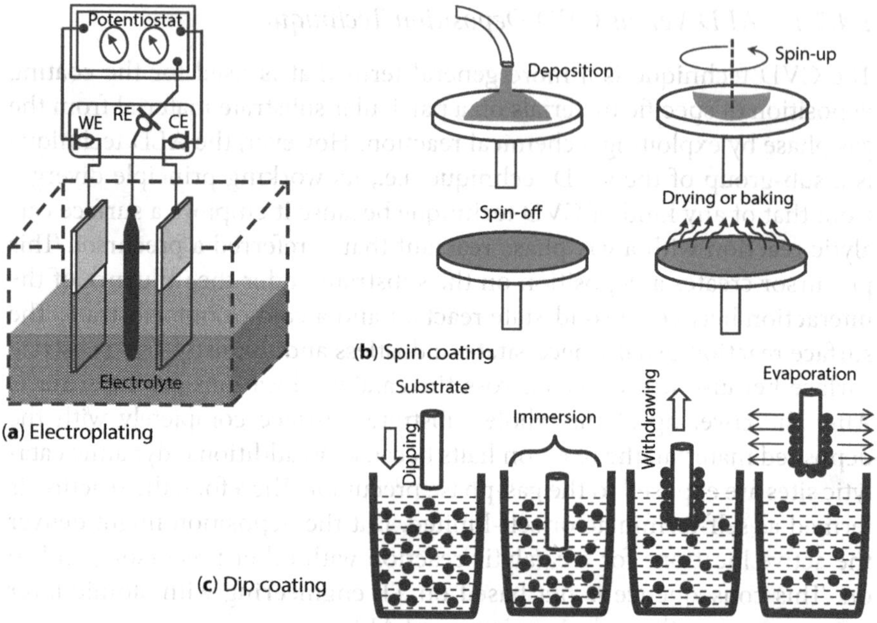

Figure 1.13 Representation showing the working of (a) electroplating deposition, (b) spin-coating deposition, and (c) dip-coating deposition.

All of these are linked with a device that maintains a constant potential during the process and controls the rate of electro-deposition [26]. Usually, a potentiometer is used for this purpose. Spin coating [28–30] is another simple chemical deposition technique that can be used for achieving a uniform layer of a variety of materials (mostly polymers) (see Figure 1.13b).

It is a more developed approach of fabricating a uniform thin liquid film on a specific type of substrate having a flat surface of small disc-shaped geometry. The nature of the elementary flow is unstable centrifugal drainage. In this flow, both viscous, as well as centrifugal, participate so much that a simple liquid having fixed viscosity bends toward a TF of good uniformity that grows constantly and gradually. The handling of the thickness control of TF depends on precise conditions. Spin coating is invariably used in the fabrication of photoresistors in photolithography.

The third straightforward and cost-effective technique is dip coating [29, 31] that comprises the four steps (see Figure 1.13c). In the first step, a substrate is dipped perpendicularly at a constant speed in a jar containing the coating solution. The dissolution of the substrate is classically performed vertically to achieve a symmetrical and uniform deposition. In the second step, the substrate leftovers in the solution for a chosen immersion time. In the third step, the coating deposition begins on the substrate when it is pulled out after the immersion time is passed out. The speed at which the substrate is immersed and pulled and immersion time are the key parameters that decide the thickness of the coatings. The fourth and last step includes the draining of the surplus liquid and the evaporation of the solvent from the surface of the substrate and we obtained the TF of the coating material. The multi-layer deposition is also possible in a controlled way by submerging the substrate multiple times. Such multi-layers with inconsistent layer substances may also be prepared by employing multi-vessel dip coaters with diverse dipping solutions.

1.5 Present Status and Future Scope

Currently, the coating deposition techniques are used in many applications, including microelectronics, optics, magnetic, hard resistant coatings, and micro-mechanics. Better results in these areas depend on the selective and controlled deposition. A typical plot is shown in Figure 1.14 that indicates the rapid rise in the market demand of TF-based devices based on different market surveys.

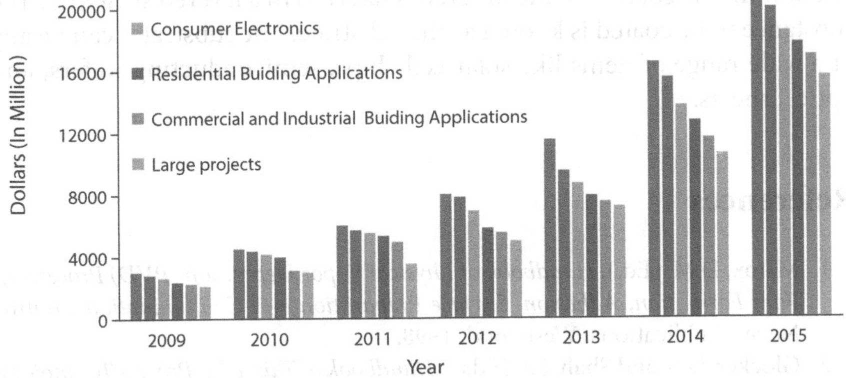

Figure 1.14 Plot showing the market survey and analysis of thin-film demand in various sectors.

The growth in the global market of TF-based devices was from \$3,000 million in 2009 to \$21,000 million in 2015 that is its compound annual growth rate was approximately 38.3% from 2009 to 2015. The comprehensive study of market size, stake, development, trends, price structure, and statistical data shows that the worldwide market of TF materials is projected to reach nearly \$31 billion by the year 2028.

1.6 Conclusion

The manufacturing processes involved in the coating deposition are at the center of modern semiconductor manufacturing units and the optical devices fabrication industries (e.g., compact disks, solar panels, integrated chips, and disk storage devices).

The deposition techniques of coatings are employed in an extensive span of applications for countless necessities. Such techniques are frequently utilized: (i) to prevent scratches on displays or to save them from environmental exposure and (ii) to generate a precise degree of reflectivity on the build layers of metallization, semiconductor wafers, and optical lenses. Some of the major markets that are aided by coating deposition techniques are optical and optoelectronic appliances, semiconducting devices, drugs and healthcare sector, and microscopical instruments.

Numerous approaches are used directly or indirectly for coating deposition with the help of vacuum techniques. These methods can be applied for the production of TFs of pure materials at the substrates of different kinds of substances. The coatings are frequently in the thickness range of angstroms to microns. The TF formed is either a single coating of a material or multiple coating of the different materials in a layered structure. The substance to be coated is known as the substrate. The substrates can be any of a wide range of items like solar cell chips, semiconducting wafers, and optical sheets.

References

1. Matox, D.M. (Ed.), *Handbook of Physical Vapor Deposition (PVD) Processing, Film Formation, Adhesion, Surface Preparation and Contamination Control*, Noyes, Publications, Westwood, 1998.
2. Glocker, D.B. and Shah, S.I. (Eds.), *Handbook of Thin Film Process Technology*, Institute of Physics Publishing, Bristol, 1995.

3. Carlsson, J.O. and Martin, P.M., Chemical Vapor Deposition, in: *Handbook of Deposition Technologies for Films and Coatings*, P.M. Martin (Ed.), pp. 314–363, Noyes Publications, William Andrew (Imprint), NY, USA, 2010.
4. Li, Z., Guo, Y., Zhao, F., Nie, C., Li, H., Shi, J., Liu, X., Jiang, J., Zuo, S., Effect of film thickness and evaporation rate on co-evaporated SnSe thin films for photovoltaic applications. *RSC Adv.*, 10, 16749, 2020.
5. Holland, L. and Steckelmacher, W., The distribution of thin films condensed on surfaces by the vacuum evaporation method. *Vacuum*, 2, 346, 1952.
6. Yoshino, H., Abbas, A., Kaminski, P.M., Smith, R., Walls, J.M., Mansfield, D., Measurement of thin film interfacial surface roughness by coherence scanning interferometry. *J. Appl. Phys.*, 121, 105303, 2017.
7. Faraday, M., Experimental relations of gold (and other metals) to light. *Philos. Trans. R. Soc. Lond.*, 147, 145, 1857.
8. M. von Pirani, Production of homogeneous bodies from tantalum or other metals, US Patent 848600, 1907.
9. R. Ruhle, Verfahren zur Bedampfung im Vakuum, Deutsche Reich Patent 764927, 1939.
10. Brown, I.G., Cathodic arc deposition of films. *Annu. Rev. Mater. Sci.*, 28, 243, 1998.
11. Sanders, D.M. and Anders, A., Review of cathodic arc deposition technology at the start of the new millennium. *Surf. Coat. Tech.*, 133–134, 78, 2000.
12. Farrow, R.F.C., Recent developments in molecular beam epitaxy. *J. Cryst. Growth*, 104, 556, 1990.
13. Joyce, B.A., Molecular beam epitaxy. *Rep. Prog. Phys.*, 48, 1637, 1985.
14. Arthur, J.R., Molecular beam epitaxy. *Surf. Sci.*, 500, 189, 2002.
15. Powell, C.F., Chemical Vapor Deposition, in: *Vapor Deposition*, C.F. Powell, J.H. Oxley, J.M. Blocher (Eds.), pp. 249–276, John Wiley & Sons, New York, 1966.
16. Manasevit, H.M., Single-crystal gallium arsenide on insulating substrates. *Appl. Phys. Lett.*, 12, 156, 1968.
17. Manasevit, H.M., The use of metalorganics in the preparation of semiconductor materials: Growth on insulating substrates. *J. Cryst. Growth*, 13–14, 306, 1972.
18. Manasevit, H.M. and Simpson, W.I., The Use of Metal-Organics in the Preparation of Semiconductor Materials: I. Epitaxial Gallium-V Compounds. *J. Electrochem. Soc.*, 116, 1725, 1969.
19. Ready, J.R. (Ed.), *Industrial applications of lasers*, Academic Press, New York, 1978.
20. Roy, S.K., Laser chemical vapour deposition. *Bull. Mater. Sci.*, 11, 129, 1988.
21. Beauerle, D. (Ed.), *Laser Processing and Chemistry*, Springer, Berlin, 2011.
22. Conde, O. and Silvestre, A.J., Laser-assisted deposition of thin films from photoexcited vapour phases. *Appl. Phys. A: Mater. Sci. Process.*, 79, 489, 2004.
23. Leyendecker, G., Beauerle, D., Geittner, P., Lydtin, H., Laser induced chemical vapor deposition of carbon. *Appl. Phys. Lett.*, 39, 921, 1981.
24. Deutsch, T.F., Ehrlich, D.J., Osgood, R.M., Laser photodeposition of metal films with microscopic features. *Appl. Phys. Lett.*, 35, 175, 1979.
25. Suntola, T., Atomic layer epitaxy. *Mater. Sci. Rep.*, 4, 261, 1989.

26. Besra, L. and Liu, M., A review on fundamentals and applications of electrophoretic deposition (EPD). *Prog. Mater. Sci.*, 52, 1, 2007.
27. Ojo, A.A. and Dharmadasa, I.M., Electroplating of Semiconductor Materials for Applications in Large Area Electronics: A Review. *Coatings*, 8, 262, 2018.
28. Sahu, N., Parija, B., Panigrahi, S., Fundamental understanding and modeling of spin coating process: A review. *Indian J. Phys.*, 83, 493, 2009.
29. Scriven, L., Physics and Applications of DIP Coating and Spin Coating. *MRS Proc.*, 121, 717, 1988.
30. Birnie, D.P., Spin Coating Technique, in: *Sol-Gel Technologies for Glass Producers and Users*, M.A. Aegerter and M. Mennig (Eds.), pp. 49–55, Springer, Boston, MA, 2004.
31. Puetz, J. and Aegerter, M.A., Dip Coating Technique, in: *Sol-Gel Technologies for Glass Producers and Users*, M.A. Aegerter and M. Mennig (Eds.), pp. 37–48, Springer, Boston, MA, 2004.

2

Tribological Investigation of Developed Nanocomposite Mos_2-Tio_2-Zro_2 Coating Material

Avinash V. Borgaonkar[1], Deepak Rai[2] and Ismail Syed[1]*

[1]*Dept. of Mechanical Engineering, National Institute of Technology Warangal, Warangal, Telangana, India*
[2]*Collins Aerospace, Bengaluru, Karnataka, India*

Abstract

The bonded MoS_2 is widely used as solid lubricant in different applications due to their superior tribological properties. However, the pure MoS_2 coating exhibits poor bond strength. In the present work, an effort has been made in order to improve the bond strength and endurance life of the pure MoS_2 coating by reinforcing different elements such as TiO_2 and ZrO_2 into the MoS_2 base matrix. The composite MoS_2-TiO_2-ZrO_2 coatings were developed with different wt. % of the compounding elements. Before deposition of the coating film, the surface of the substrate was pre-treated by employing phosphating process which enhances the bond strength. The composite MoS_2-TiO_2-ZrO_2 coating bonded onto a pre-treated steel substrate employing sodium silicate binder. The experimental study was performed on the pin-on-disc friction and wear test rig at various loads and speeds. The effect of different wt. % addition of the compounding elements on the frictional and wear behavior of the composite MoS_2-TiO_2-ZrO_2 coating has been studied. The results obtained for composite MoS_2-TiO_2-ZrO_2 coating were compared with pure MoS_2 coating for the same operating conditions. The test results shown that coating with sample B (i.e., MoS_2 - 60% and TiO_2-ZrO_2 - 20% each) depicts lowest wear rate and COF of around 0.12 (mm^3/m) and 0.52 respectively. The endurance life of the composite coating has also been studied. The test results exhibit that composite coating withstands up to a sliding distance of 7,400 m.

**Corresponding author*: syedismail7@nitw.ac.in

Sarfraj Ahmed and Vinayak S. Dakre (eds.) Tribology and Characterization of Surface Coatings, (33–48) © 2022 Scrivener Publishing LLC

***Keywords*:** Composite, coating, MoS_2, solid lubricant, TiO_2, tribological properties, ZrO_2

2.1 Introduction

There is a growing interest on usage of solid lubricants, which exhibit superior tribological performance in critical and adverse environmental conditions like high temperature and high load, ultralow temperature and vacuum, clean and dry lubrication, protection against contamination, and corrosion [1, 2]. The solid lubricant gets shear out easily due to which it offers better lubrication and possess high load carrying capacity. The various types of dry lubricants include layered inorganic compounds (MoS_2, graphite, and HBN), soft metals (In, Ag, Sn, Pb, and Au) and polymers (Teflon) [3]. The molybdenum disulphide (MoS_2) offers low friction because of its anisotropic layered structure, i.e., weak Van der Waal forces and covalent bonding between the subsequent lamellae as shown in Figure 2.1. Due to this, for many years, MoS_2 has been considered as a popular solid lubricant. Its natural shearing ability and lamellar structure provides better tribological performance [4]. As per the literature many researchers have used MoS_2 a solid lubricant over the past four decades [5] and still many researchers interested to work out on it, for space applications [6–8].

Most commonly, the solid lubricants, especially MoS_2, have been deposited employing various methods like bonding, burnishing, particle embodiment, and thin-film coating [9, 10]. In the recent, several researchers have studied the performance of the bonded solid lubricants (BSL). The BSL are cheaper and easy to deposit and exhibit excellent tribological properties. A significant amount of reduction in friction coefficients of the novolac-epoxy bonded MoS_2 compared to uncoated specimen have been reported by Luo *et al.* [11] under fretting wear test. Vadiraj *et al.* [12] carried out frictional and wear analysis of various solid lubricants such as graphite, MoS_2, boric acid, and TiO_2 using pin-on-disc friction and wear test rig. The experimental results shown that in case of MoS_2 and graphite about 30% to 50% reduction in wear rate was noted in comparison with remaining solid lubricants at same operating conditions. In case of boric acid, higher COF and wear rate were observed due to transformation of boric acid into erosive boric oxide. Asmoro *et al.* [13] investigated the effect of various solid lubricants like graphite and MoS_2 on brake lining composite at different operating parameters such as contact pressure and sliding speeds. The test result shows that MoS_2 provide excellent tribological performance in comparison with the graphite at different load and speed conditions. Savan *et al.*

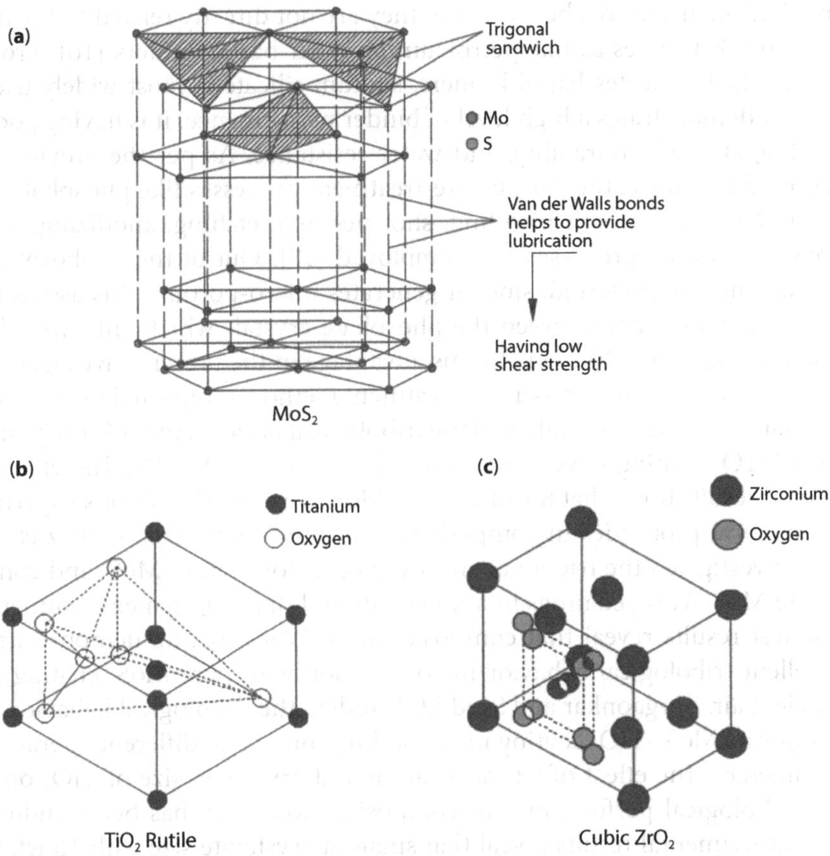

Figure 2.1 The crystallographic structure of (a) MoS_2, (b) TiO_2, and (c) ZrO_2.

[14] demonstrated from their experimental study that application of multi-layered coating assists to enhance adhesion of the soft coating. The tribological study reveals that, due to application soft MoS_2 coating contacting surfaces exhibits lower COF as compared to bare surface [15].

As the direct deposition of the solid lubricant onto the contacting surface lead to weak adhesion. Numerous researchers started to employ a bonding agent and/or pre-treatment onto substrate surface, which enhances the bonding between coating and substrate surface. The materials used as bonding agents are organic as well as inorganic binding agents, resins, and silicates. From the past few decades, soluble silicates have been mostly used as binders in many applications. Since problem-related to handling, safety and environmental issues which are associated with nitrogen oxides (NOX) and volatile organic compounds (VOC) are not in case of soluble silicates.

In addition, these are cheaper since they are not directly related to fluctuating market prices as like petroleum or sugar-based binders [16]. From the available silicates based binders, sodium silicate is most widely used since it demonstrates a high level of binder performance. It is having good binding strength, durability, and water resistance. As per the previously reported literature, the various pre-treatment processes like phosphating, sand-blasting, salt-bath nitriding, shot peening, etching, anodizing, and abrasive blasting processes were employed [17]. Out of these, phosphating was mostly preferred, since it generates micro-porous. This assists to entrap the lubricant between the phosphate crystals which enhances the bond strength [18–20]. Due to this advantage in the present investigation phosphating was chosen as a pre-treatment method as reported in [21–24].

Shankara *et al.* [25] evaluated the tribological performance of composite MoS_2/ZrO_2 coating developed with different wt. % of ZrO_2. The experimental result shows that 8% of ZrO_2 in MoS_2 base matrix exhibits superior tribological properties in comparison with pure MoS_2. Dominguez *et al.* [26] investigated the frictional and wear behavior of pure MoS_2 and composite MoS_2-WSe_2 coatings in ambient air and dry nitrogen environment. The test results reveal that composite MoS_2-WSe_2 coating demonstrated excellent tribological behavior in comparison with pure MoS_2 coating in ambient air. Borgaonkar and Syed [27] studied the tribological behavior of composite MoS_2-TiO_2 coating in dry sliding contacts at different operating parameters. The effect of wt. % addition and crystallite size of TiO_2 onto the tribological performance of composite MoS_2-TiO_2 has been studied. The experimental results reveal that smallest crystallite size with 15 wt. % addition of TiO_2 improves the tribological performance significantly in comparison with pure MoS_2 coating. Further, the effect of temperature has been studied on the performance of composite MoS_2-TiO_2 coating [28]. The test results showed that at higher temperature, i.e., at 400°C, the COF has been reduced drastically but at the same time wear rate also increased. This may be due to the oxidation process.

Recent developments in the area of tribological evaluation of the MoS_2 coating with various compositions of metals (MoS_2/Pb-Ti and MoS_2-Ti), selenide (MoS_2-WSe_2), and oxides (MoS_2-ZnO and MoS_2-ZrO_2) have been carried out. There is very limited study available on the tribological investigation of composite coated contacting surfaces (with different wt. % addition of reinforcement material into MoS_2 base matrix) operating at various loads and sliding speeds, which need to further investigate. In this study, a tribological investigation of composite MoS_2-TiO_2-ZrO_2 coating material (with different wt. % addition of TiO_2 and ZrO_2 into MoS_2 base matrix) was conducted using pin-on-disc tribometer.

2.2 Materials and Methods

AISI 52100 steel is used as a substrate pin material (having dimensions of 12-mm diameter and 25-mm length). Before experimentation, the specimens were polished using SiC paper (from 200 to 800 grit size), cleaned with acetone using ultra-sonication and further cured in ambient condition. The phosphating was employed as a pre-treatment for pin samples to improve the bond strength between substrate and coating. For the development of pure MoS_2 coating, MoS_2 (70–90 nm) powder was bought from Himedia Laboratories, Mumbai, India. For preparing composite coating, TiO_2 powder (with size 30–80 nm), and ZrO_2 powder (with size 40–70 nm) were procured from Swastik Chemicals Ltd., Nagpur, India. In the present study, sodium silicate (Na_2SiO_3) is used as a binder to hold these powders together. The gel type mixture of powders and binder was bonded onto the pre-treated steel substrate samples by bonding technique [25]. The different steps involved in the preparation of coated specimens are depicted in Figure 2.2. The wt. % of TiO_2 and ZrO_2 changed from 10% to 30% into the MoS_2 matrix and accordingly the proportion of Na_2SiO_3 was varied. Further, these specimens were cured at 170°C for 2 hours in the furnace.

The coating samples prepared with different compositions are depicted in Table 2.1.

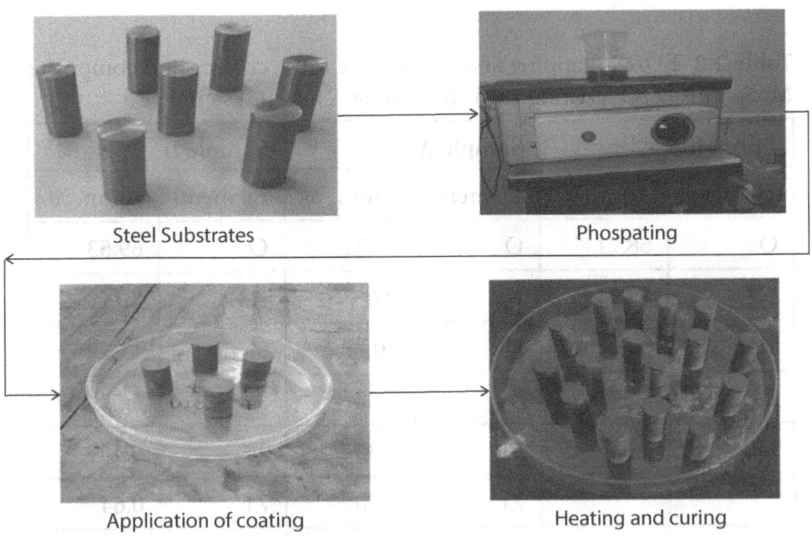

Figure 2.2 Steps for preparing the coated specimens.

Table 2.1 Coating samples with different compositions.

Sample	Weight %		
	MoS_2	TiO_2	ZrO_2
0	100	0	0
A	80	10	10
B	60	20	20
C	40	30	30

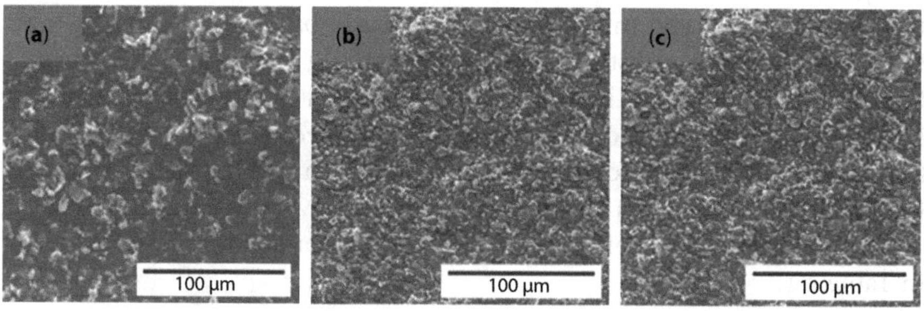

Figure 2.3 (a) Pure MoS_2 coating and composite MoS_2-TiO_2-ZrO_2 coating (b) with sample A and (c) with sample B.

Table 2.2 EDAX mapping analysis of pure MoS_2 coating and composite MoS_2-TiO_2-ZrO_2 coating with (B) sample A and (C) sample B.

Pure MoS_2		Sample A		Sample B	
Element	Atom. %	Element	Atom. %	Element	Atom. %
O	68.71	O	66.12	O	69.53
Na	13.31	Na	8.05	Na	8.11
Si	9.60	Si	10.80	Si	11.79
S	5.33	S	9.04	S	5.77
Mo	3.04	Ti	0.35	Ti	0.40
		Zr	0.70	Zr	0.64
		Mo	4.94	Mo	3.76

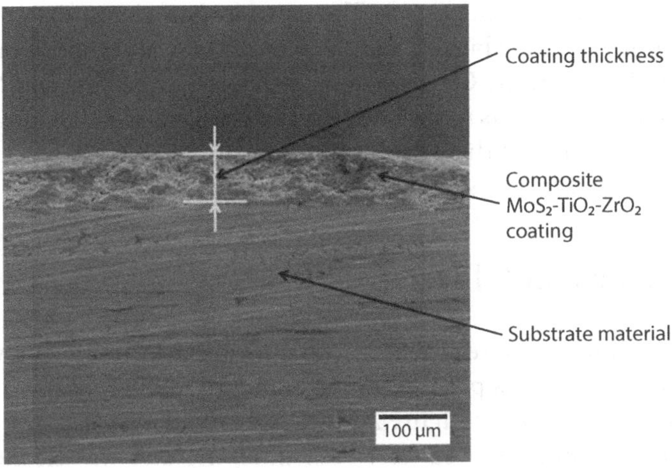

Figure 2.4 Cross-section of the composite MoS_2-TiO_2-ZrO_2 coating.

The morphology of pure MoS_2 and composite MoS_2-TiO_2 coating with different wt. % addition of TiO_2 and ZrO_2 particles was characterized by SEM, as shown in Figure 2.3. The MoS_2 particles have an average size of 90 nm and TiO_2 particles have an average size of 70 nm a spherical shape and contain a

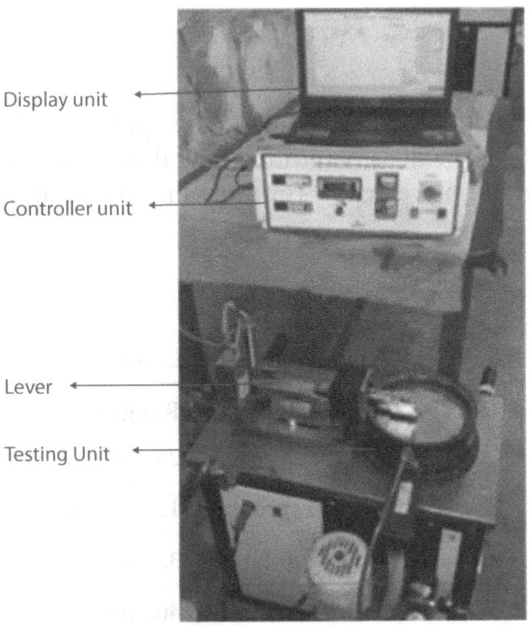

Figure 2.5 Experimental setup for tribological tests.

lot of nanocrystals in each particle. These nanocrystals intermixed with each other in sodium silicate binder. Table 2.2 shows typical compositional mappings of Na Si, Mo, S, and O elements in the pure MoS_2 coating and Mo, S, Ti, O, Na Si, and Zr elements in the composite MoS_2-TiO_2-ZrO_2 coating.

The cross-section of the deposited composite MoS_2-TiO_2-ZrO_2 coating on the substrate surface is depicted in Figure 2.4.

2.3 Experimental Details

The tribological tests were conducted on pin-on-disk friction and wear test rig (Magnum, India) as per the ASTM Standard G99-95 [29]. An experimental set up is shown in Figure 2.5.

After performing the wear test each time, the disc was cleaned using acetone. The test parameters used for tribological evaluation were mentioned in Table 2.3.

The COF and wear rates for each test were evaluated from the below mentioned equations.

$$COF = \frac{F_f}{F_n} \tag{2.1}$$

$$WR = \frac{\Delta W}{\rho \times d_s} \tag{2.2}$$

where F_f is the frictional force (N), F_n is the applied load (N), ΔW is the weight loss (g), ρ is the density (g/mm^3), and d_s is the sliding distance (m), calculated from sliding velocity v (m/min) and sliding time t (min).

Table 2.3 Tribological test condition.

Sr. no.	Parameters	Ranges
1	Load	25, 50, 75 N
2	Sliding speed	1, 2, 3 m/sec
3	Sliding distance	3,000 m
4	Track radius	40 mm

2.4 Results and Discussions

From the experimental results depicted in Figures 2.6 and 2.7, it was observed that COF and wear rate decrease with increase in both load and sliding speed. The composite coating, i.e., sample B exhibits excellent tribological properties in comparison with other samples. The pure MoS_2 coating (designated by 0 wt. % of TiO_2 and ZrO_2) demonstrated higher COF and wear rate due to poor bonding between the composite coating and the substrate. The obtained results have a number of similarities with the previous literature [25, 30, 31], which confirms that pure MoS_2 coating possesses low wear resistance causes to significant increase in wear rate.

The inclusion of TiO_2 into the MoS_2 base matrix enhances the tribological properties of the tribological pairs (i.e., samples A and B). However, the excess concentration of doping material results into inappropriate mixing between the doping material and the base matrix which results into poor bond strength between them. This, in turn, deteriorates the tribological performance of the composite coating [32].

Figure 2.8 demonstrates the pin surfaces after completion of tribological test with pure MoS_2 and composite coating with samples A, B, and C. The test results shown that with the addition of each TiO_2 and ZrO_2 upto 20 wt. %, the coating even remain exist on the substrate surfaces. The coating firmly adheres to the substrate surfaces as represented in Figures 2.8b and C. This resulted into lower magnitudes of wear rate and COF. As the wt. % of TiO_2 and ZrO_2 rises beyond 20%, the surplus quantity not well mixed into MoS_2 matrix, which results in poor bond strength between them as reported by Essa *et al.* [33]. Eventually this, results into higher wear rate of the coating as depicted in Figure 2.8d obtained from the microscope. This resulted into an increased in wear rate and COF.

The morphologies of different samples after wear test are depicted in the Figure 2.8. In case of pure MoS_2 coating and composite MoS_2-TiO_2-ZrO_2 coating sample C (refer Figures 2.8a, b, d), detachment of particles took place on the whole contact surfaces and a lot of wear debris adhered to the worn surface. The detached particles of coating trapped into wear track and also thrown away form the contacting surfaces. The oxidation transformation of MoS_2 into MoO_3 was caused by both mechanical and thermal effect, which results into the ruptures and detachments of coating from the substrate surface. Whereas in case of coating with composite MoS_2-TiO_2-ZrO_2 coating sample C (refer Figure 2.8c), experimental test results reveal that coating is still present on the substrate surface due to which it depicts lowest COF

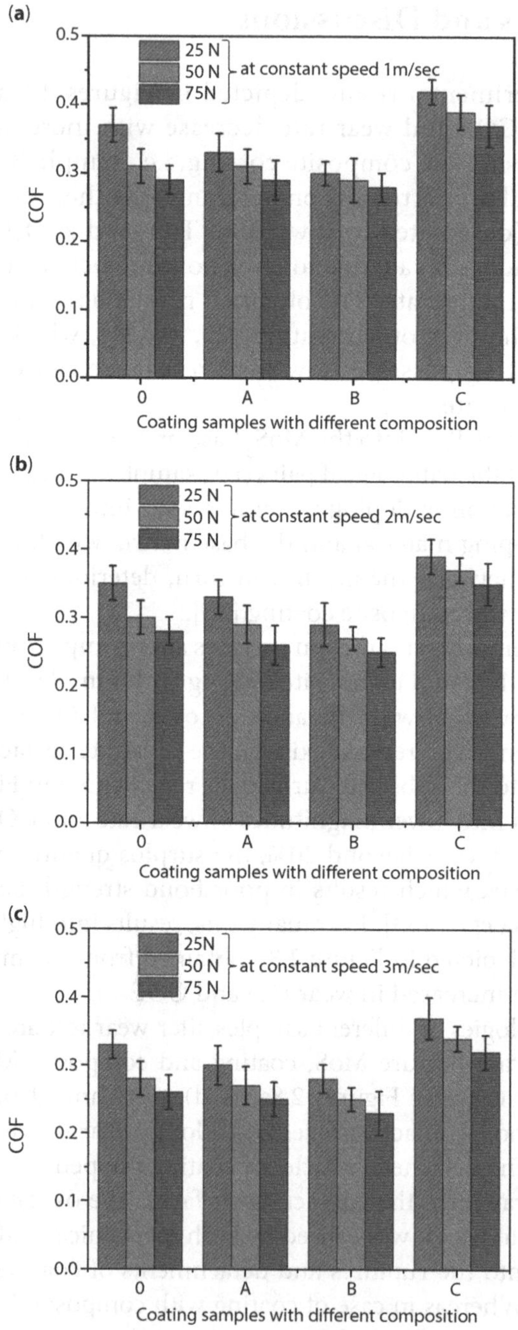

Figure 2.6 COF at load 25, 50, and 75 N at different sliding speed (a) 1, (b) 2, and (c) 3 m/sec.

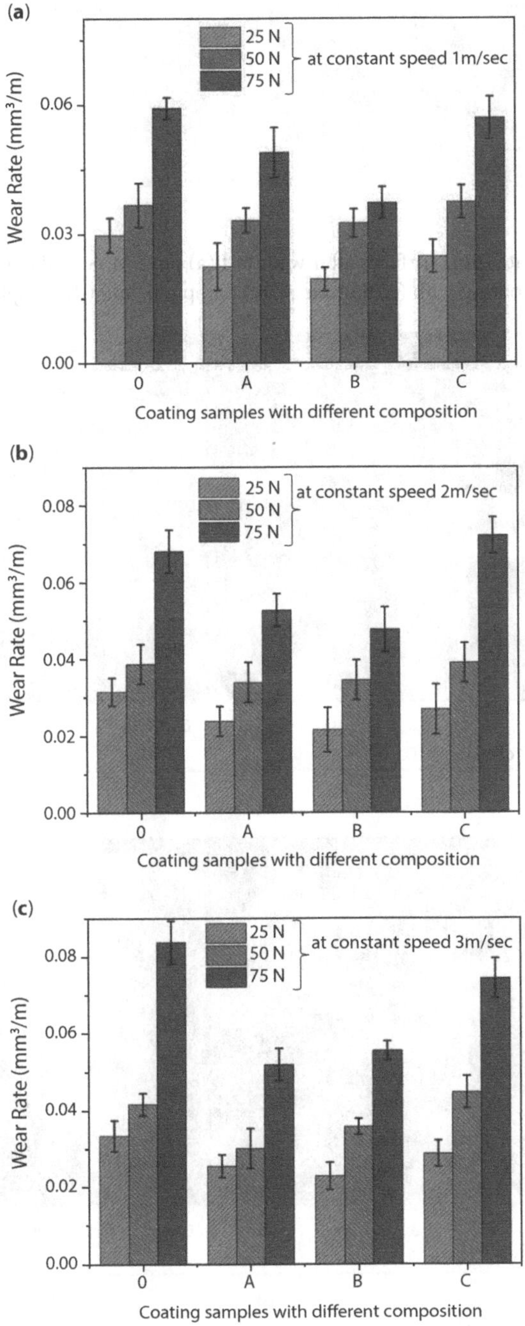

Figure 2.7 Wear rate (mm3/m) at load 25, 50, and 75 N at different sliding speed (a) 1, (b) 2, and (c) 3 m/sec.

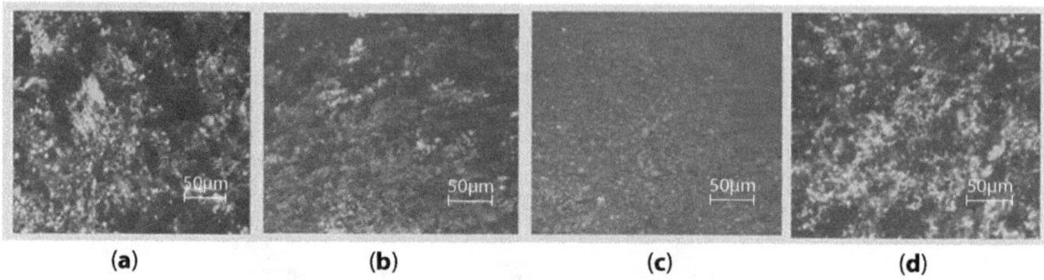

Figure 2.8 Substrate (pin) surfaces after wear test (a) pure MoS_2 and composite MoS_2-TiO_2-ZrO_2 coating with (b) Sample A, (c) Sample B, and (d) Sample C.

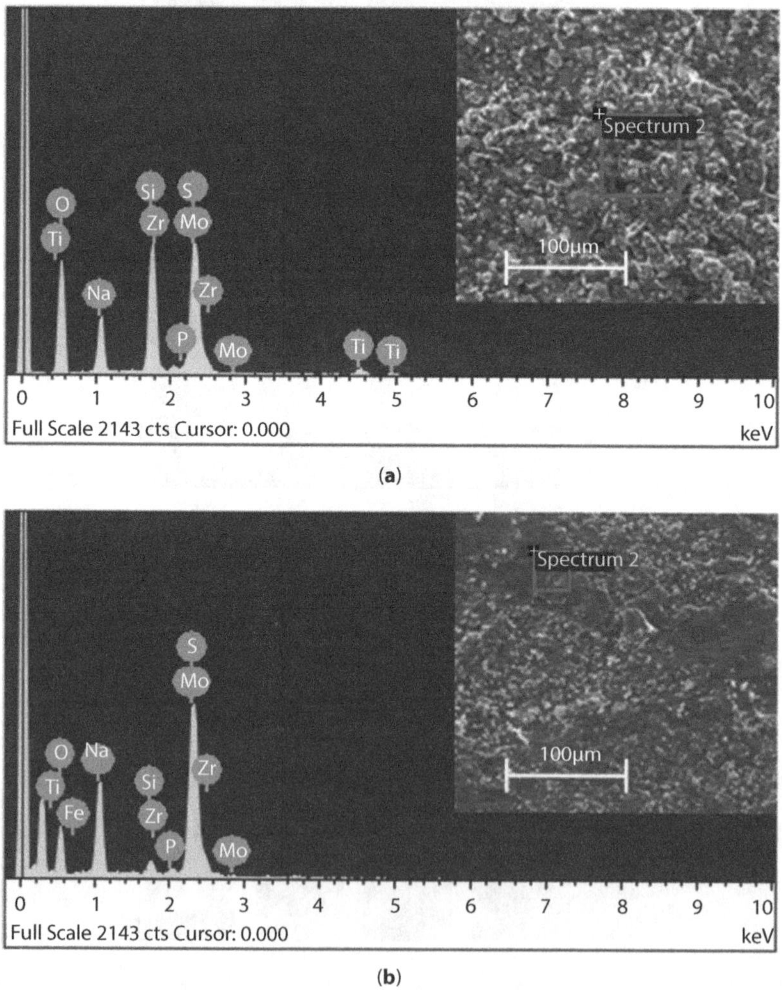

Figure 2.9 SEM/EDAX spectrums for composite MoS_2-TiO_2-ZrO_2 coating (a) before wear test and (b) after wear test.

and wear rate. Moreover, to determine the endurance life of the composite MoS_2-TiO_2-ZrO_2 coating sample C, numbers of trial tests were performed.

The endurance life of the composite coating investigated from SEM/EDAX analysis as shown in Figure 2.9. The SEM/EDAX analysis was executed before and after wear test for a coating which provide mapping of different elements present into composite MoS_2-TiO_2-ZrO_2 coating. The spectrum (a) shows various elements constituting for coatings like Mo, O, Ti, Na Zr, Si, and P, whereas the spectrum (b) represents few traces of Fe in conjunction with other listed elements.

This clearly demonstrates that composite MoS_2-TiO_2-ZrO_2 coating wears out from the substrate surface and asperity contact takes place within the contacting surfaces. The test results indicate that it has withstand upto 7,400 m before failure wear rate and COF were around 0.12 (mm^3/m) and 0.52, respectively.

2.5 Conclusions

The tribological evaluation of the composite MoS_2-TiO_2-ZrO_2 coating at various loads, sliding speed, and coating compositions has been examined. From the number of experimental investigations, the following conclusions have been drawn.

- With the increase in both applied load and sliding speed, COF and wear rate decrease.
- The introduction of TiO_2 and ZrO_2 into the MoS_2 base matrix results into improved bond strength between them without affecting the lubricating property of MoS_2. This resulted into a greater improvement in the endurance life of the composite MoS_2-TiO_2-ZrO_2 coating.
- Among all samples of the composite MoS_2-TiO_2-ZrO_2 coating, the sample B demonstrated the lower wear rate and COF values.

References

1. Bhushan, B., *Modern Tribology Handbook*, CRC Press, London, New York, Washington, D.C., 2000.
2. Ingole, S.P., Menezes, P.L., Nosonovsky, M., Lovell, M.R., Kailas, S.V., *Tribology for Scientists and Engineers: From Basics to Advanced Concepts*, Springer, New York, 2013.

3. Donnet, C. and Erdemir, A., Solid lubricant coatings: recent developments and future trends. *Tribol. Lett.*, 17, 389, 2004.
4. Ilie, F. and Tita, C., Modelling and Experimentation of Solid Lubrification with Powder MoS_2 through Self-Repairing and Self-Replenishing. *Adv. Mater. Res.*, 463, 1120, 2012.
5. Lansdown, A.R., *Molybdenum disulphide lubrication*, Elsevier Science, UK, 1999.
6. Colas, G., Saulot, A., Regis, E., Berthier, Y., Investigation of crystalline and amorphous MoS_2 based coatings: towards developing new coatings for space applications. *Wear*, 330, 448, 2015.
7. Wang, P., Qiao, L., Xu, J., Li, W., Liu, W., Erosion mechanism of MoS_2-based films exposed to atomic oxygen environments. *Appl. Mater. Interfaces*, 7, 12943, 2015.
8. Tedstone, A.A., Lewis, D.J., Hao, R., Mao, S.M., Bellon, P., Averback, R.S., Warrens, C.P., West, K.R., Howard, P., Gaemers, S., Dillon, S.J., Mechanical properties of molybdenum disulfide and the effect of doping: an *in situ* TEM study. *Appl. Mater. Interfaces*, 7, 20829, 2015.
9. Totten, G.E., Solid lubricants, in: *ASM Handbook*, ASM International, Ohio, USA, 2017.
10. Vazirisereshk, M.R., Martini, A., Strubbe, D.A., Baykara, M.Z., Solid Lubrication with MoS_2: A Review. *Lubricants*, 7, 1, 2019.
11. Luo, J., Zhu, M.H., Wang, Y.D., Zheng, J.F., Mo, J.L., Study on rotational fretting wear of bonded MoS_2 solid lubricant coating prepared on medium carbon steel. *Tribol. Int.*, 44, 1565, 2011.
12. Vadiraj, A., Kamaraj, M., Sreenivasan, V.S., Effect of solid lubricants on friction and wear behaviour of alloyed gray cast iron. *Sadhana*, 37, 569, 2012.
13. Asmoro, G., Surojo, E., Ariawan, D., Muhayat, N., Raharjo, W.W., Role of solid lubricant (MoS_2 and graphite) variations on characteristics of brake lining composite. *Mater. Sci. Eng.*, 420, 1, 2018.
14. Savan, A., Pfluger, E., Voumard, P., Schroer, A., Simmonds, M., Modern solid lubrication: recent developments and applications of MoS_2. *Lubr. Sci.*, 12, 185, 2000.
15. Borgaonkar, A.V. and Syed, I., Effect of coatings on rolling contact fatigue and tribological parameters of rolling/sliding contacts under dry/lubricated conditions: a review. *Sadhana*, 45, 1, 2020.
16. McDonald, M., Hamilton, J., Thompson, J., Recent Developments in Soluble Silicate Based Binders and Coatings. *Biennial Conference-Institute of Briquetting and Agglomeration*, vol. 30, p. 1, 2007.
17. Rovani, A.C., Kouketsu, F., da Silva, C.H., Pintaude, G., Surface Characterization of Three-Layer Organic Coating Applied on AISI 4130 Steel. *Adv. Mater. Sci. Eng.*, 1, 1, 2018.
18. Qiu, M., Lu, J., Li, Y., Lv, G., Investigation on MoS_2 and graphite coatings and their effects on the tribological properties of the radial spherical plain bearings. *Chin. J. Mech. Eng.*, 29, 844, 2016.

19. Duszczyk, J., Siuzdak, K., Klimczuk, T., Strychalska-Nowak, J., Zaleska-Medynska, A., Manganese phosphatizing coatings: the effects of preparation conditions on surface properties. *Materials*, 11, 1, 2018.
20. Rajagopal, C. and Vasu, K.I., *Conversion Coatings: A Reference for Phosphating, Chromating, and Anodizing Processes*, Tata McGraw-Hill, New Delhi, 2000.
21. Sankara Narayanan, T.S.N., Surface pre-treatment by phosphate conversion coatings-A review. *Rev. Adv. Mater. Sci.*, 9, 130, 2005.
22. Quitmeyer, J., Grasping the phosphate coating process: A properly maintained solution is key to achieving adhesion, corrosion protection, and lubrication. *Met. Finish.*, 104, 47, 2006.
23. Fouladi, M. and Amadeh, A., Effect of phosphating time and temperature on microstructure and corrosion behavior of magnesium phosphate coating. *Electrochim. Acta*, 106, 1, 2013.
24. Pokorny, P., Tej, P., Szelag, P., Discussion about magnesium phosphating. *Metalurgija*, 55, 507, 2016.
25. Shankara, A., Menezes, P.L., Simha, K.R.Y., Kailas, S.V., Study of solid lubrication with MoS_2 coating in the presence of additives using reciprocating ball-on-flat scratch tester. *Sadhana*, 33, 207, 2008.
26. Dominguez-Meister, S., Rojas, T.C., Brizuela, M., Sánchez-Lopez, J.C., Solid lubricant behavior of MoS_2 and WSe_2-based nanocomposite coatings. *Sci. Technol. Adv. Mater.*, 18, 122, 2017.
27. Borgaonkar, A. and Syed, I., Friction and wear behavior of composite MoS_2-TiO_2 coating material in dry sliding contact. *J. Braz. Soc. Mech. Sci. Eng.*, 43, 1–13, 2020, DOI: 10.1007/s40430-020-02721-8.
28. Borgaonkar, A.V. and Syed, I., Effect of temperature on the tribological performance of MoS2-TiO2 coating material, in: *Advances in Applied Mechanical Engineering. Lecture Notes in Mechanical Engineering*, H. Voruganti, K. Kumar, P. Krishna, X. Jin (Eds.), p. 611, Springer, Singapore, 2020.
29. ASTM Standard G99: Standard Test Method for Wear Testing with a Pin-on-Disk Apparatus, ASTM international, West Conshohocken, Pennsylvania, p. 1, 2010.
30. Tamilselvi, M., Kamaraj, M.A., Devikala, S., Selvi, J.A., Progress in zinc phosphate conversion coatings: a review. *Int. J. Adv. Chem. Sci. Appl.*, 3, 25, 2015.
31. Shang, K., Zheng, S., Ren, S., Pu, J., He, D., Liu, S., Improving the tribological and corrosive properties of MoS2-based coatings by dual-doping and multilayer construction. *Appl. Surf. Sci.*, 437, 233, 2018.
32. Ding, X.Z., Zeng, X.T., He, X.Y., Chen, Z., Tribological properties of Cr- and Ti-doped MoS_2 composite coatings under different humidity atmosphere. *Surf. Coat. Technol.*, 205, 224, 2010.
33. Essa, F.A., Zhang, Q., Huang, X., Ali, M.K.A., Elagouz, A., Abdelkareem, M.A., Effects of ZnO and MoS_2 solid lubricants on mechanical and tribological properties of M50-steel-based composites at high temperatures: experimental and simulation study. *Tribol. Lett.*, 65, 1, 2017.

3

Methods of Microstructural Characterization of Coatings

A. M. Kamalan Kirubaharan*, D. Dinesh Kumar, K. Gobi Saravanan and Subhenjit Hazra

Center for Nanoscience and Nanotechnology, Sathyabama Institute of Science and Technology, Jeppiaar Nagar, Chennai, India

Abstract

This chapter describes the microstructural characterization of coatings prepared through physical vapor deposition (PVD) route. Coatings prepared through PVD are extensively used to enhance the tribological performance of forming tools, cutting tools, and machine components. In such applications, the microstructure of coatings are the most important factors widely used to evaluate the product and materials property. Recently, considerable attention has given to focus on the microstructure at the micro/nano level to clarify and to study the influence of microstructure in tribological performance of the coating. The primary characterization techniques used for the microstructural analysis of the coatings are X-ray diffraction (XRD), Raman spectroscopy, scanning electron microscopy (SEM) coupled with energy dispersive X-ray spectroscopy (EDS), atomic force microscopy (AFM), transmission electron microscopy (TEM), and X-ray photoelectron spectroscopy (XPS). These techniques are particularly useful in assisting the development of novel coatings by studying their microstructure at different scales.

***Keywords*:** Microstructure, X-ray diffraction, Raman spectroscopy, tribochemistry, microscopic analysis

3.1 Introduction

The application of hard, thin coatings in tribological applications has become more extensive. There is a huge demand for increasing the durability

**Corresponding author*: kamalankiruba@gmail.com

Sarfraj Ahmed and Vinayak S. Dakre (eds.) Tribology and Characterization of Surface Coatings, (49–70) © 2022 Scrivener Publishing LLC

of tool components through physical vapor deposition (PVD) route, triggering the development of several advanced coatings in the market. In recent days, hard tribological coatings are deposited through PVD or chemical vapor deposition (CVD) are widely used to improve the wear resistance in many industrial applications. For example, titanium nitride (TiN), chromium nitride (CrN), titanium aluminium nitride (TiAlN), carbon-based coatings [diamond-like carbon (DLC) and micro/nano crystalline diamond], and multilayered/compositionally graded coating structures are currently focused in various research laboratories.

The tribological performance of the coated component is based on its microstructure, design, environment (atmosphere, vacuum, and temperature), test conditions (speed, lubrication, load, etc.), and coating architecture (single layer, multilayer, and composite layer). Thus, there is a need to understand the relationships between the microstructure, morphology, substrate/coating interface, and the tribological behavior of the coating in different environment becomes more prominent.

There are different types of coatings architectures are being used to enhance the tribological performance of the coatings. Depending on the requirement, a single layer, multilayer, multicomponent and compositionally or functionally graded coatings were studied so for. For instance, the compositionally grade coatings are designed to reduce the friction coefficient on the top coat and subsequent layers provide the abrasive wear resistance behavior. In addition, the combination of hard and soft phases (for example, ceramic-amorphous matrix) has been designed to achieve the combined wear resistant and self-lubricating property in a single layer. In these architectures, the size of crystalline phases is important for providing better tribological performance, and hence, the study of microstructure analysis using transmission electron microscopy (TEM) would provide detailed information about the coating. As another case of biological coatings, the coatings with multi-elements are designed for multifunctional properties other than tribological performance of the coatings. Thus, the understanding the optimization of chemical composition is important.

In order to predict the coating performance of a component, one must understand the microstructure of the coatings in all aspects. The coating performance and failure mechanisms are the areas where microstructural analysis can be utilized to evaluate and develop products. For example, the growth morphology of the coating shows major influence for determining the plastic deformation during mechanical interaction. It is noticed that the columnar microstructure of the coatings could be deformed plastically which can be avoided by designing lamellar type nano grain growth morphology or multilayer approach. The plastic deformation can be avoided with the help of multilayer approach and the grain size of the crystalline compound present in the coatings.

Microstructural analysis starting from simple determination of basic parameters such as thickness, grain size, growth morphology, number of layers and its interface, types of materials used in nano/microscale and coating thickness, porosity, and pore distribution of single- or multi-component systems, and the evaluation of degradation or failure mechanisms. For high-temperature applications, pin holes and porous coating limit the mechanical performance and acts as tribo-corrosive sites. A combination of characterization techniques is used to provide both physical and chemical information, with sub-micron/nanometer resolution. Hence, the primary governing factor for obtaining enhanced tribological properties is the microstructure of the coatings. Moreover, reproducibility and standard test methods for the determination of tribological properties of the coatings are not satisfactory and the effort has to be made to improve it [1].

The primary characterization techniques used for the microstructural analysis of the coatings are X-ray diffraction (XRD), scanning electron microscopy (SEM) coupled with energy dispersive X-ray spectroscopy (EDS), atomic force microscopy (AFM), TEM, electron probe microanalysis (EPMA), secondary ion mass spectroscopy (SIMS), Raman spectroscopy, and X-ray photoelectron spectroscopy (XPS). These techniques are useful for the development of advanced coatings by studying the microstructural features of them at different scale.

Important characterization of coatings to obtain the microstructural properties of wear resistant coatings is as follows:

Property of coatings	**Characterization techniques**
Microstructure	• X-ray Diffraction (XRD) • Raman spectroscopy • Transmission electron microscopy (TEM)
Surface studies	• Light optical microscopy (OM) • Scanning electron microscopy (SEM) • Atomic force microscopy (AFM) • Transmission electron microscopy
Residual stress	• X-ray diffraction • Stylus profilometry
Thickness	• Contact Stylus profilometry • Cross-sectional scanning electron microscopy • X-ray fluorescence (XRF)

Adhesion	• Scratch tester
Chemical composition	• Energy dispersive X-ray spectroscopy (EDS) • Wavelength dispersive X-ray Spectroscopy (WDS) • X-ray photoelectron spectroscopy • Auger electron spectroscopy • Rutherford backscattering spectrometry (RBS), Glow discharge optical emission spectroscopy (GDOES), Secondary ion mass spectroscopy (SIMS)
Hardness	• Indentation

The contribution of this chapter is to brief and clarify the application of selected number of characterization methods to obtain a good illustration of microstructural analysis of the coatings/films prepared through PVD route.

3.2 X-Ray Diffraction

The most commonly used standard technique for the analysis of structural properties of materials is XRD. This technique is used to identify the unknown crystalline information, unit cell dimension, and phases present in the coating, purity, identification of crystallite size, and strain. The Gaussian fitting was used to extract full width at half maximum (FWHM) values from the peak reflections. The d-spacing value can be calculated using Bragg's equation [Equation (3.1)] [2].

$$n\lambda = 2dsin\theta \tag{3.1}$$

where n is the order of reflection, λ is called the wavelength of X-ray used, d is the interatomic distance between the two atomic planes, and θ is the angle of diffraction. The 2θ positions will be compared with JCPDS files of the respective materials published with star quality. From the diffraction angle and Miller indices, accurate lattice parameters can be derived using regression analysis combined with the nonlinear least-square fitting with the help of "Unit cell program" [3]. The lattice parameter also calculated from d-values for cubic, tetragonal and monoclinic phases and are given below in Equations (3.2), (3.3), and (3.4), respectively [4].

$$\frac{1}{d^2} = \frac{(h^2 + k^2 + l^2)}{a^2} \tag{3.2}$$

$$\frac{1}{d^2}=\frac{(h^2+k^2)}{a^2}+\frac{l^2}{c^2} \tag{3.3}$$

$$\frac{1}{d^2}=\frac{1}{sin^2\beta}\left\{\frac{h^2}{a^2}+\frac{k^2 sin^2\beta}{b^2}+\frac{l^2}{c^2}-\frac{2hlcos\beta}{ac}\right\} \tag{3.4}$$

where a, b, and c are called as lattice parameters; h, k, and *l* are called as miller indices; and β is the angle between a and c axes. The crystallite size of a phase present in the coating is calculated from the FWHM due to peak broadening. XRD can be effectively used to measure the crystallite size in the range 10 to 100 nm [5]. From the diffraction angle and FWHM value, the crystallite size (D) can be calculated using Scherrer equation [Equation (3.5)].

$$D=\frac{k\lambda}{\beta cos\theta} \tag{3.5}$$

where θ is the angle and β is the FWHM obtained from the film after subtracting instrumental broadening, $\beta=\sqrt{B^2-\beta_0^2}$, where β_o is the measure of instrumental broadening obtained from standard reference material, and B is the broadening obtained from the sample. The instrumental broadening is calculated using reference material XRD pattern and the spectrum must be recorded under similar conditions [5]. While calculating the crystallite size, broadening caused by strain would not be considered.

The presence of crystallite size and strain present in the coatings are calculated by using Williamson-Hall relation [Equation (3.6)] [6]:

$$\frac{\beta cos\theta}{\lambda}=\frac{1}{D_v}+2\varepsilon\left(\frac{2sin\theta}{\lambda}\right) \tag{3.6}$$

The plot of $\left(\frac{\beta cos\theta}{\lambda}\right)$ versus $\left(\frac{2sin\theta}{\lambda}\right)$ gives you the value of strain from the slope value. Dislocation density (δ) is calculated from Equation (3.7):

$$\delta=\frac{1}{D^2} \tag{3.7}$$

The crystal structure of the hard coating can be analyzed using XRD on the as-deposited condition and then rubbed surfaces coated onto a steel substrate. The crystallinity of the as-coated and rubbed region can be analyzed based on the presence of the peaks and their intensity. No crystalline peaks in the XRD pattern depicts the amorphous nature of the coating. XRD analysis on the wear tracks helps to identify the formation of new phases or the increment of existing phases during the wear test. Yang *et al.* observed the increment of γ-Fe phase in wear tracks than as-coated surface due to stress induced phase transformations [7].

Advantage of XRD technique is that, this is a powerful method to for the identification of unknown material and minimum sample preparation is sufficient. The data interpretation is relatively easy and mineral determination is possible. The detection limit is ~2% of sample and complexity of indexing of non-isometric crystal system are the limitation of this method.

3.2.1 Coating Texture

Qualitative information on preferred orientation (texture coefficient, T_c) is calculated from the intensity ratio of first four strongest reflections of a particular phase present in the XRD pattern [8]. The texture coefficient (T_c) can be calculated using the following relation [Equation (3.8)]:

$$T_C = \frac{\frac{I(hkl)}{I_0(hkl)}}{\frac{1}{n}\sum \frac{I(hkl)}{I_0(hkl)}} \tag{3.8}$$

where $I_{(hkl)}$ is the intensity of the specific plane and $I_{o(hkl)}$ is the intensity of the particular plane of the same phase obtained from JCPDS card data.

The tribological performance of the coating could be enhanced to some extent through surface texturing by either reducing friction or increasing wear resistance. A proper textured surface can be acts as a lubricant reservoir and provide lubricant when necessary [9]. It also used to capture wear debris in order to prevent abrasion from third bodies. Chen *et al.* studied the tribological performance of the textured coating and shows less friction than untextured coatings [10]. Zhang *et al.* studied surface-textured DLC films and observed the significant improvement in the hydrodynamic lubrication performance through texture [11]. The micro-textured DLC coating deposited on titanium alloy shows low friction coefficient (80.6% lesser) than titanium alloy substrate and the wear volume also decreased upto 97.5% [12].

3.2.2 Residual Stress Measurement

Residual stress arises because of individual particle stress. Residual stress analysis of sputtered coatings is mostly related to the microstructure of the coating prepared through PVD route. Residual stress accumulation during hard thin film/coating deposition process results in microcracking and plastic deformation, which affects the tribological properties of the coated components. Thus, it is necessary to understand if the residual stress present in the coating.

Many determinant methods are available to measure the stress within the film, and they are categorized as destructive and nondestructive. There are two non-destructive methods which are widely used for the determination of residual stress within the films. One is to measure the lattice dimension through XRD [13] and the other called stylus profiler which measures the surface curvature of the substrate before and after the deposition [14]. From the literature, the standard technique for the measurement of residual stress determination is through $\sin^2\psi$ technique, which is derived from XRD data to obtain stress values [15]. Valvoda *et al.* [16] proposed a modified $\sin^2\psi$ technique using Seeman-Bohlin geometry [17]. From the XRD data, the d-spacing of all the peaks with respect to miller indices was measured using grazing incidence XRD method. Through this method, the film thickness lesser than 100 nm also can be measured. The depth resolution of less than 1 nm can be achieved by varying the angle of incidence of x-rays which is a great advantage of this technique than other methods available [18].

Another alternative method is to measure the deflection of the disc by using surface profilometry. Surface scan profile of the coated component is recorded before and after the coating deposition and the stress is calculated using Stoney equation [Equation (3.9)].

$$\text{Residual Stress}(\sigma) = \frac{E_s}{6(1-\upsilon_s)} \frac{t_s^2}{t_f} \left(\frac{1}{R_1} - \frac{1}{R_o} \right) \tag{3.9}$$

where E_s called substrate biaxial modulus, R_0 and R_1 are the radius of surface curvature before and after coating deposition, and t_s and t_f are the thickness of the substrate and film, respectively. This method uses relatively larger scan length (up to 10 cm) to measure the substrate surface curvature. The substrate curvature is proportional to the residual stress and thickness of the coating.

The wear resistance of the coating that depends on microstructure, porosity, hardness, and residual stress exists in the coating that is induced by thermal gradient and variation in thermal expansion coefficient of coating/substrate materials and lattice mismatch. Luo *et al.* reported that the presence of compressive stress in the plasma sprayed WC-Co coating possesses a positive impact in the wear behavior whereas for tensile stress, it displayed a negative effect [19]. Higher hardness is observed when the coating consists of smaller grain size and compressive residual stress [20]. Lower residual stress and dense microstructure design of hard wear resistant coatings are vital in order to reach better performance of the coatings during service.

3.3 Atomic Force Microscopy

AFM generally used to analyze the as-coated and wear track topographic and roughness information. Unlike SEM, this technique does not require any special sample preparation step to make it electron conducting [21]. In addition, AFM study can be performed at ambient condition. Three different modes are available in AFM instrument, namely, tapping mode, contact mode, and non-contact mode [22]. The tip radius determines the resolution limit of the instrument. The surface topography as well as root mean square (RMS) surface roughness of the coatings can be investigated in this study.

AFM can be used to study the nanoscale tribology for MEMS applications as well. This is derived from the capability of performing experiments using nanometer sized stylus in the AFM over the surface of the nano coating. Although conventional tribology testing is not carried out with the help of AFM tip, there are many applications are possible as listed below.

1. It can be used to visualize three-dimensional (3D) view of wear tracks, cracks, or debris on the surface.
2. It helps to measure the thickness of the coating in nanometer scale.
3. Provide detailed information on the surface texture, roughness, and topographical information,
4. Determination of nanoscale mechanical properties such as plastic deformation, elasticity fracture, and hardness.
5. The analysis can be done in a vacuum or liquid environment.

AFM provides 3D visualization of wear tracks. The image of AFM can be displayed in either 2D or 3D projection. From the image, direct

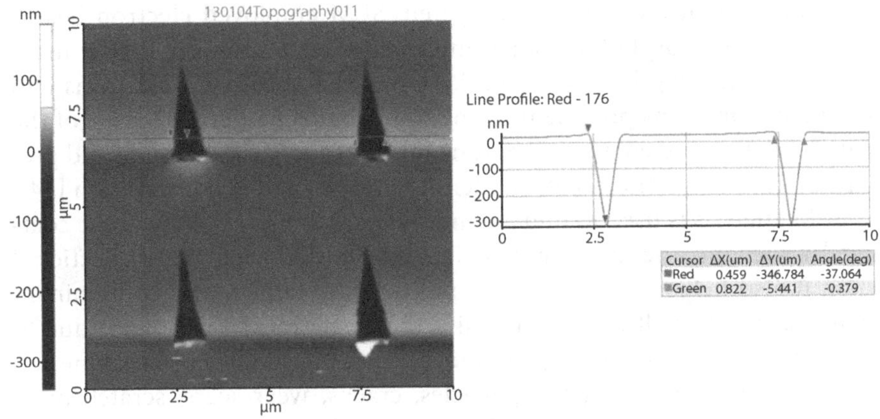

Figure 3.1 AFM image of Berkovich diamond indenter onto TiN coating with line profile information.

measurement of depth profile of wear track is possible through the line profile obtained from the AFM image. AFM provides high contrast images of surfaces at the nanoscale. Hence, optical and electron microscopy cannot distinguish surface texture easily which can be done using AFM.

AFM can be used to study the hardness at nanoscale by applying AFM probe onto the sample's surface. The advantage of this method is that we can select the region of indentation after visualizing the coating surface at the nanoscale. One can choose the grain or grain boundary region to find out the hardness values. AFM provides 3D topographical images of the indentations and depth information through which one can understand the fracture toughness and material behavior at nanoscale. AFM images of the residual intents used by Berkovich diamond indenter on to TiN coating prepared through RF magnetron sputtering on the silicon substrate at 400°C are shown in Figure 3.1. The maximum depth of indentation can be visually seen from the line profile data which is about 350 nm. Also, from the AFM indentation images, resistance to plastic deformation is being clearly observed.

3.4 Scanning Electron Microscopy

Electron interaction with samples has been extensively studied for characterizing the materials. Secondary electrons, backscattered electrons, and X-rays produced by the electron interactions carry information about the samples that are being investigated. SEM employed electron beams to witness the morphology of a sample at higher resolution, higher magnification, and depth of focus [23]. Herein, backscattered electrons and secondary electrons provide information about the microstructure of the sample. X-rays are energetic photons produced from the sample and give information about the chemical identity, structure, and composition [24].

SEM analysis is used to understand the growth morphology of the coating by viewing the substrate/coating interface through cross-sectional view. It also helps to identify wear mechanism, failure, and contaminant analysis. Additionally, it can magnify the wear tracks takes place due to both contact bodies (coating/ball indenter) at the sliding surface. One can observe the patches, trapped particles, cracks, wear scars, scratch morphology, and adhesive layers that present in the wear tracks (Figure 3.2). Oxidation process during wear could stimulate removal of material from the coating surface due to change in the properties of the material and grain rearrangement on the coated surface [26]. SEM images ZrN/Cu multilayer hard coatings prepared through PVD method are shown in Figure 3.2 which indicates the presence of wear scars and adhesive layers in the wear tracks [25].

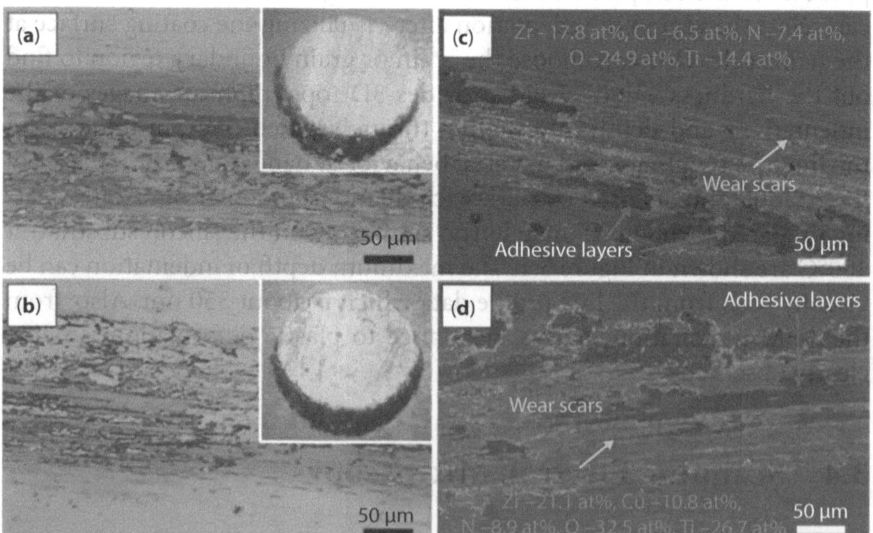

Figure 3.2 Wear morphology of ZrN/Cu multilayer coating deposited on (a) and (c) 316 L SS and (b) and (d) Ti substrates under dry conditions [25].

3.5 Energy Dispersive X-Ray Spectroscopy

SEM coupled with EDS analysis is a powerful tool that combines with high-power microscope with an energy dispersive spectrometer, which makes it possible to analyze the wear mechanism. EDS analysis is used to analyze the tribochemical composition and wear behavior of the wear tracks of the samples tested in wear conditions. Figure 3.3 shows a typical example of elemental mapping of TaN coating using 100Cr6 SS ball on the sliding region. Elemental mapping is performed for the elements such as Ta, N, Fe, O, and Cr and their distribution quantitatively seen from the spectrum. From the elemental mapping, one can understand the possible mechanism of the wear behavior of the coatings.

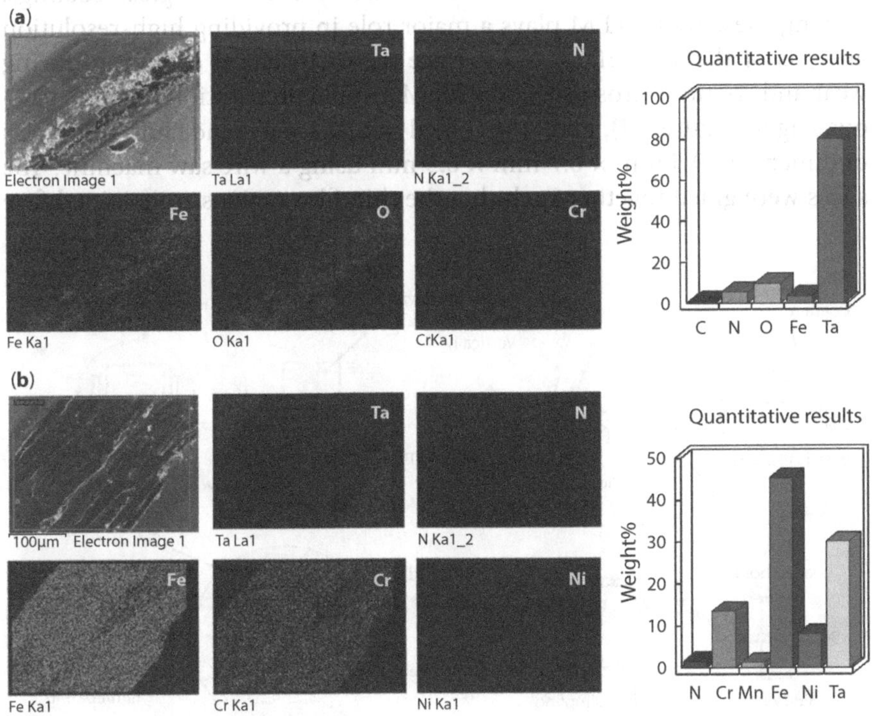

Figure 3.3 Elemental mapping of TaN coating against 100Cr6 SS ball under (a) ambient and (b) high-vacuum sliding conditions [27].

3.6 Transmission Electron Microscopy

TEM is the leading tool to understand the internal microstructure of any materials at nanometer level. The resolution can reach up to few tenths of nanometers, depending up on the imaging criterion, and can acquire electron diffraction pattern from the specific regions in the image as small as 1 nm [28]. The high-resolution TEM (HRTEM), an interference image creates by the uses both transmitted and diffracted beams. This phase contrast image can be as small as the unit cell of the crystal system. TEM has been comprehensively and effectively used to analyze crystal structures as well as the imperfections of lattice in numerous kinds of advanced functional materials in atomic resolution scale. This analytical tool can be used for the depiction of point defects, dislocations, stacking faults, and precipitates at grain boundaries. However, the contrast in TEM images is strongly depending up on the sample thickness.

Various microstructural characterization tools were employed to scrutinize the friction and wear characteristic of the tribological coatings. Among these tools, TEM plays a major role in providing high-resolution analysis of the subsurface and surface nano-structural evolution during wear and friction. Cross-sectional TEM sample preparation involves various stages (Figure 3.4). First, the coated samples were made into thin slices of dimension 0.5 mm × 0.7 mm × 0.5 mm using a wire saw machine. The slices were glued together such that the thin film regions were joined face

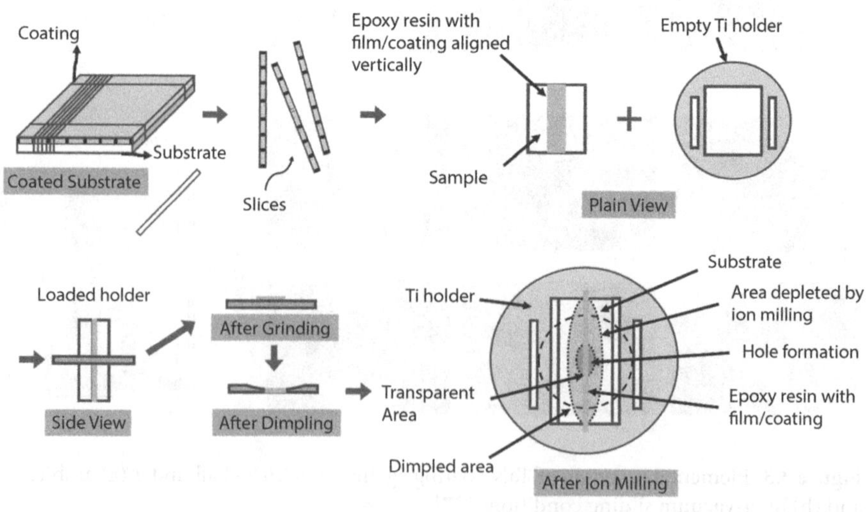

Figure 3.4 Steps involved in cross-sectional TEM specimen preparation.

to face. The pasted slices were hot pressed at 423 K for 2 h, and then, it is made to fit into titanium grid of slot size 1.2 mm × 1.7 mm followed by crimping. The specimen was ground up to a thickness of ~100 μm using SiC grit papers of 1,200, 2,400, and 4,000 particles per unit area in successive steps. Dimpling process was carried out to obtain ~15 μm thickness by employing any dimple grinder and then introduced into ion beam miller for further thinning such that electron transparency is achieved. Care must be taken in order to avoid the milling of glue line at the interface region by setting a stage rotation defined by an angular range.

This technique is more prominent when the coating consists of more than one layer. The removed fragments which are chipped away from the wear track can be taken out and the sample has to be thinned using focused ion beam technique in order to obtain a better results. TEM bright field image and SAED pattern provide the crystallographic information of the coating along with phase composition which is very useful to understand the wear mechanism. Ti/TiN coating with silver nanoparticle microstructure performed through STEM technique gives a detailed information on wear mechanism of the coating (Figure 3.5) [29]. The analysis provides a

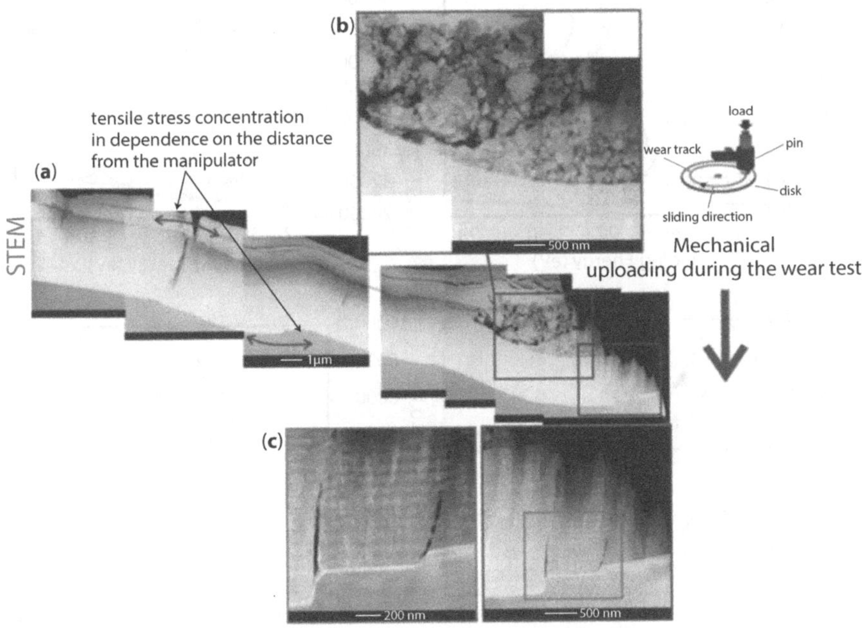

Figure 3.5 Microstructural characterization of Ti/TiN coating after wear test: (a) wear mechanism at low magnification, (b) layer by layer removal at high magnification, and (c) cracking mechanism at higher magnification [29].

detailed information on layer by layer removal process and cracking mechanism of the coating during wear process.

3.7 X-Ray Photoelectron Spectroscopy

XPS is a key spectroscopic non-destructive technique for the surface characterization and analysis of the material [30, 31]. This technique is also termed as Electron Spectroscopy for Chemical Analysis (ESCA), and it offers complete elemental analysis of the samples except hydrogen and helium. XPS is surface sensitive (10–200 Å) and it also provides elemental and chemical bonding information more quantitatively. In XPS, first, the sample surface is irradiated by soft X-rays, and simultaneously, the ejected photoelectrons from the sample are being analyzed. The difference between binding energy of the electrons at their initial atomic energy levels and the

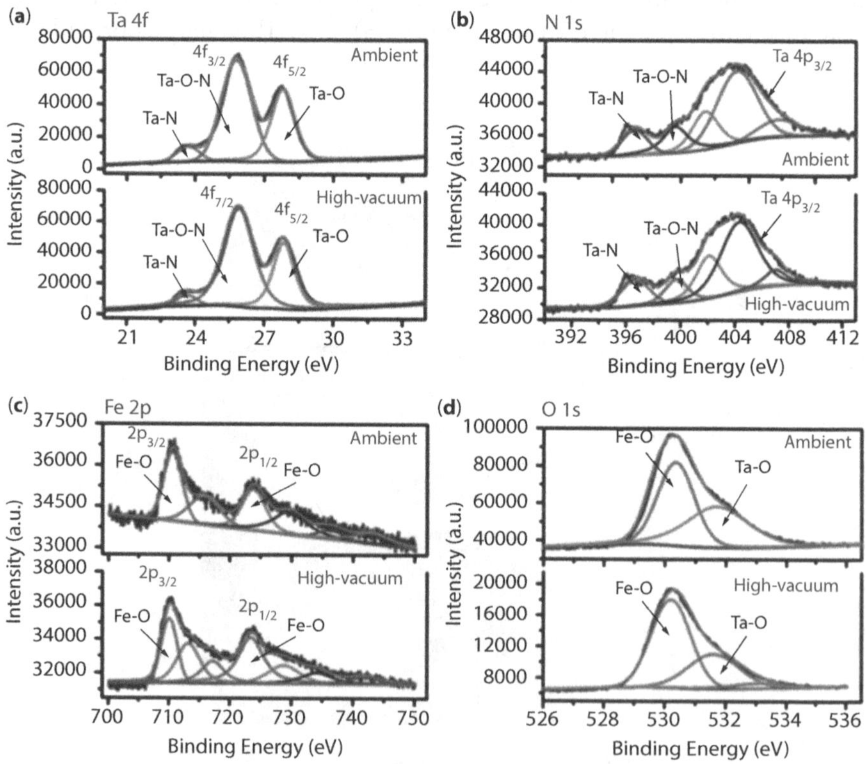

Figure 3.6 High resolution XPS spectra of (a) Ta 4f, (b) N 1s, (c) Fe 2p and (d) O 1s from the wear track of TaN coating under ambient and high-vacuum conditions [27].

energy of the irradiated X-ray is equivalent to the energy of the emitted photoelectrons, and this energy is characteristic for each individual element. By measuring the photoelectron energy with intensity, the elements and their abundances are recorded.

Chemical bonding characteristics of as-deposited coatings and on wear profile can be analyzed employing XPS technique. The survey spectrum provide information about the elements present in the scanning region. Before starting the analysis on the coating surface, pre-sputtering on the surface of about 10 nm is recommended to remove surface oxides and other contaminant due to atmospheric conditions. The influence of atmospheric condition, the formation of oxide layer and precipitate analysis on the wear profile will provide the tribo-chemical information of the wear track and the surface of the coatings. Figure 3.6 shows XPS spectra of TAN coating obtained from the wear track after tribology test under ambient and high vacuum condition. From the survey spectrum, the elements of Ta, N, O, and Fe were analyzed to find the formation of oxide layer and counter body elements present on the wear profile.

3.8 Electron Probe Microanalysis

EMPA is an analytical non-destructive technique that is used to analyze the composition of a specimen. The importance of an EPMA analysis is the ability to acquire accurate quantitative elemental investigation at very small scan area (as little as 1 to 2 μm), primarily by wavelength-dispersive spectroscopy (WDS) [32]. Except H, He, and Li, rest of the elements can be detected because the energy and wavelength of X-rays are small. This analytical technique possesses a high spatial and depth resolution, and each analysis requires only a minute or two. EPMA also provide image capabilities allowing mapping of composition contrast.

EPMA is useful to analyze the elemental composition of multilayer/compositionally graded coating design. The inward/outward diffusion of one element to another and interlayer diffusion can be clearly seen quantitatively through elemental mapping. Figure 3.7 shows electron image of Ni/Si multilayer coating after heat treatment [33]. The reaction between Ni and Si during heat treatment cause the formation of intermetallics at the interfaces can be identified using this technique. In addition to that, the variation in the concentration of Ni and Si around the interface can be derived from the compositional mapping. Figure 3.7 shows the variation in silicon and nickel concentration at different intermediate level which will give the information about the elemental distribution and precipitate formation.

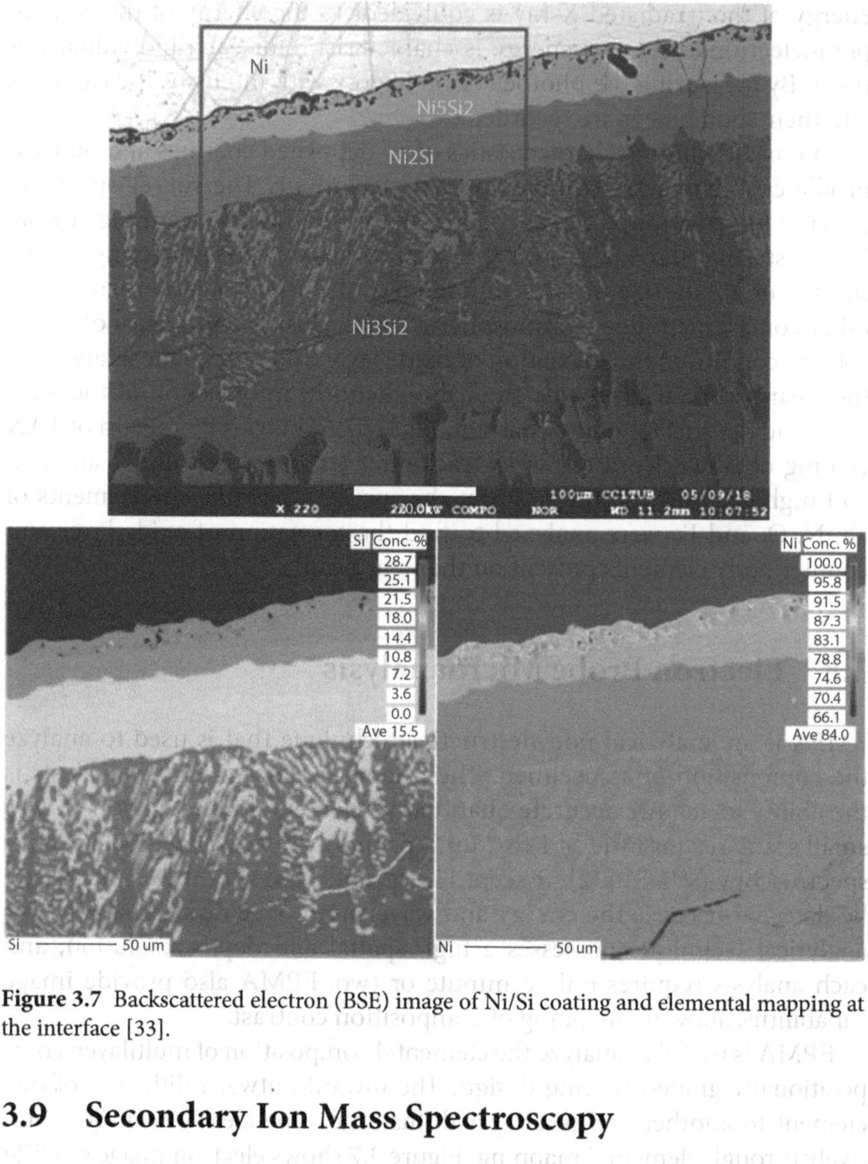

Figure 3.7 Backscattered electron (BSE) image of Ni/Si coating and elemental mapping at the interface [33].

3.9 Secondary Ion Mass Spectroscopy

The mechanical properties of the hard coatings depend on atomic concentration and elemental distribution across the thickness of the coating. Thus, the depth profile of the hard coatings with sub nanometer accuracy is required for better understanding of microstructure of the coatings. SIMS technique provides an exceptional combination of enormously high sensitivity for all elements starting from hydrogen to uranium and high lateral resolution imaging [34]. This technique is a destructive by its

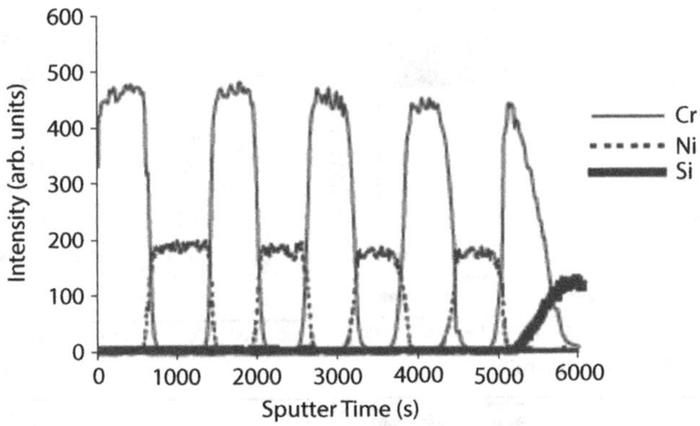

Figure 3.8 Depth profile of Ni (66 nm per layer)/Cr (53 nm per layer) multilayer coating on Si(100) substrate [35].

process because analysis takes place by sputtering the material. It can be applied to any kind of materials (insulators, semiconductors, and metals) that can sustain under vacuum. The limit of detection is as low as few parts per billion (ppb) levels for many elements. When a sample is being sputtered by primary ions, secondary ions are emitted from the target. SIMS involves to analyze those emitted secondary ions with a mass spectrometer. Secondary ions emitted by a solid sample provide information about the element, isotopic, as well as molecular composition of its atomic layers. Figure 3.8 shows SIMS investigation on the multilayer coating of Ni/Cr on Si(100) substrate [35]. The depth of the sputtered region can be correlated with the sputtered time with the help of stylus profilometer [36], and hence, the thickness of the layer can be calculated. Layer interface and diffusion of the coating during heat treatment can be identified across the coating thickness.

3.10 Raman Spectroscopy

Raman spectroscopy is a spectroscopic technique that used to define the vibrational modes of molecules present in the system. Raman spectroscopy provides information to identify molecules, intramolecular bonds and also to study the nature of chemical bonding. The spectrum provides a fingerprint data to identify molecules. In solid-state physics, Raman spectroscopy technique is used to characterized molecules and crystallographic orientation of materials. This is a commonly applied technique for the study of tribo-chemical changes on the wear tracks to understand the wear

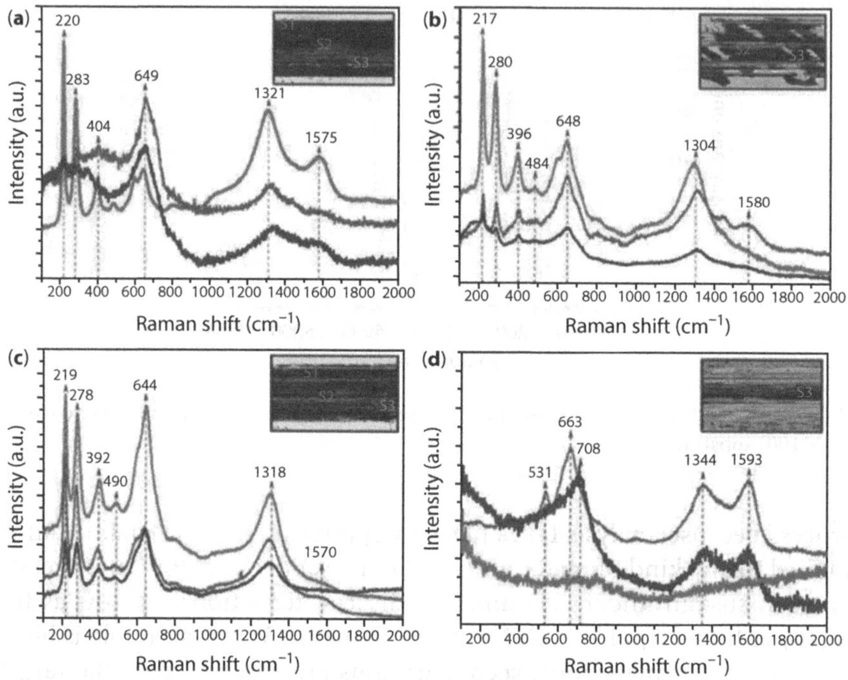

Figure 3.9 Raman spectra obtained from the wears tracks of (a) TiC, (b) CrC, (c) ZrC, and (d) WC coatings [37].

mechanism of the coating. Figure 3.9 shows the Raman spectra obtained from the wear tracks of TiC, CrC, ZrC, and WC coatings [37].

From the peak position and their intensity, we can identify the bonding characteristics and their chemical state. The identification of carbon and its state will be useful to understand the wear behavior of the coating. Shift in the peak is normally due to lattice strain of the phases. Raman spectroscopy reveals the presence of amorphous carbon present in the system which will enhance the lubricity of the coating. In general, the improved crystalline phase, microstructure and a-C matrix in the coating system will improve the tribo-mechanical properties of the carbide-based coating system [38].

3.11 Conclusion

This chapter briefly discusses the microstructural characterization techniques required for the tribological coatings prepared through PVD route. A relative importance had given to wide range of techniques such as XRD, AFM, SEM, TEM, EDS, Raman, and XPS were discussed. The properly controlled crystallographic phases, textured coatings, or multilayer structures provide improved wear resistance to the coated component. The type of bonding, lattice mismatch, surface topography, chemical composition, interfacial porosity, as well as the coating growth morphology will influence the performance of the coating. Hence, tailoring of coating microstructure and their detailed analysis could help to achieve the development of novel hard coatings with superior tribological performance.

References

1. Hogmark, S. and Hedenqvist, P., Tribological characterization of thin, hard coatings. *Wear*, 179, 147–154, 1994.
2. Patterson, A.L., The Scherrer formula for X-ray particle size determination. *Phys. Rev.*, 56, 978–982, 1939.
3. Holland, T.J.B. and Redfern, S.A.T., Unit cell refinement from powder diffraction data : The use of regression diagnostics. *Mineral. Mag.*, 61, 65–77, 1997.
4. Suryanarayan, C. and Grant Norton, M., *X-ray diffraction : A practical approach*, 1st ed., Springer US, New York, USA, 1998.
5. Cullity, B.D., *Elements Of X-ray diffraction, I*, Addison-Wesley Metallurgy Series, Massachusetts, USA, 1956.
6. Williamson, G.K. and Hall, W.H., X-ray line broadening from filed aluminium and wolfram. *Acta Metall.*, 1, 22–31, 1953.
7. Yang, S., Li, X., Renevier, N.M., Teer, D.G., Tribological properties and wear mechanism of sputtered C/Cr coating. *Surf. Coat. Technol.*, 142–144, 85–93, 2001.
8. Singh, A., Kuppusami, P., Thirumurugesan, R., Ramaseshan, R., Kamruddin, M., Dash, S., Ganesan, V., Mohandas, E., Study of microstructure and nanomechanical properties of Zr films prepared by pulsed magnetron sputtering. *Appl. Surf. Sci.*, 257, 9909–9914, 2011.
9. Niemczewska-Wójcik, M. and Piekoszewski, W., The surface texture and its influence on the tribological characteristics of a friction pair: metal–polymer. *Arch. Civ. Mech. Eng.*, 17, 344–353, 2017.
10. Chen, J., Zeng, L., Wan, Q., Chen, K., Lu, Y., Effects of texture geometry on the tribological performance of brass with a TiN coating under lubricated rotation. *AIP Adv.*, 9, 105116, 2019.
11. Zhang, Y., Fu, H., Wang, X., Liang, H., Puoza, J.C., Ji, J., Hua, X., Xu, X., Fu, Y., Additional tribological effect of laser surface texturing and diamond-like

carbon coating for medium carbon steel at near room temperature. *Coatings*, 10, 1–22, 2020.
12. Peng, R., Zhang, P., Tian, Z., Zhu, D., Chen, C., Yin, B., Hua, X., Effect of textured DLC coatings on tribological properties of titanium alloy under grease lubrication. *Mater. Res. Express*, 7, 066408, 2020.
13. Welzel, U., Ligot, J., Lamparter, P., Vermeulen, A.C., Mittemeijer, E.J., Stress analysis of polycrystalline thin films and surface regions by X-ray diffraction. *J. Appl. Crystallogr.*, 38, 1–29, 2005.
14. Feng, X., Huang, Y., Jiang, H., Ngo, D., Rosakis, A.J., The effect of thin film/substrate radii on the Stoney formula for thin film/substrate subjected to nonuniform axisymmetric misfit strain and temperature. *J. Mech. Mater. Struct.*, 1, 1041–1053, 2006.
15. Amirtharaj Moses, K.K., Parasuraman, K., Chakravarty, S., Rabel, A.M., Selvaraj, A.J., Singh, A., Residual stress measurements in electron beam evaporated yttria doped zirconia films deposited on Si (111) substrates. *J. Vac. Sci. Technol. A Vac. Surf. Films*, 36, 021504, 2018.
16. Valvoda, V., Kuzlel, R., Cerny, R., Rafaja, D., Musil, J., Kadlec, S., Perry, A.J., Structural analysis of tin films by Seemann-Bohlin X-ray diffraction. *Thin Solid Films*, 193–194, 401–408, 1990.
17. Haase, E.L., The determination of lattice parameters and strains in stressed thin films using X-ray Diffraction with Seemann-Bohlin focusing. *Thin Solid Films*, 124, 283–291, 1985.
18. Vourlias, G., Application of X-rays diffraction for identifying thin oxide surface layers on zinc coatings. *Coatings*, 10, 1–14, 2020.
19. Luo, W., Selvadurai, U., Tillmann, W., Effect of Residual Stress on the Wear Resistance of Thermal Spray Coatings. *J. Therm. Spray Technol.*, 25, 321–330, 2016.
20. Xi, Y., Bai, Y., Gao, K., Pang, X., Yang, H., Yan, L., Volinsky, A.A., Residual stress and microstructure effects on mechanical, tribological and electrical properties of TiN coatings on 304 stainless steel. *Ceram. Int.*, 44, 15851–15858, 2018.
21. Meyer, E., Atomic force microscopy. *Prog. Surf. Sci.*, 41, 3–49, 1992.
22. Jalili, N. and Laxminarayana, K., A review of atomic force microscopy imaging systems: application to molecular metrology and biological sciences. *Mechatronics*, 14, 907–945, 2004.
23. Tanaka, K. and Maeda, K., Scanning electron microscopy. *J. Health Med. Informat.*, 05, 1000167, 2014.
24. Gotsch, T., Wallisch, W., Stoger-Pollach, M., Klotzer, B., Penner, S., From zirconia to yttria: Sampling the YSZ phase diagram using sputter-deposited thin films. *AIP Adv.*, 6, 025119, 2016.
25. Kumar, D.D. and Kaliaraj, G.S., Multifunctional zirconium nitride/copper multilayer coatings on medical grade 316L SS and titanium substrates for biomedical applications. *J. Mech. Behav. Biomed. Mater.*, 77, 106–115, 2018.

26. Gheisari, R. and Polycarpou, A.A., Tribology of Selected Hard Coatings for Oil and Gas Applications Up to 450 °C. *Tribol. Lett.*, 67, 1–16, 2019.
27. Kumar, D.D., Rani, R., Kumar, N., Panda, K., Kirubaharan, A.M.K., Kuppusami, P., Baskaran, R., Tribochemistry of TaN, TiAlN and TaAlN coatings under ambient atmosphere and high-vacuum sliding conditions. *Appl. Surf. Sci.*, 499, 143989, 2020.
28. Williams, D.B. and Carter, C.B., The transmission electron microscope, in: *Transm. Electron Microsc*, pp. 3–22, Springer US, Boston, MA, 2009.
29. Major, L., Lackner, J.M., Kot, M., Major, B., Nanostructural aspects of the wear process of multilayer tribological coatings. *Tribol. Int.*, 151, 106411, 2020.
30. Sarma, D.D., Santra, P.K., Mukherjee, S., Nag, A., X-ray photoelectron spectroscopy: A unique tool to determine the internal heterostructure of nanoparticles. *Chem. Mater.*, 25, 1222–1232, 2013.
31. Majumdar, D. and Chatterjee, D., Xray photoelectron spectroscopic studies on yttria, zirconia, and yttria stabilized zirconia. *J. Appl. Phys.*, 70, 988–992, 1991.
32. Falcone, R., Sommariva, G., Verità, M., WDXRF, EPMA and SEM/EDX quantitative chemical analyses of small glass samples. *Microchim. Acta*, 155, 137–140, 2006.
33. Llovet, X., Electron Probe Microanalysis, in: *Encycl. Anal. Sci.*, pp. 30–38, Elsevier, Netherlands, 2019.
34. Fearne, S., *An introduction to time-of-flight secondary ion mass spectrometry (ToF-SIMS) and its application to materials science*, IOP Publishing, CA, USA, 2015.
35. Sostarecz, A.G., Sun, S., Szakal, C., Wucher, A., Winograd, N., Depth profiling studies of multilayer films with a C 60+ ion source. *Appl. Surf. Sci.*, 231–232, 179–182, 2004.
36. Kirubaharan, A.M.K., Kuppusami, P., Priya, R., Divakar, R., Gupta, M., Pandit, D., Ningshen, S., Synthesis, microstructure and corrosion behavior of compositionally graded Ni-YSZ diffusion barrier coatings on inconel-690 for applications in high temperature environments. *Corros. Sci.*, 135, 243–254, 2018.
37. Kumar, D.D., Kumar, N., Kalaiselvam, S., Radhika, R., Maximus Rabel, A., Jayavel, R., Tribo-mechanical properties of reactive magnetron sputtered transition metal carbide coatings. *Tribol. Int.*, 114, 234–244, 2017.
38. Kumar, D.D., Kumar, N., Kalaiselvam, S., Radhika, R., Dash, S., Tyagi, A.K., Jayavel, R., Reactive magnetron sputtered wear resistant multilayer transition metal carbide coatings: microstructure and tribo-mechanical properties. *RSC Adv.*, 5, 81790–81801, 2015.

4

Wear Mechanisms and Methods of Wear Testing

Ahmed Abdelbary

Chief Engineer and Tribology Consultant Egyptian Government, Alexandria, Egypt

Abstract

Wear is defined as the progressive removal of the material from a solid surface and manifested by a change in the geometry of the surface. Accordingly, it is extensively supposed that wear is one of the major causes of failure in many engineering applications. Most of the wear mechanisms are built around concepts such as brittle fracture, plastic deformation, fatigue, and cohesive and adhesive failures in bonded structures.

Since there are several mechanisms of wear, we should expect no unique or typical form or size of wear debris. The complexity of wear is related to the fact that, in many cases, the involved mechanism may act in such a way as to affect the others. Besides, wear behavior is associated with operating conditions. The significance of these conditions (or factors) tends to change with the type of the wear mechanism. The approach of wear-regime maps is introduced as a graphical technique used to characterize various aspects of wear behavior in terms of operating conditions. Finally, diverse test methods used in evaluating wear are reviewed in the last section. The main purpose of conducting wear tests is to provide required data for tribologists to be used for selecting the best material for the design.

Keywords: Wear mechanisms, wear curve, three-body wear, wear test

Email: elbary1972@yahoo.com

Sarfraj Ahmed and Vinayak S. Dakre (eds.) Tribology and Characterization of Surface Coatings, (71–98) © 2022 Scrivener Publishing LLC

4.1 Introduction

Wear, as a natural phenomenon, usually arises once two surfaces with a relative motion interact with each other. It is a progressive loss of material from the solid surface as a consequence of mechanical action [1]. It is usually considered a vital reason for failure in many machine parts. Opposing to what is previously believed, wear phenomenon is not wholly destructive, in some situations such as polishing, grinding, and many other forming operations, controlled wear is accepted.

Many researchers have recognized four foremost mechanisms (or types) of wear: adhesive, abrasive, fatigue, and corrosive wear. Additionally, erosion, galling, cavitation, thermal, impact, and fretting wear are minor types of wear. Most of these mechanisms are built around concepts such as brittle fracture, plastic deformation, fatigue, and cohesive and adhesive failures in bonded structures. The difficulty of wear is that the involved mechanism, in many cases, may act in such a way as to affect the others. Furthermore, wear behavior is highly sensitive to operating conditions. Therefore, preliminary validation of machines and machine parts by laboratory testing is required before integrating a new solution into the system. The main purpose of testing wear in laboratories is to provide required data for tribologists for selecting the best material for the design.

4.2 Mechanisms of Wear

Wear processes, generally, involve four main groups of mechanisms; adhesive, abrasive, fatigue, and corrosive wear. Each of these mechanisms is administrated by its laws. However, on many occasions, this mechanism may act to affect the other mechanisms [2]. It is significant to emphasize that it is not always easy to differentiate between these types because they are interrelated and rarely occur separately. Other types of wear such as erosive, fretting, galling, impact, thermal, cavitation, impact, biocorrosion, and electro-arcing are also considered [3].

4.2.1 Adhesion

Wear by adhesion is a communal type in mechanical systems that occurs whenever two counterparts are in contact with a relative motion, whether lubricated or not. Adhesion arises at junctions of surface asperity as a micro spot welding. Consequently, the tips of the softer asperities are sheared plastically. As a result, a loss of material in the form of minor fragments which adhere to the counterpart surface, as in Figure 4.1. With continuous sliding, the transferred wear debris may detach from the surface on which they are adhered and be transferred back to the original surface, or separate as debris particles. In this wear type, wear debris are characterized by relatively large particles composed of two surfaces.

Archard [4] suggested that if the required force to shear through the interface detachment of the materials is higher than the required force to shear through one of the materials, then the shearing will take place at the original surface. It may occur also in the weakest region in one of the two materials, as in Figure 4.2.

In utmost cases of dissimilar material combinations, wear fragments are formed from the softer material. However, fragments of the harder counterpart are also formed. This suggests the existence of local regions of low strength within the harder material. Also, fragments of the material with a harder surface may result from the fatigue process which takes place after some repeated stresses. Wear debris particles produced by adhesive wear usually show a fragment pulled off one surface and adhered to the other, as represented in Figure 4.3 [5, 6].

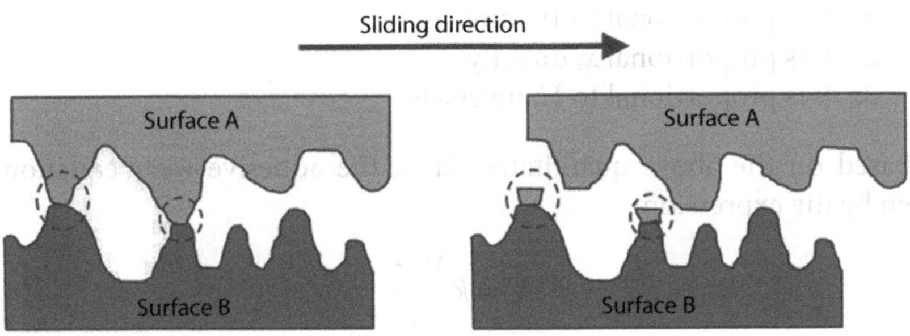

Figure 4.1 Mechanism of adhesive wear.

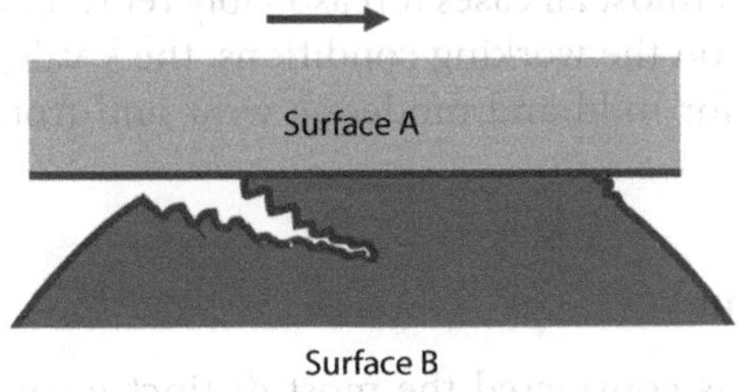

Figure 4.2 Archard proposal for adhesive wear.

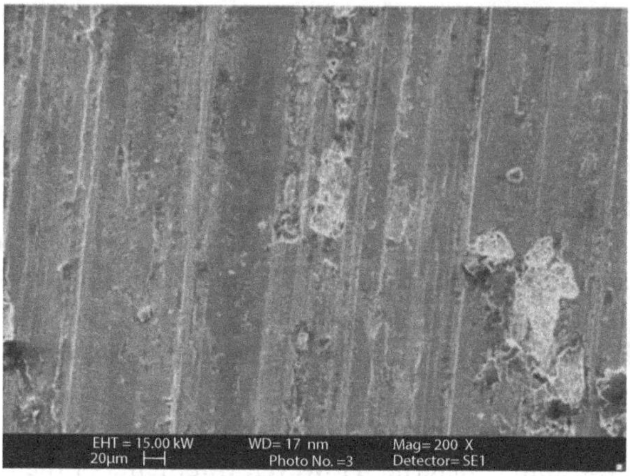

Figure 4.3 Typical morphology of adhesive wear. Adapted from Dwivedi, D. K. [5] with permission.

Experimental results for unlubricated sliding of similar and different metal combinations suggested the following quantitative laws of the adhesive wear [7] to find a relation among the worn volume (V) and both load (L), hardness (H), and distance (x):

1. V is proportional to W, directly.
2. V is proportional x, directly.
3. V is proportional to H, inversely.

Based on the above quantitative laws, the adhesive wear equation is given by the expression:

$$V = k\frac{Wx}{H} \tag{4.1}$$

where k is a non-dimensional wear coefficient dependent on the materials in contact and their cleanliness. The above expression is not obeyed at all times, however, in almost all cases it reasonably represents the experimental data [8]. Based on the working conditions, the k value typically ranges from 10^{-8} to 10^{-4} for mild and moderate wear and from 10^{-4} to 10^{-2} for severe wear [3].

4.2.2 Abrasion

Wear by abrasion is considered the most distinct form of wear in many mechanical applications. This wear type mostly occurs in tribo-systems of two rubbing surfaces which are different in hardness. The obvious surface topography feature of abrasive wear is that plowing occurs during sliding resulting in abrading of surface material. Consequently, long abrasive grooves are cut on the softer (ductile) surface which usually runs in the direction of motion, as presented in Figure 4.4.

In brittle materials with low fracture toughness, wear arises by brittle fracture [3]. In such cases, the worn zone includes significant cracking.

The "abrasive wear" term refers to two distinct wear mechanisms: *(1) two-body abrasion* and *(2) three-body abrasion*. Both situations are conjugated with the plowing-out of the softer material surface by a harder one, as described in Figure 4.5.

(1) Tow-Body Abrasion
It occurs under two conditions: (1) one of the mating surfaces must be harder compared to the other surface and (2) the harder surface must be rough. In such a case, hard asperities of the rough surface slide against the

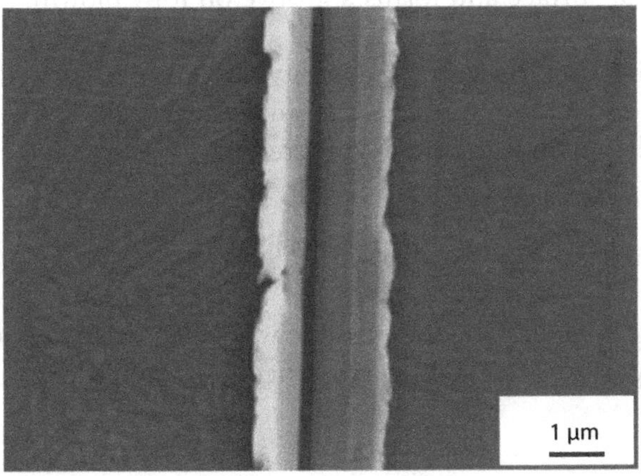

Figure 4.4 Single abrasive wear track of ultrahigh strength bainitic steel. Adapted from Balji N. [9], with permission.

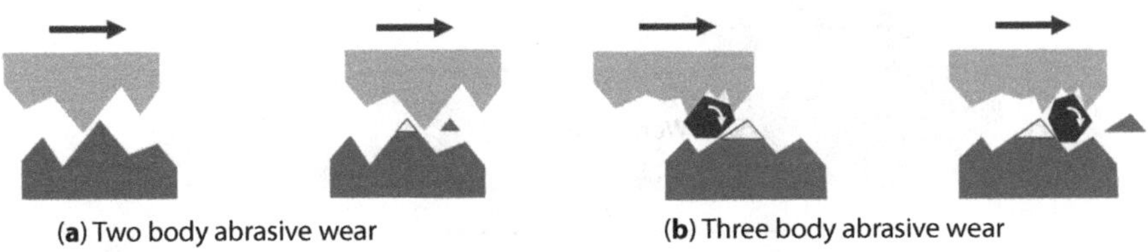

Figure 4.5 A schematic diagram for (a) two-body and (b) three-body wear.

softer surface, dig into it and damage the interface by plastic distortion or fracture. Examples of this wear mechanism are found in various machining operations, such as grinding and metal cutting.

(2) Three-Body Abrasion

This wear mechanism arises when hard particles (third body) are trapped among two contact surfaces in relative motion and abrade either one of the two surfaces or both of them. This form also includes abrasion by wear debris generated by other wear mechanisms. The two conditions for this mechanism to occur are as follows:

(1) The abrasive fragment must have considerable dimensions comparable to the space between sliding surfaces.
(2) The abrasive fragment must be harder than the mating surfaces.

To deduce a quantitative law for abrasive wear, we assume a model of a single contact point where a hard asperity of conical shape is indented against the flat surface and forms a groove on it by plowing, as illustrates in Figure 4.6.

Considering that the abrasive asperity has a projected area in the horizontal plane equal to πr^2 and H is the softer surface hardness. If the material has yielded under the load dW; therefore,

$$dW = \frac{1}{2}\pi r^2 H \tag{4.2}$$

According to Figure 4.6, the possible wear volume dV, which is plowed by harder asperities after sliding a distance x, is given by

$$dV = x\, r^2 \cdot \tan\theta \tag{4.3}$$

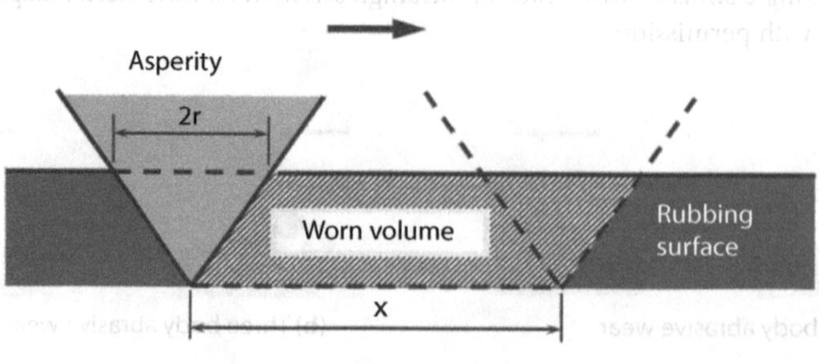

Figure 4.6 A simple model for abrasive wear.

Equations (4.2) and (4.3), altogether

$$dV = \frac{2dWx \cdot \tan\theta}{\pi H} \tag{4.4}$$

Emerging Equation (4.4) for contributions of all asperities, we will have

$$V = \frac{2\,\overline{\tan\theta}}{\pi} \cdot \frac{Wx}{H} \tag{4.5}$$

where roughness factor $\overline{\tan\theta}$ is an average of the slope values of all the individual asperities.

This relation is quite similar to Archard Equation (4.6) for adhesive wear, except that the factor $k_c/3$ is replaced by the term $\frac{2\,\overline{\tan\theta}}{\pi}$.

$$V = \frac{k_c}{3}\frac{Wx}{H} \tag{4.6}$$

To accommodate abrasion in terms of Equation (4.6), an asperity geometry parameter k_{abr} is introduced in Equation (4.5) as a nondimensional abrasive wear coefficient. Finally, the removed volume by abrasive wear is

$$V = k_{abr}\frac{Wx}{H} \tag{4.7}$$

Typical experimental results proposed that the values of k_{abr} in two-body situations are about 10^{-2} to 10^{-1}, whereas in three-body situations, they are about 10^{-3} to 10^{-2} [8]. Besides, the coefficient of friction during three-body abrasion is generally less than that in two-body abrasion, by as much as a factor of two. Based on this, two important remarks are drawn:

a) The abrasive grain in a three-body situation tends to roll rather than sliding and wearing the surface.
b) Generally, the rate of abrasive wear is larger than the adhesive wear rate by 2 to 3 orders of magnitude.

4.2.3 Surface Fatigue

Surface fatigue wear mechanism is closely associated with the general fatigue phenomenon of materials where a correlation between the contact stress and the cycles-to-failure number exists. However, there are two differences between wear due to surface fatigue and ordinary fatigue; first, the high fluctuations in the life in fatigue wear applications compared to ordinary fatigue cases. Second, there is no fatigue limit stress in fatigue wear as in ordinary fatigue testing.

Practically, there are two mutual situations of wear by fatigue fracture: in rolling elements (e.g., rolling bearings, gears, and cams) and in sliding on brittle surfaces (e.g., ceramic materials), as presented in Figure 4.7.

In such applications, during operation, under the normal loading, the contacting components are elastically deformed rather than plastically. Subsequently, stressing and un-stressing the rolling surface or sliding track continues to occur. The repeated load cycles may induce the subsurface or surface cracks formation and formation of large fragments. Thereafter, surface deterioration by spalling or flaking becomes rapid, as presented in Figure 4.8.

The existence of subsurface cracks is possible according to the fact that the point with maximum shear stress, and hereafter, the point with the maximum tendency for plastic yielding is at a small distance below the surface [11].

In rolling contact, fatigue wear is distinguished by large fragments formation after a certain number of cycles. Accordingly, the typical experimental investigations of fatigue wear have considered running rolling

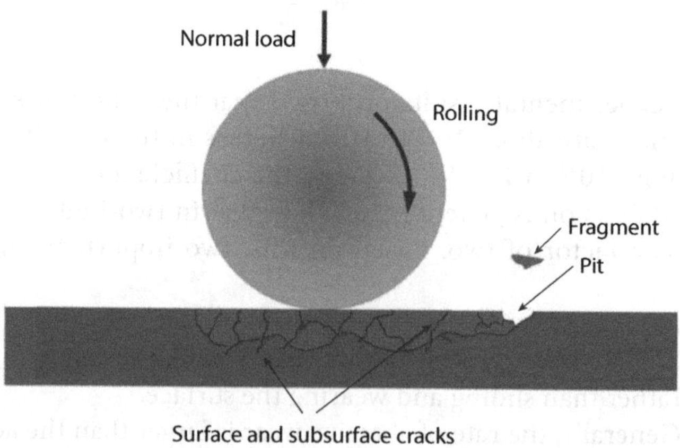

Figure 4.7 Fatigue fracture in a rolling motion.

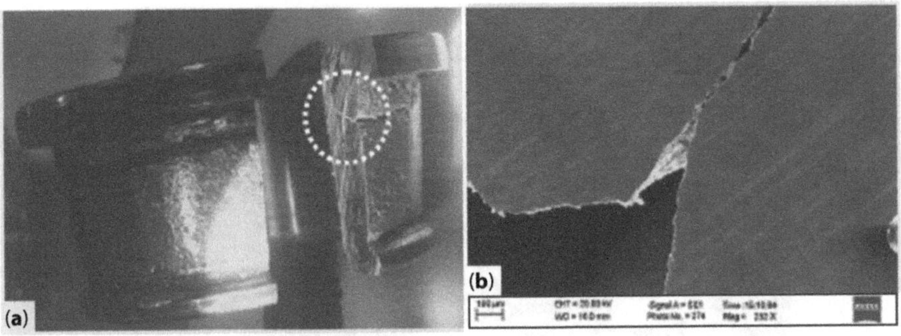

Figure 4.8 A section of bearing inner cone: (a) crack starts at the deep groove and propagates toward the surface; (b) SEM image showing the nucleation of the crack. Adapted from Gurumoorthy [10], with permission.

contact devices until spalling occurs, and noting the rolling cycles (time) to spalling [8]. Testing of several rolling elements has postulated that the critical number of high-cycle fatigue N_f, as defined above is given by the following empirical relation [12, 13]:

$$N_f = bW^{-n} \tag{4.8}$$

where

W	normal load,
b and n	experimental constants.

The n value depends on the rolling element shape. In rolling ball bearings, $n \cong 3$.

In sliding, fatigue wear takes another form probably resulting from repeated stress cycles applied to the surface with asperity interactions. During sliding, asperities can make contact without adhering and pass leaving them plastically deformed. After several contacts, the resultant cyclic plastic shear deformation leads to subsurface crack initiation at the surface [14].

4.2.4 Corrosion

Wear by corrosion (or corrosive wear) is categorized by material degradation and it is considered a dangerous form of corrosion because the failure can occur without clear warning [15–17].

Corrosive wear arises in conditions in which the environment chemically interacts with the rubbing surface. Or it can arise when sliding arises in corrosive media, as shown in Figure 4.9.

Figure 4.9 Corrosion wear in helical gear. Adopted from http://metroforensics with permission.

The mechanism of corrosive wear is extremely complex and not well understood. It involves the contacting surfaces' behaviors, the contact mechanics, and the conditions of corrosion. In the first stage, the tribo-chemical reaction produces a film (e.g., oxides) naturally less than a micrometer thick on the surface of the material, which would inhibit or delay the corrosion action. If such films strongly adhere to the surface and behave like the bulk material, then the mechanism of wear would be the same as that of the bulk material. However, practically, the action of sliding abrades the chemical film resulting in a virgin surface, so that the chemical attack can continue. Thus, chemical wear involves both corrosion and rubbing.

4.2.5 Erosion

The term *erosion* may designate abrasive wear in a case that only one surface is involved. The erosion process refers to the advanced loss of a surface resulting from mechanical contact between the surface and a gas or liquid fluid. Abrasive erosion, slurry erosion, cavitation erosion, fluid erosion, and spark erosion are the common types of erosive wear. They are produced by the impact of solid particles, liquid droplets, bubbles, or electrical sparks [18, 19].

Erosion by abrasion is defined as "the process of metal removal due to impingement of sharp solid particles on a surface", as illustrated in Figure 4.10.

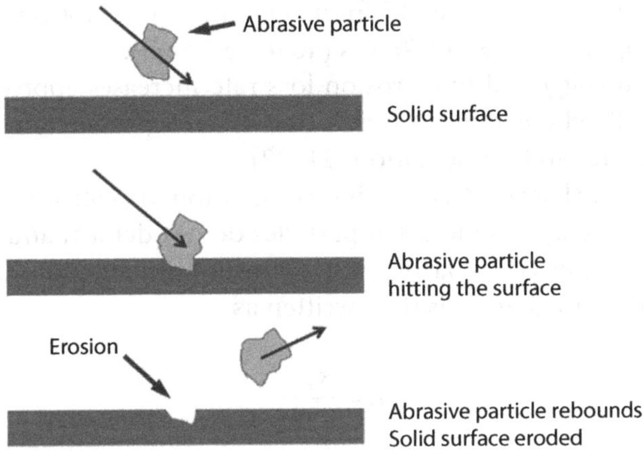

Figure 4.10 Schematic of erosive wear showing abrasive particle hitting a surface.

Erosive wear is a mixture of impact and abrasive wear mechanisms. The main differences are as follows:

(1) The surface roughness produced may become relatively great, because an impinging particle may readily material remove from surface valleys.
(2) The contact stress in erosion arises from the kinetic energy of particles flowing in an air or liquid stream as they encounter a surface.

Generally, erosive wear is an undesired type of wear in many situations such as gas turbine blades and refractories in electric arc melting furnaces. Nevertheless, it is deliberated in sandblasting, abrasive deburring, and erosive drilling of hard materials.

Since this type of wear is ruled by the kinetic energy of the particles impinging on surfaces, it is found that the erosion ratio (E), the mass of removed material to the mass of erosive particles striking the surface, depends on some factors contributing to the impact process [20]. These factors include impact angle, impact velocity, and relative material hardness, and incident particles; therefore,

$$E = cv^n f(\alpha) \tag{4.9}$$

where v is the particle velocity, α is the impact angle with respect to the material surface, and the coefficient n varies from ductile to brittle materials.

In general, n ranges from 2 to 2.5 for metals and 2.5 to 3 for ceramics, while v usually ranges from 15 to 170 m/s (40 to 150 mph).

It was also suggested that erosion loss rate increases approximately by (particle size)^m where $m \approx 3$. It is reported that n is proportional to particle size, perhaps due to fragmentation [21, 22].

Hutchings [23] derived a quantitative equation to determine the erosion ratio E given the assumption that particles do not deform and the surface deformation is perfectly plastic with a constant indentation hardness, H. The erosion wear equation is then written as

$$E = \frac{k\rho v^2}{2H} \tag{4.10}$$

where ρ is the eroded material density, v is the particle velocity, and k is the proportion of the displaced material resulting as wear debris.

4.2.6 Fretting

The word "*fretting*" was first introduced by Tomlinson [24] to refer to a small amplitude of oscillation ranging from a fraction of μm to hundreds of μm. Since fretting action is conjugated to vibration, it is frequently found in most machinery, bolt joints, splined couplings, shrink fits, transformer

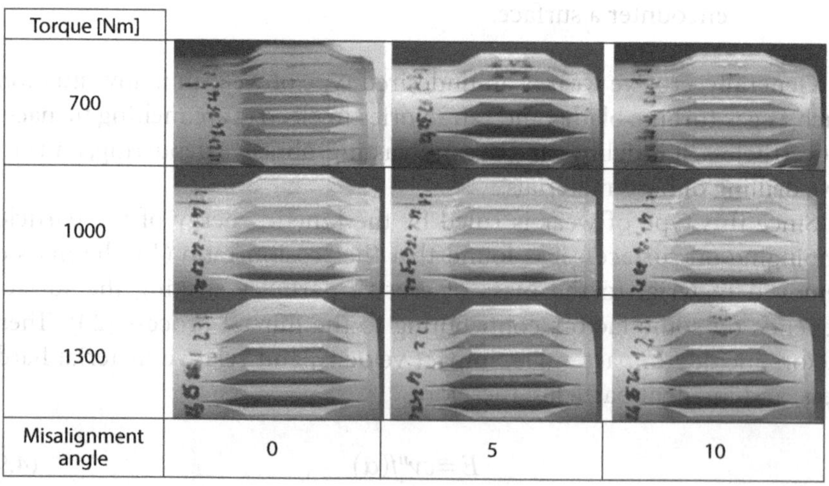

Figure 4.11 Spline coupling after fretting wear test. Adapted from F. Curà F. *et al.* [26] with permission.

cores, and in many other cases. Ultimately, fretting is a type of adhesive or abrasive wear where the applied load results in adhesion between asperities, while the low-amplitude oscillatory motion causes rupture of the formed joints [1, 3, 25]. Though, fretting can be considered as a special case of fatigue wear at the surface [18].

Figure 4.11 shows fretting wear damage appearing on splined couplings because of the virtual motion between teeth, primarily due to angular misalignments of these components [26].

4.2.7 Impact

Wear by impact (or percussive wear) is repeated impacts that result in continuous loss of material. It arises by a combination of the following mechanisms: adhesive, abrasive, surface fatigue, fracture, and tribo-chemical wear. This form of wear can be found, for example, in jaw crusher plates, railway points, and excavator buckets. Although this type of wear is a source of failure in numerous industrial situations, it has generally not been considered as extensively as other wear mechanisms.

Failure of tools used for drilling rocks and other media raises cost concerns, not only associated with the need for frequent replacement of parts but also in the downtime incurred. Impact motion also widely exists in industrial machinery applied in the mining industry, metallurgy, and electric power [3, 27–29]. To find some wear-resistant materials or structures by adapting some special physical and chemical methods, numerous researches were introduced. Probably, heat treatment and surface coating are the most effective methods that can enhance the impact and abrasion resistance with some considerations.

Many studies reflect the situation of the contact conditions in dry contact and there have been fewer investigations on lubricated impact wear. In industry, the lubricated shaft or rolling bearings suffered failure because impact wear is very common [27]. In dry contact, spherical particles are formed inside the impact dent. On the other side, in lubricated dents, especially in those with high viscosity oil, direct contact occurs where the spherical particles appear in clusters and different sizes.

An expression identical to Archard's law, Equation (4.6), can be obtained by recognizing that the term Wx represents sliding energy E_s. Consider n identical particles of mass m impacting a surface with the same velocity v. Their individual kinetic energy E_k is $1/2mv^2$ of the total kinetic energy of the particles. Only a fraction β results in material removal. Therefore, the impacting energy E_i resulting in wear is given by

$$E_i = \beta E_k = \beta \frac{1}{2} m v^2 n \tag{4.11}$$

Therefore, substituting this impact energy instead of the sliding energy into Archard's equation yields

$$V = k_i \frac{E_k}{H} \tag{4.12}$$

In this case, the wear coefficient k_i can then be interpreted as the fraction of all impact events that result in material removal.

4.2.8 Thermal

Thermal wear is not considered as a typical mechanism of wear in general tribo-systems. However, it is generated by unexpected contact conditions, such as hard inclusions at the contact interface or sudden overloading due to vibration. The thermal wear is ruled by surface melting results mainly from frictional heating. *Diffusive wear* is also included in the term "thermal wear" since it becomes noticeable only at elevated temperatures [30]. Quantitative estimation of the worn volume may not be required. Even though the original material properties of a contact surface are degraded by losing significant chemical compositions, and consequently, the wear rate can still be increased by enhancing other wear modes, such as adhesive or abrasive wear [31, 32]. E. Usui *et al.* [32] proposed that the rate of wear increases linearly with $\exp(-\Delta E/kT)$, where k is the wear coefficient, T is the temperature, and E is the modulus of elasticity.

4.3 Methods of Wear Testing

When wear is equated with material removal, volume is the fundamental measured for wear. Nevertheless, in many investigations, mass loss is used instead of volume loss. Engineering designers, on the other hand, maybe more anxious about the change of a dimension due to volume loss, V. Dimensional changing frequently increases clearance or results in a change in a contour. Besides the former measures for wear, there is a variation of other operational ones that may be used such as crazing, vibration level, noise level, life, and appearance of the surface.

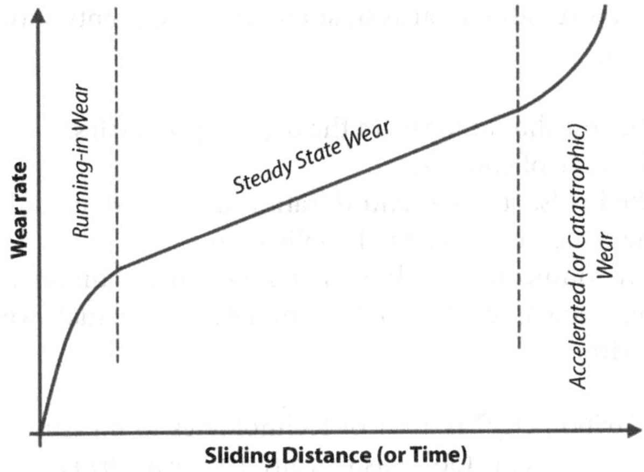

Figure 4.12 A typical wear curve.

Wear is usually designated by "Wear Rate" *WR*, which is calculated by dividing the volume of material removed (*V*) by the sliding distance (*x*), or in some cases per unit time. Wear rate is generally not constant, it is a multifaceted function of time [33].

A typical wear curve is illustrated in Figure 4.12. At the initial stage, wear usually starts low and then rises. After a time, the wear rate becomes constant for a period. The first stage during which wear rate changes is called running-in wear. During this stage, wear depends on the initial material structure and properties and surface conditions such as surface finish and the nature of any contaminated films present. Following the running-in phase, the wear goes through a steady-state stage which is characterized by a constant wear rate. Sometimes, after a certain sliding distance, the accelerated (or catastrophic) wear stage starts due to surface fatigue.

Wear could be also presented in terms of a "Wear Factor" *WF* (or specific wear rate). Wear factor is the wear volume per unit applied load (*F*) and per unit sliding distance (*x*), it can be calculated by Equation (4.12):

$$WF = \frac{V}{F \cdot x} \tag{4.13}$$

There are several types of wear including sliding, scratch, abrasion, and others mechanisms. Laboratories associated with wear testing possess practice a variety of tribometers and procedures to address each specific wear type. The choice of the proper tribometer is very important.

Thus, to select a suitable wear test, some intrinsic points should be considered [34, 35]:

i) The test should measure the desired properties.
ii) The type of contact.
iii) The loads, stresses, and duration in the test should be in the same range of actual application.
iv) Environmental conditions (temperature, humidity, sliding media, etc.) should simulate the actual working environment.

The term "tribometer" is a set of technologies to measure the friction and wear of tribosystems [36]. Historically, the term "*tribometer*" was first introduced in the 1700s by Goldsmith to refer to a device to measure friction, but in current usage, the scope of the tribometer has been broadened to mean an instrument used to measure friction, wear, interface temperature, and other *tribo-parameters*.

4.3.1 Sliding Wear Test

Generally, the apparatuses used in the sliding wear tests, and their features, indicate that these tribometers have the same general feature of the applications. The test rigs accommodate test specimens and provide different forms of wear paths: unidirectional, oscillating, spiral, and others. The nature of wear situations has typically resulted in the development of unique apparatuses and test methods to simulate these situations and to provide the needed data. ASTM G99 [37] and ASTM G133 [38] are the standards for these tests, which do not specify specific values for the parameters but allows those to be selected by the researchers to provide a simulation model.

The most famous tests are as follows:

a) Unidirectional Sliding

In unidirectional sliding applications, the pin-on-disc, Figure 4.13, is used extensively in the wear test process. It has been also employed in many other applications including material wear and friction properties at elevated temperatures, controlled atmospheres, and lubrication.

The basic configuration of this apparatus consists of a pin sliding against a rotating counterface disk. The motion is generally unidirectional at a constant speed. The test parameters that have been used include the size and the shape of the pin, load, speed, and material pairs.

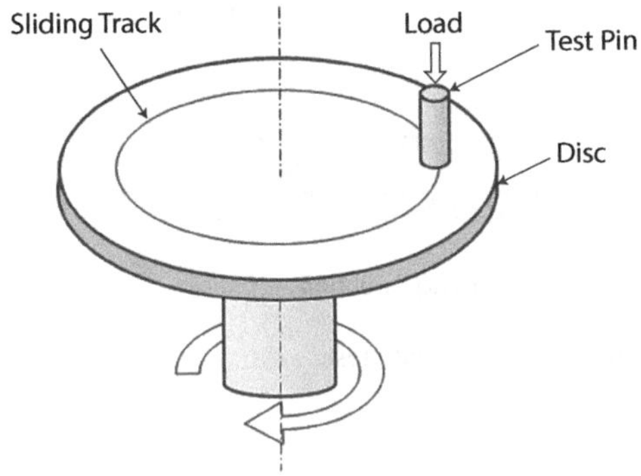

Figure 4.13 Schematic of a pin-on-disc sliding wear test.

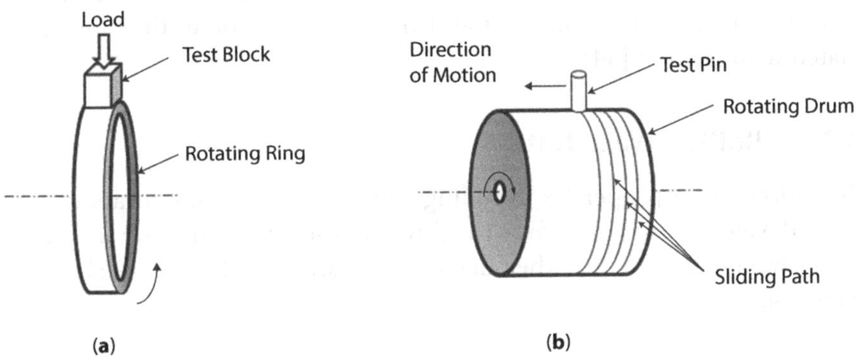

Figure 4.14 (a) Block-on-ring. (b) pin-on-drum.

In many studies, modelling of sliding aspects were conducted using block-on-ring, ASTM G137 [39], pin-on-drum, ASTM G65 [40], ball-on-flat [41, 42], as shown in Figure 4.14.

b) Reciprocating Sliding

Usually, pin-on-plate wear tests are used to evaluate wear and friction during the reciprocating sliding. Unlike material combinations, wear may occur at different rates depending on which material is the pin (or ball) and which is the flat plate. The generic configuration of the pin-on-plate tribo-tester is shown in Figure 4.15 [43]. Normally, the mode of rubbing is simulated by a radius-tipped, or flat-ended, test pin pressed against a flat counterpart that is moving backward and forward. Alternate configurations in which the specimen moves while the flat plate is fixed may be used.

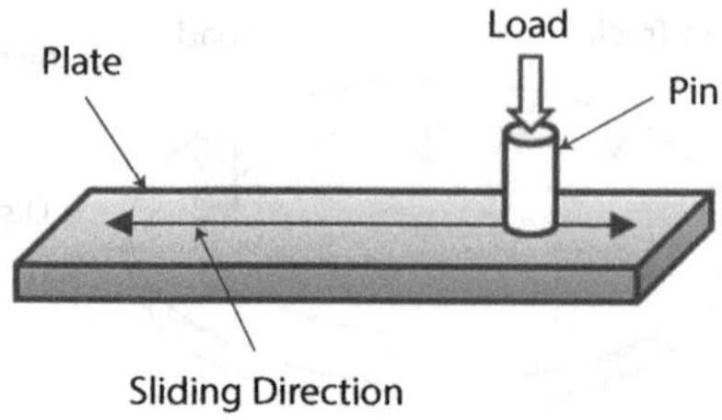

Figure 4.15 Diagram of the pin-on-plate test.

The generic feature of the change in direction of motion is that each cycle contains acceleration and deceleration portions. The sliding velocity profile tends to vary with different rigs and depends on the drive mechanism used. The reciprocating sliding test can be used for the simulation of fretting situations, by reducing the stroke of the motion to the range associated with fretting [44].

4.3.2 Rolling Wear Test

To address wear properties in rolling contact, several configurations have been developed [45]. The four (or three) ball-on-plate rigs are utilized to study wear behavior in either dry or lubricated conditions, as shown in Figure 4.16.

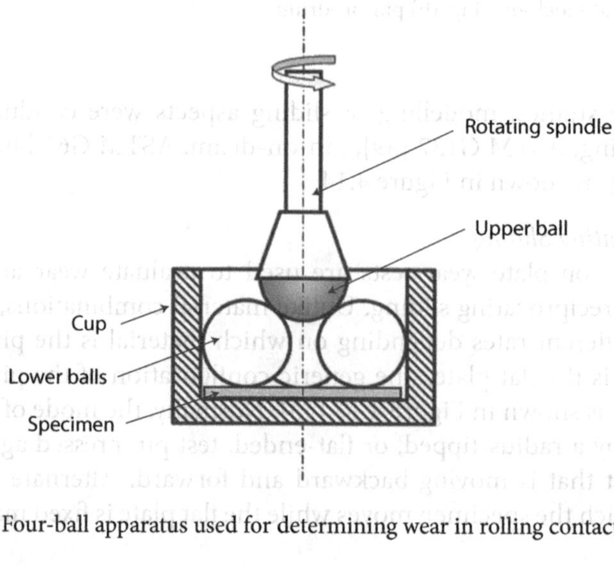

Figure 4.16 Four-ball apparatus used for determining wear in rolling contact.

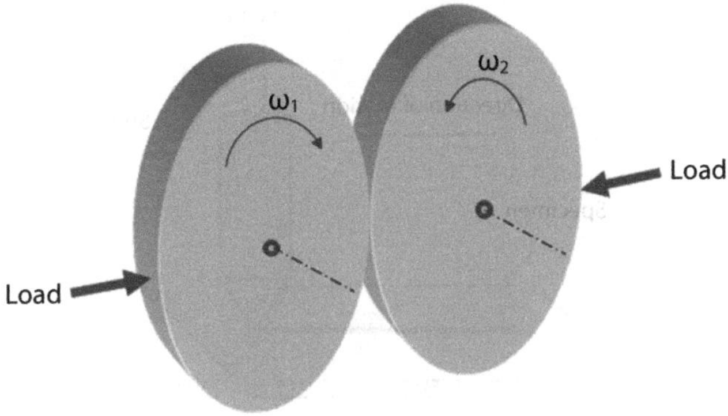

Figure 4.17 The basic configuration of a twin-disc apparatus used in rolling wear tests.

The motion of steel balls on the specimen plate is similar to the motion in a thrust bearing [46]. Also, a twin-disc test rig can be used to examine the impact of wear in rolling contact. In fact, for modeling the gear with a slip ratio (close to the pitch point), a twin-disc setup is the appropriate solution [47]. It consists of two driven cylinders or discs pressed against one another, as presented in Figure 4.17. Two motors drive the shafts so that different rotational speeds can be applied. Thus, a great advantage of this test rig is the possibility to apply relative slip of one surface on another.

4.3.3 Scratch Wear Test

Meanwhile, a roughly linear relationship exists between material hardness and wear resistance, and scratch test, ASTM G 171 [48], has often been utilized to give a guide to abrasive wear resistance. This is applied to metals, ceramics, polymers, and coated surfaces. The basic principle for all scratch tests is, nearly, the same. It involves producing a scratch on a solid surface by moving a hard, sharp stylus of specified geometry along a specified path. The normal force magnitude and the motion speed are constant, as shown in Figure 4.18.

Besides, scratch speed, scratch load, loading rates, number of scratches, and scratch length can be changed to give enough flexibility to define the desired test. Usually, a scratch test is performed under dry conditions. Nevertheless, it is possible to perform it under lubricated and selected temperature conditions.

Two shapes of stylus indenters are used in scratch tests: square-base pyramid, and circular cross-sections, Figure 4.19. It is performed to determine the wear amount that takes place under specified conditions. The wear

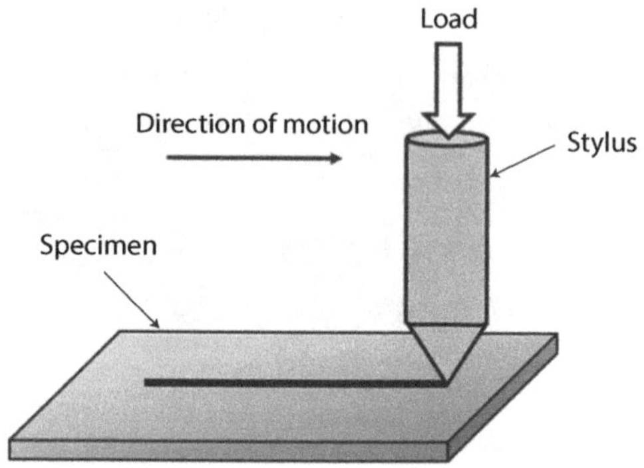

Figure 4.18 Illustration of the scratch test.

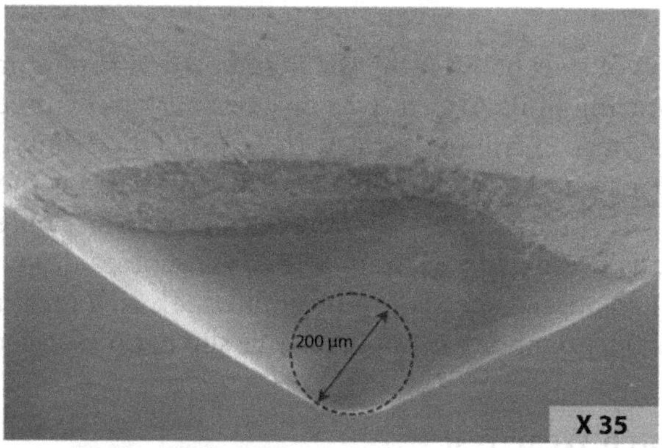

Figure 4.19 Image of the circular cross-section indenter (Rockwell diamond tip). Adapted from Ghosh, G. *et al.* [51], with permission.

mechanism, which takes place under each asperity in the nano-scale, can be magnified to a micrometer and millimeter-scale and reproduced. This gives researchers a good idea about what goes on under each asperity. These values can be applied for multiple asperities using simulations and can recreate the whole wear mechanism taking place on the solid surface [49, 50].

Effect of melting and microstructure on the microscale friction of silver-bismuth alloys, 2010, Tribology letters 38(3):275–282.

4.3.4 Abrasion Wear Test

In general, the main testing methods of abrasion wear are two and three-body abrasion. Abrasion of two-body can be imitated by utilizing material specimens. This pin specimen is loaded against a rotated relatively rough counterface (Figure 4.20). The applied load results in indentations and scratching of the wearing surfaces. It is postulated that high-stress abrasion enables the removal of material by cutting, plastic deformation, and surface fracture [52].

A three-body abrasion test was developed to simulate wear conditions in which low-stress scratching abrasion is the primary wear stage. Scratching wear arises by abrasive particles under light load. On the microscopic scale, these particles move and impose across the abraded surface, plowing and cutting. A dry sand/rubber wheel ASTM G65 is usually utilized to study the impact of various parameters on this mode of wear, such as abrasive particle size and shape and material parameters [53–55]. The test configuration is presented in Figure 4.21.

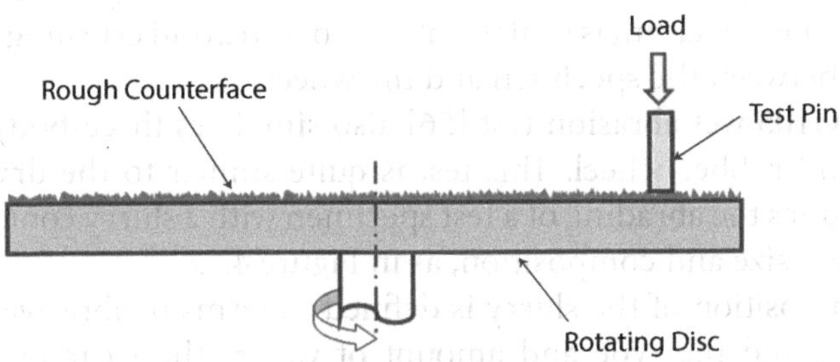

Figure 4.20 Tow-body abrasion wear test apparatus.

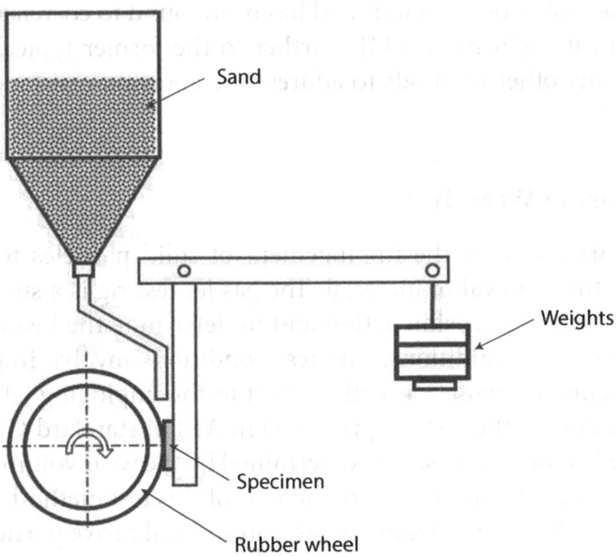

Figure 4.21 Schematic illustration of the sand/rubber wear test apparatus.

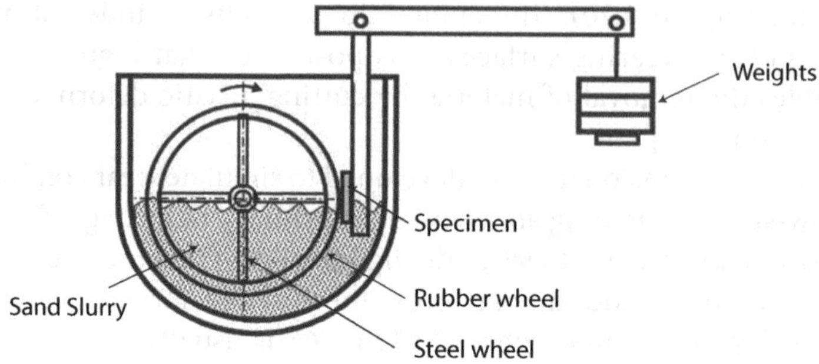

Figure 4.22 Wet sand/rubber wheel abrasion wear test apparatus.

The abrasive particles are trapped between the rotating wheel and a test specimen. The wheel rim is coated with rubber to avoid crushing the grains in the nip between the specimen and the wheel.

ASTM G105 wet abrasion test [56] also simulates three-body abrasion using a sand/rubber wheel. This test is quite similar to the dry abrasion test. It includes the abrading of a test specimen with a slurry containing grit of controlled size and composition, as in Figure 4.22.

The composition of the slurry is defined in terms of abrasive size, type and source, and the type and amount of water. The wear in both tests is determined by weight loss and converted to volume loss for material rankings. Wet abrasion test has been also extensively used to investigate scratching abrasion of polymers, and has been found to correlate well with many practical applications [34]. Further to the former typical methods, there are many other methods to address the abrasion resistance in some special cases.

4.3.5 Erosion Wear Test

Erosion is arising from the impingement of solid particles to a surface resulting in the removal of material. The gas jet test rig is a suitable apparatus to simulate the erosion action and to determine the loss of material under controllable conditions. The test conditions involve impact speed and the specimen orientation with respect to the impinging jet. The standard conditions for the test are presented in ASTM standard G76-05 [57]. The volume loss due to erosion is determined by the wear volume per gram of abraded material [54]. The fundamentals of this test method comprise a small nozzle delivering a stream of gas-entrained abrasive particles, which hits the test specimen surface (Figure 4.23).

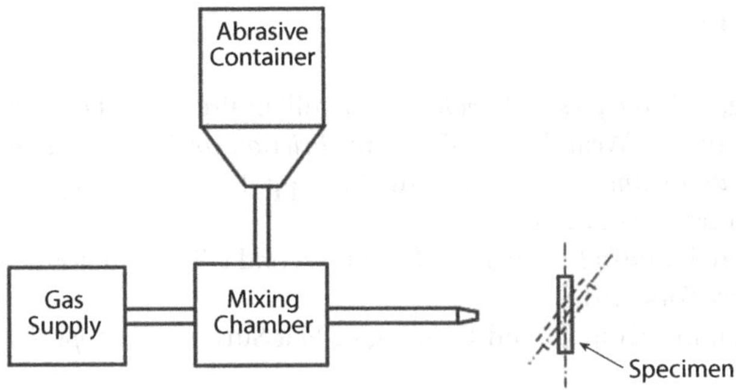

Figure 4.23 Erosion test apparatus.

4.3.6 Impact Wear Test

Impact wear has been identified as one of the dominant types of wear in many mechanical applications, especially in valves and seats. Although wear by an impact is possible to arise in these circumstances, there are no broadly used tests for this type of wear. However, several methods and mechanisms are suggested to determine impact wear and to compare the resistance of materials to this wear mode [58, 59]. Where there is a high impact velocity and therefore high impact energy, plastic deformation occurs in the soft material in contact with ductile extrusion away from the contact. Conversely, if impacts occur on a brittle specimen, then there is little plastic deformation and fracture occurs [60]. Generally, apparatuses used in impact wear tests can be grouped into two categories, pure impact, and compound impact. A hammer-type impact wear test rig is usually applied for pure or normal impact. In such cases, there is no sliding or adhesion involved, while the flat specimen is stationary. For compound impact, that is combined impact and sliding, the flat member rotates or oscillates beneath the hammer or projectile. It is important to mention that there is no fixed number of impacts used in this method for evaluating materials for engineering applications, but in some cases, it was selected somewhere around 100 impacts. Similar to testing methods and techniques used for sliding wear, resistance of materials to impact wear can be characterized using weight (or volume) loss after a fixed number of repeated impacts.

References

1. Halling, J., *Principles of Tribology*, Macmillan Press Ltd, London, 1979.
2. Straffelini, G., Wear Mechanisms, in: *Friction and Wear: Methodologies for Design and Control*, G. Straffelini (Ed.), pp. 85–113, Springer International Publishing, Cham, 2015.
3. Bhushan, B., *Introduction to Tribology*, second edition ed, John Wiley & Sons Inc, New York, 2013.
4. Archard, J.F., Contact and Rubbing of Flat Surfaces. *J. Appl. Phys.*, 24, 981–988, 1953.
5. Dwivedi, D.K., Adhesive wear behaviour of cast aluminium–silicon alloys: Overview. *Mater. Des.*, 31, 5, 2517–2531, 2010.
6. Abdelbary, A., *Extreme Tribology*, CRC Press, Boca Raton, 2020.
7. Holm, R., *Electric Contacts*, H. Gerbers (Ed.), Stockholm, Sweden, Springer-Verlag Berlin Heidelberg, 1946.
8. Rabinowicz, E., *Friction and Wear of Materials*, second edition ed, John Wiley and Sons, New York, 1995.
9. Narayanaswamy, B. *et al.*, The Impact of Retained Austenite Characteristics on the Two-Body Abrasive Wear Behavior of Ultrahigh Strength Bainitic Steels. *Metall. Mater. Trans. A*, 47, 10, 4883–4895, 2016.
10. Gurumoorthy, K. and Ghosh, A., Failure investigation of a taper roller bearing: A case study. *Case Stud. Eng. Fail. Anal.*, 1, 2, 110–114, 2013.
11. Davies, R.M., The determination of static and dynamic yield stresses using steel ball. *Proc. R. Soc.*, 197, 1050, 416–432, 1949.
12. Lundberg, G. and Palmgren, A., Dynamic capacity of roller bearings. Ingenicorsrenten-skapsakademiens, Acta Polytechnica Scandinavica. Mechanical Engineering Series, 2, 4, 1952.
13. Verma, S.K. and Tiwari, R., Robust optimum design of tapered roller bearings based on maximization of fatigue life using evolutionary algorithm. *Mech. Mach. Theory*, 152, 103894, 2020.
14. Suh, N.P., Update on the delamination theory of wear, Fundamentals of friction and wear of materials, in: *ASM Materials Science Seminar*, Rigeny, D.A. (Ed.), pp. 43–71, 1980.
15. Anaee, R.A.M. and Abdulmajeed, M.H., Tribocorrosion, in: *Advances in Tribology Pranav*, H. Darji (Ed.), IntechOpen, London, UK, 2016.
16. Robert, J.K.W., Tribo-corrosion of coatings: a review. *J. Phys. D: Appl. Phys.*, 40, 18, 5502, 2007.
17. Hua, N. *et al.*, Corrosive wear behaviors and mechanisms of a biocompatible Fe-based bulk metallic glass. *J. Non-Cryst. Solids*, 542, 120088, 2020.
18. Stachowiak, G.W., *Wear – Materials, Mechanisms and Practice*, WILEY, U.K, 2005.
19. Mohan Babu, P. *et al.*, Spark erosion machining behaviour of coconut shell ash reinforced silicon metal matrix. *Mater. Today: Proc.*, 33, 4602–4604, 2020.

20. Divakar, M., Agarwal, V.K., Singh, S.N., Effect of the material surface hardness on the erosion of AISI316. *Wear*, 259, 1, 110–117, 2005.
21. Ludema, K.C., *Friction Wear and Lubrication*, CRC Press, Boca Raton, 1996.
22. Paulo Davim, J., *Tribology for Engineers*, Woodhead Publishing Limited, Cambridge, UK, 2011.
23. Hutchings, I. and Shipway, P., 6 - Wear by hard particles, in: *Tribology*, Second Edition, I. Hutchings and P. Shipway (Eds.), pp. 165–236, Butterworth-Heinemann, Oxford, UK, 2017.
24. Tomlinson, G.A., The rusting of steel surfaces in contact. *Proc. R. Soc., London*, 115, 771, 472–482, 1927.
25. Li, D. *et al.*, Fretting wear of bolted joint interfaces. *Wear*, 458–459, 203411, 2020.
26. Curà, F. and Mura, A., Evaluation of the fretting wear damage on crowned splined couplings. *Proc. Struct. Integrity*, **5**, 1393–1400, 2017.
27. Wang, J., Lu, L., Ji, Z., A Study on the Impact Wear Behaviors of 40Cr Steel. *Tribol. Online*, 10, 4, 273–281, 2015.
28. Sarker, A.D., *Wear of Metals*, Pergamon Press, U.K, 1976.
29. Katiyar, P.K. *et al.*, Modes of failure of cemented tungsten carbide tool bits (WC/Co): A study of wear parts. *Int. J. Refract. Met. Hard Mater.*, 54, 27–38, 2016.
30. Kato, K., Classification of wear mechanisms/models. *Proc. Inst. Mech. Eng., Part J: J. Eng. Tribol.*, 216, 6, 349–355, 2002.
31. Lim, S.C. and Ashby, M.F., Wear Mechanism Maps. *Acta Metall.*, **35**, 1–24, 1987.
32. Usui, E., Shirakashi, T., Kitagawa, T., Analytical prediction of cutting tool wear. *Wear*, 100, 1, 129–151, 1984.
33. Bhushan, B., *Principles and applications of tribology*, 2nd edition ed, John Wiley & Sons, Inc, New York, 2013.
34. Kennedy, D.M. and Hashmi, M.S.J., Methods of wear testing for advanced surface coatings and bulk materials. *J. Mater. Process. Technol.*, 77, 1, 246–253, 1998.
35. Abdelbary, A., 7 - Methodology of testing in wear, in: *Wear of Polymers and Composites*, A. Abdelbary (Ed.), pp. 159–183, Woodhead Publishing, Oxford, 2014.
36. Richard, C., Manivasagam, G., Chen, Y.-M., Chapter 8: Measurement of Wear and Friction Resistance of Bulk and Coated Materials, In: *Materials Characterization*, Naryanaswami (Mohan) Ranganathan (Ed.), Jenny Stanford Publishing, New York, 259-287, 2015.
37. International, A., ASTM Standard G99, in: *Standard Test Method for Wear Testing with a Pin-on-Disk Apparatus*, ASTM International, West Conshohocken, PA, 2004.
38. International, A., ASTM Standard G133, in: *Standard Test Method for Linearly Reciprocating Ball-on-Flat Sliding Wear*, ASTM International, West Conshohocken, PA, 2005.

39. International, A., ASTM G137-97, in: *Standard Test Method for Ranking Resistance of Plastic Materials to Sliding Wear Using a Block-On-Ring Configuration*, ASTM International, West Conshohocken, PA, 2017.
40. International, A., ASTM Standard G65, in: *Standard Test Method for Measuring Abrasion Using the Dry Sand/Rubber Wheel Apparatus*, ASTM International, West Conshohocken, PA, 2004.
41. Kelley, J.E., Stiglich, J.J., Sheldon, G.L., Methods of characterization of tribological properties of coatings, in: *Surface Modification Technologies, in 1 st International Conference of Surface Modification, 25-28, January 1988*, T.S. Sudarshan and D.G. Bhat (Eds.), p. 169, The Metallurgical Society, Phoenix, AZ, USA, 1988.
42. Roshon, D.D., Testing machine for evaluating wear by paper. *Wear*, 30, 1, 93–103, 1974.
43. Ayerdi, J.J. *et al.*, Ball-on-flat linear reciprocating tests: Critical assessment of wear volume determination methods and suggested improvements for ASTM D7755 standard. *Wear*, 470–471, 203620, 2021.
44. Wayson, A.R., A study of fretting on steel. *Wear*, 7, 5, 435–450, 1964.
45. Harrass, M., Friedrich, K., Almajid, A.A., Tribological behavior of selected engineering polymers under rolling contact. *Tribol. Int.*, 43, 3, 635–646, 2010.
46. Stolarski, T.A., Hosseini, S.M., Tobe, S., Surface fatigue of polymers in rolling contact. *Wear*, 214, 2, 271–278, 1998.
47. Sukumaran, J. *et al.*, Modelling gear contact with twin-disc setup. *Tribol. Int.*, 49, 1–7, 2012.
48. International, A., ASTM Standard G 171 – 03, in: *Standard Test Method for Scratch Hardness of Materials Using a Diamond Stylus*, ASTM International, West Conshohocken, PA, 2003.
49. Sinha, S.K., Chong, W.L.M., Lim, S.-C., Scratching of polymers—Modeling abrasive wear. *Wear*, 262, 9, 1038–1047, 2007.
50. Rajendhran, N. *et al.*, Single-point scratch testing for understanding particle engagement in abrasion of multiphase materials. *Wear*, 476, 203689, 2021.
51. He, B. *et al.*, Effect of Melting and Microstructure on the Microscale Friction of Silver–Bismuth Alloys. *Tribol. Lett.*, 38, 3, 275–282, 2010.
52. Unal, H., Sen, U., Mimaroglu, A., Abrasive wear behaviour of polymeric materials. *Mater. Des.*, 26,8, 705–710, 2005.
53. Harsha, A.P., Tewari, U.S., Venkatraman, B., Three-body abrasive wear behaviour of polyaryletherketone composites. *Wear*, 254, 7, 680–692, 2003.
54. Bijwe, J. *et al.*, Friction and wear behavior of polyetherimide composites in various wear modes. *Wear*, 249, 8, 715–726, 2001.
55. Curley, M. and Joseph, T.G., Cyclic loading and a modified ASTM G65 abrasion test. *Wear*, 390–391, 346–354, 2017.
56. International, A., ASTM Standard G105, in: *Standard Test Method for Conducting Wet Sand/Rubber Wheel Abrasion Tests*, ASTM International, West Conshohocken, PA, 2002.

57. International, A., ASTM Standard G76, in: *Standard Test Method for Conducting Erosion Tests by Solid Particle Impingement Using Gas Jets*, ASTM International, West Conshohocken, PA, 2005.
58. Bayer, R.G., Engel, P.A., Sirico, J.L., Impact wear testing machine. *Wear*, 19, 3, 343–354, 1972.
59. Mahoney, N.J., Grieve, R.J., Ellis, T., A simple experimental method for studying the impact wear of materials. *Wear*, 98, 79–87, 1984.
60. Slatter, T., Lewis, R., Jones, A.H., The influence of cryogenic processing on wear on the impact wear resistance of low carbon steel and lamellar graphite cast iron. *Wear*, 271, 9, 1481–1489, 2011.

5

Mechanical Behavior of Thermal Spray Coatings

Jayant Gopal Thakare

School of Mechanical Engineering, Vellore Institute of Technology, Vellore, Tamil Nadu, India

Abstract

Thermal spray coating as any other industrial product must be tested for quality assurance after deposition. Coating properties vary with the deposition technique and is strongly dependent on spraying conditions and feedstock material. The mechanical behavior of the coating is a combined response of coating microstructure and its constituents. Thermal spray coatings are tested at research and development stage to ascertain the adhesion-cohesion strength, bending strength, wear resistance, and corrosion resistance. The tests are conducted as per the service condition requirements, most tests focus on the mechanical behavior of the coating and performance in working environment. The aim of this chapter is to inform the readers about various methods that can be used to analyze the mechanical behavior of the coating. The focus of the chapter is to successively described fracture toughness of coatings, effect of microstructure on fracture toughness, evaluation of residual stress (RS) and its effect on properties of the coating, Young's modulus and hardness of the coating and its effect on coating properties, effect of coating microstructure on mechanical behavior, and the effect of various reinforcements on the mechanical behavior of the coating.

Keywords: Coating, residual stress, fracture toughness, thermal spray, TBC

5.1 Introduction

The performance of a coating in a working environment depends on few specific properties of the coating. These mechanical properties include

Email: jayantgopal.thakare@vit.ac.in

Sarfraj Ahmed and Vinayak S. Dakre (eds.) Tribology and Characterization of Surface Coatings, (99–114) © 2022 Scrivener Publishing LLC

elastic-plastic behavior, adhesive-cohesive strength, thermomechanical behavior, fracture behavior, and friction properties. Coating properties mainly depend on torch design, feedstock composition, deposition parameters that govern a coating's microstructure, and mechanical properties [1]. However, the high-temperature performance of the layer depends mainly on feedstock composition (coating constituents). The coating microstructure is heterogeneous, anisotropic, and its properties are affected by measurement technique. Mechanical properties of thermal spray coatings can be measured by conventional bend test, adhesion test, and tensile test. Indentation method can also be used to measure coating properties effectively [2]. The coating properties are either measured for freestanding coating or coated specimens, properties will be different for coated specimen and free-standing coating when measure by conventional methods.

Thermal spray coatings are generally subjected to thermal and mechanical fatigue which leads to failure of the coating. This fatigue loading facilitates crack propagation and hence cause the coating to delaminate. The elastic modulus of coating varies with the method of measurement. The elastic modulus of a ceramic coating in a thermal barrier coating was measured to 200, 10, and 100 GPa, and when measured through nanoindentation, bending, and microindentation, respectively [3, 4]. The properties of a thermally sprayed coating are also affected by post processing. High-temperature sintering closes the microdefects and facilitates grain growth, reducing inter-splat defects [5, 6].

Plastic properties (temperature dependent) of metallic coatings are vital in understanding the performance of the coating. The plastic properties of a metallic coating play a vital role in initiating the brittle failures of the coating. In a thermal barrier coating, a higher creep strength of the bond coat avoids negative effect of compressive stresses induced as a result of thermally grown oxide which is predominantly Al_2O_3 in case of NiCrAlY bond coat [7]. A less ductile and weak bond coat results in an unstable TBC.

Residual stress (RS) generated in a coating can accelerate failure of the coating. The failure can be either due to cracking or spallation. The RS in a thermal spray coating should be evaluated before commissioning the coated component. The initial stress state can be detrimental during the cyclic thermal or isothermal conditions. The RS when combined with service stress will results in amplification of the stress experienced by the coating. The presence of RS results in formation of segmentation cracks which can be seen along the cross section of the coating [8, 9]. RS in a thermal spray coating is generated when molten, semi-melted, and heated particles impinge on a cooler substrate or a pre-existing lamella.

The RS is introduced in the coating from two main sources. Firstly, quenching stresses are introduced due to sudden cooling or impact of the heated feedstock particles. The second major cause of RS is the cooling of the coating substrate system to room temperature referred as secondary cooling stresses [10]. The quenching stress are tensile in nature and the secondary stresses can be tensile or compressive depending on the difference between the coefficient of thermal expansion (CTE) of substrate and the CTE of coating [11]. RS in the coating is also affected by the inflight velocity of the feedstock particles. The presence of compressive stresses inhibits crack growth whereas presence of tensile RS facilitates the same.

RS in a coating can be estimated by destructive as well as nondestructive methods. The X-ray diffraction technique and neutron diffraction technique are few examples of nondestructive methods of determining RS. Incremental hole drilling, curvature method, and layer removal method are few of the destructive techniques of measuring RS in the coating [12, 13]. Nondestructive testing methods estimate RS by comparing the lattice distortion of the stressed material to the unstressed one. The processing parameters of the coating invariably affect its RS and the elastic modulus. The elastic modulus of the coating increases with increase in deposition temperature as increase in deposition temperature increases adhesion between the splats.

The fracture toughness of a ceramic coating depends on the state of RS in the coating, and presence of compressive RS in the coating reduces the critical strain required to generate a transverse crack. The increase in compressive RS improves the fracture toughness of the coating.

5.2 Elastic Modulus and Fracture Toughness

The fracture toughness of a ceramic coating can be determined using indentation method which relates indentation hardness and elastic modulus to crack length. The hardness is measured either using nanoindentation or microindentation. A nanoindentation measurement involves the use of a Berkovich or Vickers indenter. In nanoindentation measurement, a depth to area correlation is established by calibrating the indenter using a standard sample. Each indentation cycle involves a loading, dwell, and unloading part; a load-displacement plot can then be used to calculate the Young's modulus and hardness of the coating. The modulus value obtained from the load-displacement properties is overestimated as it includes the effect of both the indenter material and the coating. Hence, reduced modulus is calculated using the relation given in Equation (5.1).

$$\frac{1}{Er} = \frac{1 - vc}{E_c} + \frac{1 - vi}{E_i} \tag{5.1}$$

ν_s, ν_i, E_s, and E_i are Poisson's ratio for coating, indenter, Young's modulus for coating, and indenter, respectively. The reduced Young's modulus is related to the slope of the elastic portion of the load vs. displacement plot using the relation given in Equation (5.2). The hardness is directly calculated from the projected area which is a function of depth. The hardness measured using nanoindentation will also be overestimated as compared to microindentation, because the penetration depth is affected by strain accumulation [14].

$$E = 0.5\left(\frac{\pi}{A}\right)^{0.5}\left(\frac{dP}{dh}\right) \tag{5.2}$$

The fracture toughness measurement can be carried out on the coating cross section, as the microstructure of a thermal spray coating is lamellar and interlamellar cracks are favorable points for initiation of delaminating cracks [15]. The coating fracture toughness can be calculated using the Antis relation shown in Equation (5.3) [16].

$$Kic = 0.016\left(\frac{E}{H}\right)^{0.5}\frac{F}{C^{1.5}} \tag{5.3}$$

E and H are the Young's modulus and the hardness in GPa, respectively, F is the load applied, and C is the crack length measured from the center of the diagonal, as shown in the Figure 5.1.

Fracture toughness can be calculated from the load that is required to initiate the radial cracks, cause chipping, and initial signs of delamination in the coating. Fracture toughness K_{IC} and the surface RS can be calculated by the measuring the length of a radial crack as a function of indentation load, as shown in Equation (5.4) [17].

$$\chi_r \frac{P}{C^{1.5}} = K_{IC} - 2\sigma\left(\frac{c}{\pi}\right)^{0.5} \tag{5.4}$$

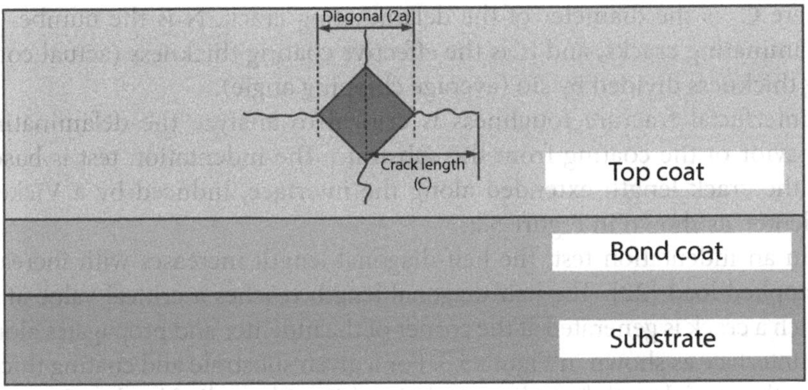

Figure 5.1 Schematic showing fracture toughness measurement location.

where χ_r depends on the type of indenter; for Berkovich type indenter, it is given by

$$\chi_r = .016\frac{E^{0.5}}{H} \tag{5.5}$$

The above analytical relation was initially developed for bulk material but has been successfully used for coatings [18–20]. The indentation depth causing radial cracks exceeds 10 % of the coating thickness. Hence, an elastic plastic zone might extend to the substrate, and to take this factor into consideration, an method based on energy dissipated during chipping was developed by lee and Bhushan [21]. The discontinuities in the loading curve are associated with chipping of the coating. The fracture toughness is determined at a point where the crack proceeds through the coating and the coating chips off. In this method, a line is extrapolated from the starting point of the discontinuity to the same indentation depth as measured at the end of the discontinuity. The dissipation energy is calculated as a difference between the extrapolated and the measured load displacement curve. The fracture toughness can be calculated by the relation given in Equation (5.6).

$$K_{IC} = \left[\left(\frac{E}{N(1-\vartheta^2)2\pi C_d}\right)\left(\frac{U}{h'}\right)\right]^{0.5} \tag{5.6}$$

where C_d is the diameter of the delaminating crack, N is the number of delaminating cracks, and h′ is the effective coating thickness (actual coating thickness divided by sin (average chipping angle).

Interfacial fracture toughness is critical to analyze the delaminating behavior of the coating from the substrate. The indentation test is based on the crack length extended along the interface, induced by a Vickers indenter, as shown in Figure 5.2.

In an indentation test, the half-diagonal length increases with increase in applied load [22]. The half-diagonal length reaches a critical value after which a crack is generated at the corner of the indenter and propagates along the interface as shown in Figure 5.3. For a given substrate and coating thickness, the crack length has a linear relationship with applied load as shown in Figure 5.4. The crack initiation point can be identified by the coordinates of intersection of critical indentation load and the corresponding crack length

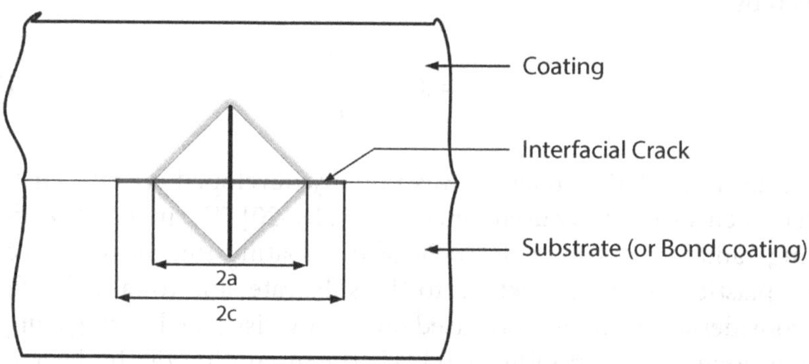

Figure 5.2 Inter-facial crack geometry induced by indentation testing.

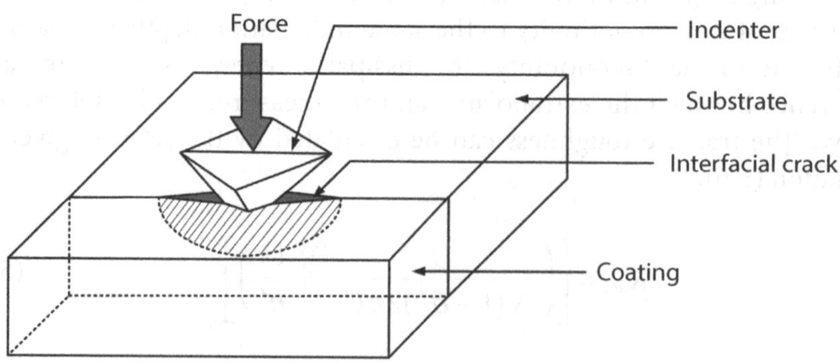

Figure 5.3 Schematic showing the generation of inter-facial crack.

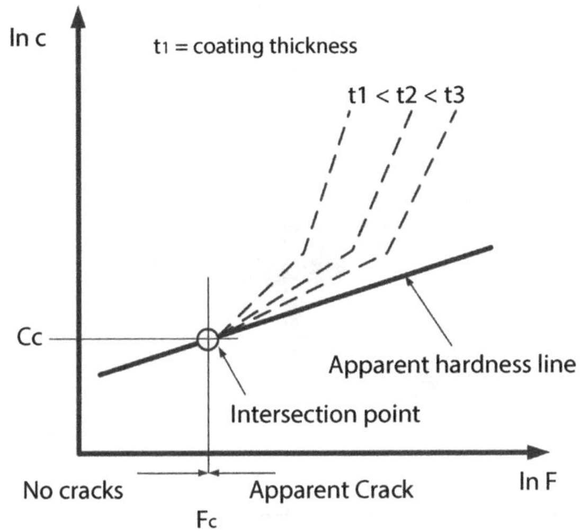

Figure 5.4 Relation between crack length (c) and test force (F) on a logarithmic scale [26].

[23–26]. The interfacial fracture toughness referred as apparent fracture toughness can be determined using the relation given below.

$$K_{ca} = 0.015\left(\frac{E}{H}\right)_I^{\frac{1}{2}} \frac{F_C}{C_c^{\frac{3}{2}}} \tag{5.7}$$

The ratio $\left(\frac{E}{H}\right)_I^{0.5}$ is ratio of the respective properties at the interface expressed as

$$\left(E/H\right)_I^{0.5} = \frac{\left(E_s/H_s\right)^{0.5}}{1+\left(H_s/H_c\right)^{0.5}} + \frac{\left(E_c/H_c\right)^{0.5}}{1+\left(H_c/H_s\right)^{0.5}} \tag{5.8}$$

Real-time monitoring of elastic modulus and fracture toughness can be carried out via digital image correlation (DIC) system. To analyze the effect of temperature on coating properties, *in situ* monitoring is highly advantageous. The monitoring is carried out on a three-point bend test sample and analysis is done using the DIC system which relates the digitized image to the Young's modulus value and the fracture toughness. The same system can also be used to analyze the *in situ* stress pattern of the coating system. Figure 5.5 shows schematic of *in situ* three-point bend test setup with DIC system. The three-point bend test specimen is clamped such that the clamp

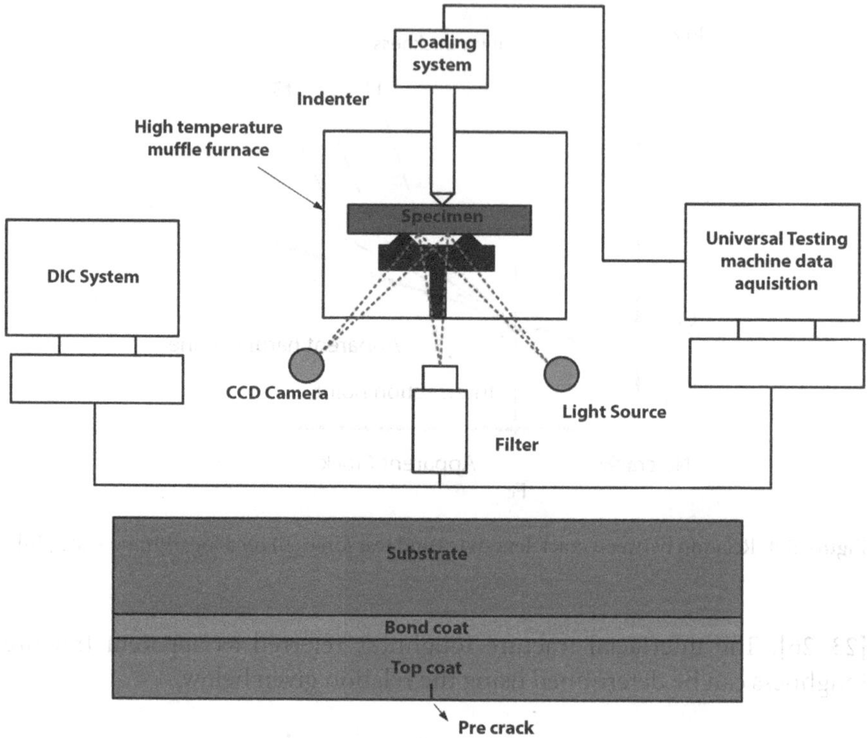

Figure 5.5 Schematic of high temperature three-point bend test with DIC.

and the center of the specimen should be in straight line. A muffle furnace is used to create uniform heating environment, and to ensure uniform heating of the specimen, thermocouples are placed near the specimen. The specimen is heated to a target temperature and dwelled to ensure thermal equilibrium is attained. A CCD camera with an inbuilt infrared filter is placed in front of the coating cross section. The filter is used to eliminate the effect of infrared light on image quality; a sampling rate of one image per second should be maintained. The composite beam theory can then be applied to calculate Young's modulus from observed deflection.

A coated specimen is a kind of composite beam, and depending on the number of deposited layers, composite beam theory can be applied to calculate the elastic modulus of the coating. A thermal barrier coating specimen is a composite beam and the bond coat can be considered as an extension of the substrate as the properties of the metallic bond coat is very close to the substrate properties. In the composite beam theory, the elastic modulus of the ceramic coating is given by the relation below.

$$E_c I_c + E_s I_s = \frac{FL^3}{48\delta} \tag{5.9}$$

The moment of inertia for the coating and the substrate is given by [27]

$$I_c = b\int_{-l-l_c}^{-l} y^2\,dy \tag{5.10}$$

$$I_s = b\int_{-l}^{l_s-l} y^2\,dy \tag{5.11}$$

where *lc* and *ls* are the thickness of the ceramic coating and the substrate and b is the width of the specimen. *l* is the distance from the neutral axis to the interface of the bond coat and the ceramic top coat is given in Equation (5.12).

$$l = \frac{E_s l_s^{0.5} - E_c l_c^{0.5}}{2E_s l_s + 2E_c l_c} \tag{5.12}$$

The elastic modulus of the coating can be evaluated from Equation (5.9), and the required Young's modulus of the substrate can be calculated from the bend test by Equation (5.13).

$$E_s = \frac{FL^3}{4bh^3\delta} \tag{5.13}$$

L is the length of the span, b and h are the width and the thickness of the substrate, respectively, and F and δ are the load and displacement, respectively.

The fracture toughness of the ceramic coating can also be evaluated during the above-mentioned three-point bend test. A groove of 0.5 mm is provided in the middle of the ceramic top coat as a through width pre-crack to capture the crack growth during the three-point bend test as shown in Figure 5.5. The crack initiates and propagates under the effect of applied load and this growth is monitored by using digital image correlation. The fracture toughness is determined by the relation given in Equation (5.14).

$$K_{TC} = Y\sigma_{cr}\sqrt{\pi a_o} \tag{5.14}$$

The left-hand side of the equation designates the fracture toughness of the ceramic top coat, and Y is the geometry factor which is "1" in case of bending test [28, 29]. The value of a_o is 0.5 mm, and the critical stress for crack prorogation can be calculated from Equation (5.15)

$$\sigma_{cr} = \frac{E_C y}{\rho} + \sigma_r \tag{5.15}$$

σ_r is the RS, and Y and ρ are the distance of the neutral axis from the crack tip and the curvature radius of the neutral axis, respectively.

$$y = l + \frac{l_c}{2} \tag{5.16}$$

$$\rho = \frac{\left[(L/2)^2 + \delta_{cr}^2\right]}{2\delta_{cr}} \tag{5.17}$$

5.2.1 Effect of Microstructure on Fracture Toughness

Thermal spray coating is an amalgamation of randomly impinging particles resulting in porous, and microcracked structure. The microstructure of a thermal spray coating is affected by the type of powder morphology. Hollow sphere morphology and fused-crushed morphology is most commonly used for thermal spraying. Yttria-stabilized zirconia coating deposited using hollow sphere morphology has lower fracture toughness as compared to the one with fused and crushed powder particles. The coating deposited using hollow sphere produced large number of interfacial defects, whereas that produced with fused and crushed have lower porosities with better interfacial bonding [30].

Eshelby inclusion model is employed to investigate the effect of microstructure on the coating mechanical properties [31, 32]. This model is used to investigate the effect of a phase on the variations in Young's modulus of the coating. The increase in volume fraction of β phase in a NiCrAlY coating decreases the Young's modulus of the coating, as predicted by the Eshelby inclusion model. The increase in porosity decreases the Young's modulus of the coating, also the type of porosity expressed in terms of the aspect ratio has an impact on the Young's modulus. The nature of porosity has effects on the Young's modulus, and oblate spheroids with their axes perpendicular to the interfacial plane have prominent effect as compared to the prolate spheroids with similar axes orientation [33]. Annealed ceramic coatings exhibit higher fracture toughness as compared to the as sprayed coating.

Various reinforcements are added in the feedstock of plasma sprayed coating with an aim to improve the fracture toughness of the APS sprayed ceramic coating. Thakare *et al.* reported an increase in fracture toughness of the plasma sprayed composite coating after carbon nanotubes (CNT) reinforcement [34]. The addition of CNT increased the fracture toughness by 15% which varied with the weight percentage of the reinforcement. The addition of reinforcement increases the fracture toughness by crack deflection, crack bridging, and CNT pull out. Mahato *et al.* [35] reported increase in fracture toughness of 8YSZ to 4.34 MPa $m^{1/2}$ after CNT reinforcement compared to the non-reinforced coating (3.90 MPa $m^{1/2}$).

5.3 Evaluation of Residual Stress

The magnitude of RS in coating affects the performance of the coating. High magnitude of tensile RS results in cracking, whereas higher compressive RS results in spallation and delamination. The initial state of RS in the coating must be evaluated to estimate the performance of the coating under cyclic and isothermal loading conditions.

The RS in a thermal spray coating can be evaluated using destructive as well as non-destructive techniques. The destructive testing methods rely on the measurement of strain values obtained by relaxation process. Incremental hole drilling method is one of the most commonly used stress relaxation methods [28, 29, 36–38]. The hole drilling procedure involves drilling a hole at the center of a strain Rosset. The drilling is done in equal steps and the relieved strain is recorded after every step. The stresses are then calculated from the relieved strain based on the assumptions that the strain relaxed after every increment is entirely due to the stress existing before the step. However, the assumption might not be initially true as the relaxed strain is a combined effect of change in hole geometry and the new hole depth increment [38, 39]. The integral method takes into consideration the effect of all relaxed strains at all depths and also estimates the stress at every step. The integral method is more adequate when the stress varies rapidly along the depth. The stress value are calculated on the basis of equation given below Equation 5.18 in the integral method [38, 40].

$$\sum_{j=1}^{j=i} \overline{A_{ij}}\, P_j = P_i$$

$$\begin{aligned} &\Sigma_{j=1}^{j=1} \overline{A_{ij}}\, P_j = P_i \\ &\Sigma_{j=1}^{j=1} \overline{B_{ij}}\, Q_j = q_i \qquad 1 \le j \le i \le n_c \\ &\Sigma_{j=1}^{j=1} \overline{B_{ij}}\, T_j = t_i \end{aligned} \tag{5.18}$$

$$P_j = \frac{(\sigma_{1j} + \sigma_{3j})}{2}; Q_j = \frac{(\sigma_{1j} - \sigma_{3j})}{2}; T_j = \tau_{13j}$$

$$p_i = \frac{(\varepsilon_{1i} + \varepsilon_{3i})}{2}; q_i = \frac{(\varepsilon_{1i} - \varepsilon_{3i})}{2}; t_i$$

$$= \frac{\varepsilon_{3i} + \varepsilon_{1i} - 2\varepsilon_{2i}}{2}$$

P_j and p_i are the RS component and volumetric strain component, respectively. Qj and Tj are the shear stress, respectively, whereas q_i and t_i are the shear strain component, respectively. ε_1, ε_2, and ε_3 represent the relaxed strain measurement, and σ_1, σ_3, and τ_{13} are the stress fields corresponding to the strains.

The problem can be solved by determining the calibration coefficients A_{ij} and B_{ij}. The coefficient can be analytically determined in case of through thickness holes, whereas in case of blind hole, FEM simulation should be used to determine the coefficients.

The RS in the ceramics and metallic coating can also be determined using curvature method. The curvature method estimates RS based on the change in the curvature resulting from the force and moment balance. The substrate is flat before deposition, the curvature changes when the coating is deposited on the substrate. The curvature changes can be recorded using optical microscopy. The sample is placed under the objective lens and moved horizontal (along X-axis). The sample is focused at each position, and the change in focus is measured as vertical height "h". The curvature can be obtained by the fitting the X-Z plot using circle function. The RS in a coating substrate system can be calculated using the modified Stoney's equation [41].

$$\sigma = \frac{1}{6h_c}\left[\left(\frac{E_C}{1-\vartheta_C}\right)\frac{h_c^3}{h_c+h_s}+\left(\frac{E_S}{1-\vartheta_s}\right)\frac{h_s^3}{h_c+h_s}\right]\Delta k \tag{5.19}$$

In case of bi-layered coating, the first layer (commonly metallic) is assumed to be one part of the substrate (an equivalent layer). The elastic modulus of the equivalent layer can be calculated by

$$E_{eq} = \frac{E_s h_s + E_{BC} h_{BC}}{h_s + h_{BC}} \tag{5.20}$$

The change in curvature of the sample is also affected by various heat treatments, and hence, curvature method can also be used to estimate the effect of various heat treatment procedures on the RS in the coating.

5.4 Conclusion

The present chapter highlights the criticality of mechanical properties of a thermal spray coating, its significance, and different methods that are used to estimate the mechanical properties of the coating. The mechanical properties primarily include Young's modulus, fracture toughness, and residuals stress in the coating.

The coating properties and its performance are affected by presence of nano-reinforcement. The chapter highlights different methods to estimate various properties of a thermally sprayed coating. The effect of properties on performance of the coatings can be understood from the variation in fracture toughness of the coating. The interfacial fracture toughness can be measured to understand the effect of thermal cycling on the coating. The *in situ* measurement of RS and Young's modulus methods mentioned in the chapter can be used to monitor the effect of process parameters on properties.

References

1. Taylor, R., Brandon, J.R., Morrell, P., Microstructure, composition and property relationships of plasma-sprayed thermal barrier coatings*. *Surf. Coat. Technol.*, 50, 141–149, 1992.
2. Kim, H.J. and Kweon, Y.G., Elastic modulus of plasma-sprayed coatings determined by indentation and bend tests. *Thin Solid Films*, 342, 201–206, 1999.

3. Harsha, S., Dwivedi, D.K., Agrawal, A., Influence of WC addition in Co-Cr-W-Ni-C flame sprayed coatings on microstructure, microhardness and wear behaviour. *Surf. Coat. Technol.*, 201, 5766–5775, 2007.
4. Li, S., An, Y., Zhou, H., Chen, J., Plasma sprayed YSZ coatings deposited at different deposition temperatures, part 1: Splats, microstructures, mechanical properties and residual stress. *Surf. Coat. Technol.*, 350, 712–721, 2018.
5. Chawla, V., Jayaganthan, R., Chandra, R., Structural characterizations of magnetron sputtered nanocrystalline TiN thin films. *Mater. Charact.*, 59, 1015–1020, 2008.
6. Kamal, S., Jayaganthan, R., Prakash, S., Evaluation of cyclic hot corrosion behaviour of detonation gun sprayed Cr3C2-25%NiCr coatings on nickel- and iron-based superalloys. *Surf. Coat. Technol.*, 203, 1004–1013, 2009.
7. Cook, L.S. and A.W., Temperature dependence of dynamic Young's modulus and internal friction in LPPS NiCrAIY. *J. Mat. Sci.*, 29, 5104–5108, 1994.
8. Kuroda, T. and Kobayashi, A., Adhesion characteristics of zirconia-alumina composite coatings by gas tunnel type plasma spraying. *Vacuum*, 73, 635–641, 2004.
9. Kuroda, S. and Clyne, T.W., The quenching stress in thermally sprayed coatings. *Thin Solid Films*, 200, 49–66, 1991.
10. Thakare, J.G., Pandey, C., Mahapatra, M.M., Mulik, R.S., Thermal Barrier Coatings—A State of the Art Review. *Met. Mater. Int.*, 27, 1–22, 2020.
11. Goswami, C., Bhat, I.K., Patnaik, A., Singh, T., Fekete, G., Fabrication of Ceramic Hip Implant Composites: Influence of Silicon Nitride on Physical, Mechanical and Wear Properties. *Silicon*, 12, 1237–45, 2019.
12. Zhang, X., Watanabe, M., Kuroda, S., Effects of residual stress on the mechanical properties of plasma-sprayed thermal barrier coatings. *Eng. Fract. Mech.*, 110, 314–327, 2013.
13. Vijigen., R.O.E. and D.J.H., Mechanical measurement of residual stress in thin PVD films. *Thin Solid Films*, 270, 264–269, 1995.
14. Wang, X., Guo, L., Peng, H., Zheng, L. *et al.*, Hot-corrosion behavior of a La2Ce2O7/YSZ thermal barrier coating exposed to Na2SO4+V2O5 or V2O5 salt at 900 oc. *Ceram. Int.*, 41, 6604–6609, 2015.
15. Movchan, M. and Rudoy, Y., Composition, structure and properties of gradient thermal barrier coatings (TBCs) produced by electron beam physical vapor deposition (EB-PVD). *Mater. Des.*, 19, 253–258, 1998.
16. Nicholls, J.R., Deakin, M.J., Rickerby, D.S., A comparison between the erosion behaviour of thermal spray and electron beam physical vapour deposition thermal barrier coatings. *Wear*, 233–235, 352–361, 1999.
17. Malzbender, J., De With, G., Den Toonder, J.M.J., Elastic modulus, indentation pressure and fracture toughness of hybrid coatings on glass. *Thin Solid Films*, 366, 139–149, 2000.

18. Thakare, J.G., Pandey, C., Mulik, R.S., Mahapatra, M.M., Mechanical property evaluation of carbon nanotubes reinforced plasma sprayed YSZ-alumina composite coating. *Ceram. Int.*, 44, 6980–6989, 2018.
19. Saremi, M., Afrasiabi, A., Kobayashi, A., Microstructural analysis of YSZ and YSZ/Al2O3 plasma sprayed thermal barrier coatings after high temperature oxidation. *Surf. Coat. Technol.*, 202, 3233–3238, 2008.
20. Vanevery, K., Krane, M.J.M., Trice, R.W., Wang, H. *et al.*, Column formation in suspension plasma-sprayed coatings and resultant thermal properties. *J. Therm. Spray Technol.*, 20, 817–828, 2011.
21. Li, X. and Bhushan, B., Measurement of fracture toughness of ultra-thin amorphous carbon films. *Thin Solid Films*, 315, 214–221, 1998.
22. Yamazaki, Y., Arai, M., Miyashita, Y., Waki, H., Suzuki, M., Determination of interfacial fracture toughness of thermal spray coatings by indentation. *J. Therm. Spray Technol.*, 22, 1358–1365, 2013.
23. Chicot, D., Démarécaux, P., Lesage, J., Apparent interface toughness of substrate and coating couples from indentation tests. *Thin Solid Films*, 283, 151–157, 1996.
24. Demarecaux, P., Chicot, D., Lesage, J., Interface indentation test for the determination of adhesive properties of thermal sprayed coatings. *J. Mater. Sci. Lett.*, 15, 1377–1380, 1996.
25. Lesage, J. and Chicot, D., Models for hardness and adhesion of coatings. *Surf. Eng.*, 15, 447–453, 1999.
26. Chicot, D., Araujo, P., Horny, N., Tricoteaux, A., Lesage, J., Application of the interfacial indentation test for adhesion toughness determination. *Surf. Coat. Technol.*, 200, 174–177, 2005.
27. Zhu, Q., He, W., Zhu, J., Zhou, Y., Chen, L., Investigation on interfacial fracture toughness of plasma-sprayed TBCs using a three-point bending method. *Surf. Coat. Technol.*, 353, 75–83, 2018.
28. Thurn, G., Schneider, G.A., Bahr, H.A., Aldinger, F., Toughness anisotropy and damage behavior of plasma sprayed ZrO2 thermal barrier coatings. *Surf. Coat. Technol.*, 123, 147–158, 2000.
29. Yang, L., Zhong, Z.C., You, J., Zhang, Q.M. *et al.*, Acoustic emission evaluation of fracture characteristics in thermal barrier coatings under bending. *Surf. Coat. Technol.*, 232, 710–718, 2013.
30. Dwivedi, G., Viswanathan, V., Sampath, S., Shyam, A., Lara-Curzio, E., Fracture toughness of plasma-sprayed thermal barrier ceramics: Influence of processing, microstructure, and thermal aging. *J. Am. Ceram. Soc.*, 97, 2736–2744, 2014.
31. Eshelby, J.D., Elastic Field Outside an Ellipsoidal Inclusion. *Am. Soc. Mech. Eng.*, 252, 561–569, 1959.
32. Eshelby, J.D., The determination of the elastic field of an ellipsoidal inclusion in an anisotronic medium. *Math. Proc. Cambridge Philos. Soc.*, 81, 283–289, 1957.

33. Saeidi, S., Voisey, K.T., McCartney, D.G., Mechanical properties and microstructure of VPS and HVOF CoNiCrAlY coatings. *J. Therm. Spray Technol.*, 20, 1231–1243, 2011.
34. Thakare, J.G., Pandey, C., Mulik, R.S., Mahapatra, M.M., Mechanical property evaluation of carbon nanotubes reinforced plasma sprayed YSZ-alumina composite coating. *Ceram. Int.*, 44, 6980–6989, 2018.
35. Mahato, N., Nisar, A., Mohapatra, P., Rawat, S. *et al.*, Effect of far-field stresses and residual stresses incorporation in predicting fracture toughness of carbon nanotube reinforced yttria stabilized zirconia. *J. Appl. Phys.*, 122, 2017.
36. Montay, G., Cherouat, A., Lu, J., Baradel, N., Bianchi, L., Development of the high-precision incremental-step hole-drilling method for the study of residual stress in multi-layer materials: Influence of temperature and substrate on ZrO2-Y2O3 8 wt% coatings. *Surf. Coat. Technol.*, 155, 152–160, 2002.
37. Sicot, O., Gong, X.L., Cherouat, A., Lu, J., Influence of residual stresses on the mechanical behavior of composite laminate materials. *Adv. Compos. Mater. Off. J. Jpn. Soc. Compos. Mater.*, 14, 319–342, 2005.
38. Valente, T., Bartuli, C., Sebastiani, M., Loreto, A., Implementation and development of the incremental hole drilling method for the measurement of residual stress in thermal spray coatings. *J. Therm. Spray Technol.*, 14, 462–470, 2005.
39. Wern, H., Finite-element solutions for mechanical drilling methods: A new integral formalism. *J. Comput. Appl. Math.*, 63, 365–372, 1995.
40. Flaman, M.T., Brief investigation of induced drilling stresses in the center-hole method of residual-stress measurement. *Exp. Mech.*, 22, 26–30, 1982.
41. McCarthy, J., Pei, Z., Becker, M., Atteridge, D., FIB micromachined submicron thickness cantilevers for the study of thin film properties. *Thin Solid Films*, 358, 146–151, 2000.

6

Tribological Properties of Carbon-Based Coatings

D. Dinesh Kumar[1*], P. Kuppusami[1†], Gobi Saravanan Kaliaraj[1], Siva Sankar Sana[2] and A.M. Kamalan Kirubaharan[1]

[1]*Center for Nanoscience and Nanotechnology, Sathyabama Institute of Science and Technology, Chennai, India*
[2]*School of Chemical Engineering and Technology, North University of China, Taiyuan, China*

Abstract
The ongoing industrial revolution demands advanced tribological coatings for enhancing the performance of mechanical components by minimizing friction and wear-related losses. Recently, carbon-based coatings have been widely studied for many tribology applications. The functionality of the carbon-based coatings can be achieved by controlled modification of sp^2/sp^3 hybridization ratio or hydrogen content during the deposition process. Amorphous carbon (a-C) coatings and hydrogenated a-C (a-C:H) or diamond-like carbon (DLC) coatings provide ultra-low friction behavior due to the enrichment of sp^2 hybridized carbon features. Similarly, hydrogen-free carbon coatings with tetrahedral sp^3 bonded networks (70%) show higher wear resistance behavior. Crystalline diamond coatings with enriched sp^3 hybridized carbon features like microcrystalline diamond (MCD) and ultra/nanocrystalline diamond (U/NCD) coatings exhibit superhardness (50–100 GPa) and remarkable wear resistance behavior. Though these coatings provide superior tribo-mechanical properties at low-temperature (<100°C), tribological properties, however, rapidly deteriorate at elevated temperatures which limit their applications in different operating conditions. To overcome these limitations, researchers have developed duplex structures of suitable metallic and/or ceramic materials doped/embedded in an a-C matrix. This chapter reviews the recent innovations that have reported on all the aspects of carbon-based coatings.

**Corresponding author*: ddinesh.tribology@gmail.com
†*Corresponding author*: pkigcar@gmail.com

Sarfraj Ahmed and Vinayak S. Dakre (eds.) Tribology and Characterization of Surface Coatings, (115–138) © 2022 Scrivener Publishing LLC

***Keywords*:** Hydrogenated and hydrogen-free DLC coatings, diamond coatings, sp^3/sp^2 hybridization, tribological properties, tribofilms, surface passivation

6.1 Introduction

Carbon is one of the most abundant elements in the earth, which can bond together in several ways and form many carbon allotropes due to its valence. Owing to the four valence electrons of carbon ($1s^2\ 2s^2\ 2p^2$), it can be able to form three major hybridization states resulting in three bonding configurations such as sp^3 bonds associated to diamond structure, sp^2 bonds related to graphite structure, and sp bonds similar to alkynes [1–3]. Diamond is a metastable form of carbon under atmospheric conditions with a cubic crystal structure that possesses fourfold sp^3 covalent bonding configurations. Graphite is a two-dimensional layered structure with sp^2 bond configurations which are interconnected by weak Van der Waals forces [4]. Apart from diamond and graphite structures, carbon exhibits in other forms including nanotubes, fullerenes, nanocrystalline graphite, and glassy carbon. Amorphous or glassy carbon has disordered nature and the structure varies based on the growth process [5, 6]. Recently, thin layers of carbon-based coatings have been widely used for a broad range of applications due to their superior optical, mechanical, and tribological properties. These coatings are applicable in various industries including cutting tools, M/NEMS, biomedical, and automotive industries. These coatings possess unique nature of adjustable properties such as superior hardness, wear resistance, chemical inertness, extremely low thermal expansion, and super-low friction which are most desirable for tribological applications [7–9]. Moreover, the tribological properties of these coatings can be tailored for specific requirements through deposition conditions. The physical properties (material characteristics) can be tailored by modifying the distribution of carbon hybridization states such as sp^1, sp^2, and sp^3 bonds, H atomic content, and the several dopants such as Si, F, N, B, and O [10].

6.1.1 Classification of Carbon-Based Coatings

Carbon-based coatings can be classified with different structures considering the hybridization states between carbon atoms (sp^1, sp^2, and sp^3), as shown in Figure 6.1. Moreover, inter bonding angles and interatomic distances have a significant influence on the determination of crystalline carbon (graphite and diamond) and amorphous carbon (a-C) coatings [11]. As seen in Figure 6.1a, diamond consists of a predominant presence of sp^3

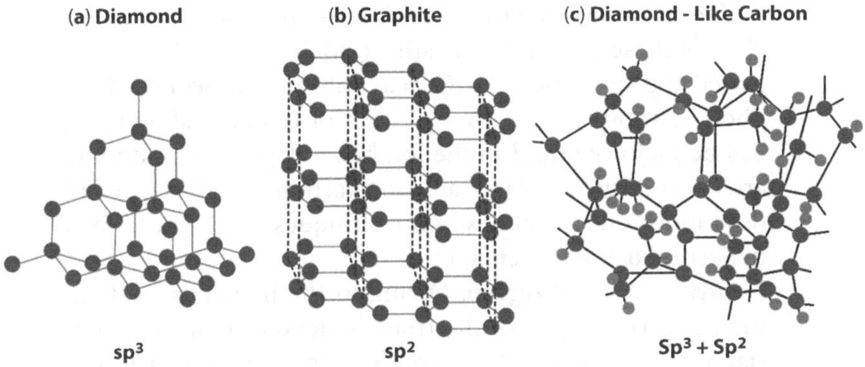

Figure 6.1 Different allotropes of carbon materials (redrawn based on Tanaka *et al.* [15]).

covalent σ-bonding between carbon atoms, and still remains as the known hardest material. In this cubic diamond structure, each carbon shares four valence electrons with other carbon atoms resulting in four single bonds [12]. The graphite structure with sp^2 bonding between carbon–carbon atoms is shown in Figure 6.1b. Graphite structure has a strong σ-bond between intra-layer atoms but weak bonding between the layers. In the case of a-C coatings, the atoms are randomly arranged by limited bonding between few atoms, and such coatings are generally termed diamond-like carbon (DLC) coatings. The DLC coatings contain mixtures of sp^2 and sp^3 hybridization states as illustrated in Figure 6.1c. DLC coatings exhibit superior mechanical properties similar to diamond, like high hardness, elastic modulus, high strength, chemical inertness, and good tribological performances. These physical properties can be altered by adjusting the sp^3/sp^2 ratios [13, 14]. In general, DLC coatings consist not only of a-C but also of hydrogenated a-C (a-C:H).

6.1.1.1 DLC Coatings

Figure 6.2 shows the ternary phase diagram of DLC coatings based on their sp^3, sp^2, and H content as first described by Jacob and Moller [16]. This phase diagram illustrates three major regions; namely, hydrogen (H) free carbon coatings along the left axis, higher H content region (right axis), and hydrogenated carbon region.

- ***H-free a-C coatings or DLC coatings***: These coatings are typically a mixture of sp^2 and sp^3 bonds without H content. Such DLC coatings are generally deposited by PVD techniques under ultra-high vacuum conditions. In order to further classify these H-free a-C coatings, McKenzie [17] has suggested the sp^3 enriched a-C materials as tetrahedral amorphous carbon (ta-C). The signature of tetrahedral bonding can be majorly found in the ion beam deposition, cathodic arc deposition, and laser ablation techniques. The detailed descriptions of the deposition techniques will be described in Section 6.3 of this chapter.

- ***Higher H-content region***: Owing to the higher H content, the right axis region of the triangle denotes that the material mostly consists of C_2H_2 and $(CH_2)n$, and these materials cannot be deposited as a coating, as illustrated in Figure 6.2.
- ***Hydrogenated carbon coatings***: These materials consist of both the signatures of tetrahedral hydrogenated a-C (ta-C:H) coatings and hydrogenated amorphous carbon (a-C:H) coatings. The ta-C:H coatings can be mostly deposited by the techniques using high-density plasmas [18] and the a-C:H can be typically deposited by chemical vapor deposition (CVD) techniques [19].

The mechanical properties (hardness and elastic modulus) of H-free a-C coatings are majorly governed by sp^3 carbons because of their higher packing density as well as the 3D bonding ability that offers higher strength

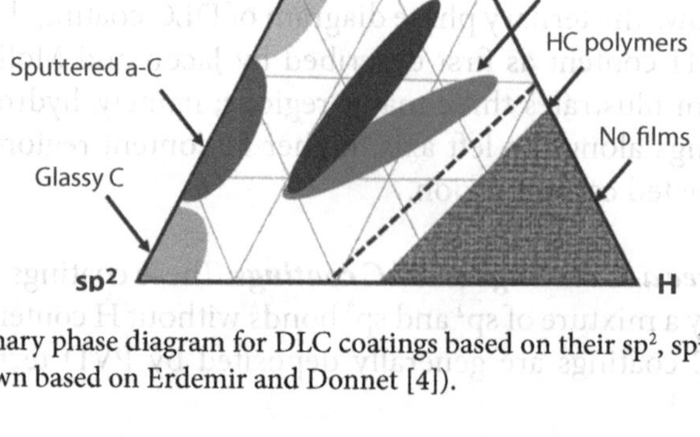

Figure 6.2 Ternary phase diagram for DLC coatings based on their sp^2, sp^3, and H contents (redrawn based on Erdemir and Donnet [4]).

to the coatings. Besides, the higher fraction of sp^3 bonds present in hydrogenated (i.e., a-C:H) coatings, the mechanical stability decreases with increasing H content. This might be due to the frequent termination of the 3D structure by the presence of higher hydrogen. Therefore, it is highly essential to design coatings for suitable tribological applications.

6.1.1.2 Diamond Coatings

Based on the research or industrial needs, the diamond coatings can be modified with different structures, such as microcrystalline diamond (MCD), nanocrystalline diamond (NCD), and ultra-nanocrystalline diamond (UNCD) coatings. Figure 6.3 illustrates the various types of carbon-based coatings based on their crystalline nature. The grain size of the diamond coatings is normally controlled by the H_2 gas concentration along with the carbon carrier gas (CH_4) and growth conditions [20]. The larger grains (≥1-μm grain size, as seen in Figure 6.3a) in MCD coatings can be mainly governed by the higher concentration of atomic H. These

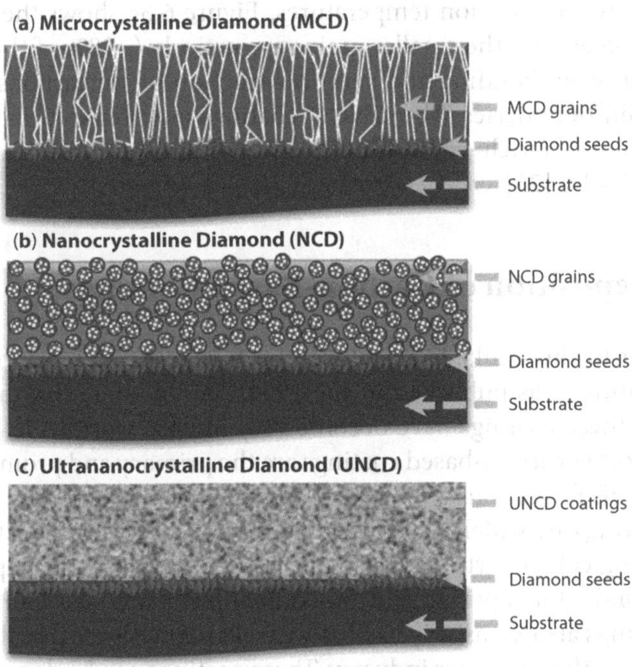

Figure 6.3 Different structures of diamond coatings (a) MCD, (b) NCD, and (c) UNCD coatings (drawn based on Narayan *et al.* [22]).

H atoms have several roles including removal of sp^2 carbons from the surface, carbon dangling bond termination, stabilizing diamond clusters and hydrogen abstraction from carbon-hydrogen bonds on the diamond surface [21]. On the other hand, deposition of suitable interlayers can also promote nucleation of the diamond as well as reducing the coating stresses. The MCD coatings have shown extremely higher mechanical properties similar to that of the bulk diamond (i.e., hardness of 100 GPa and elastic modulus over 1,000 GPa).

By controlling the parameters such as growth parameters, ratio of CH_4 in the CH_4/H_2 gas mixture, and grain size of the diamond coatings, one can control the re-nucleation density resulting in smaller grain sizes in the range of 10–100 nm [22, 23]. Such coatings are generally termed NCD coatings and the schematic of NCD coating is shown in Figure 6.3b. Such NCD coatings are relatively smoother than MCD coatings which can be further beneficial for smoother tribological contact applications. However, the roughness increases with increasing NCD coating thickness of more than few microns, and then, it behaves like MCD coating. Recently, the UNCD coatings is gaining great interest because of their smaller grain size less than 10 nm (usually 2–5 nm), lower surface roughness [24], and deposition at lower deposition temperatures. Figure 6.3c shows the schematic of UNCD coatings. The smaller grains are majorly (~98% of the material) composed of sp^3 bonding and very low fraction of sp^2 carbons is located at the grain boundaries. The UNCD coatings are prepared in hydrogen-deficient and Ar-rich environment using microwave plasma-enhanced CVD (PECVD) [25].

6.2 Deposition of Carbon-Based Coatings

According to the available literature, the first research work on carbon-based coatings was published in the 1950s, and after 1970s, the research works on these coatings have become very intense worldwide [5]. Today, DLC and other carbon-based coatings are the primary and promising coating materials for various tribological applications in numerous industries. These coatings are widely used in industries not only because of their superior tribological properties but also due to their great potential applications as semiconductor, optical, and biocompatible materials. Currently, the DLC coatings are extensively used as wear resistance layers on all magnetic hard disks in the computer industry. These coatings can be deposited using a variety of physical vapor deposition (PVD) and CVD techniques [26]. Solid carbon materials like graphite or carbon composites are used as the source material for depositing carbon-based coatings using PVD techniques. These sources are used in many PVD techniques such as ion deposition, ion-assisted sputtering, magnetron sputtering, cathodic arc deposition, pulsed laser deposition (PLD), and plasma deposition (Figure 6.4) for the deposition of a-C and DLC coatings.

6.2.1 Physical Vapor Deposition Techniques

Based on the literatures, most of the DLC and a-C coatings are deposited using various PVD techniques, and the various deposition techniques are shown schematically in Figure 6.4. Aisenberg and Chabot [27] have deposited hard a-C coatings using ion beam deposition techniques, and they have first termed these coatings as "diamond-like carbon". They found that the structure was partially crystalline with lattice constants similar to that of diamond structure. So far, the DLC coatings have been intensively investigated and applied for several industrial applications.

In the ion beam deposition (Figure 6.4a), carbon ions are generated in a discharge unit using graphite as cathode material in the presence of argon (Ar) gas [28]. Positive Ar ions and carbon ions are extracted and deposited on the negatively biased substrate by adjusting the substrate potential. Similarly, the carbon coatings can also be deposited using ion-assisted sputtering [29] by focusing the Ar ion beam on the graphite target materials as illustrated in Figure 6.4b. Sputtering is the most common industrial technique used for

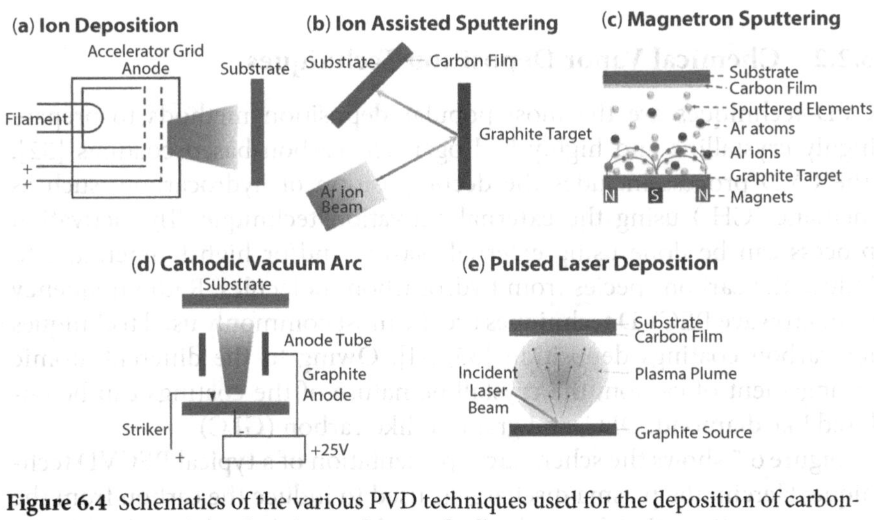

Figure 6.4 Schematics of the various PVD techniques used for the deposition of carbon-based coatings (redrawn based on Robertson [12]).

the deposition of DLC coatings by creating Ar plasma to sputter the graphite source using direct current or radio frequency power supply. Owing to the poor sputtering yield of graphite using diode sputtering, magnetron sputtering is widely used to enhance the deposition rate (Figure 6.3c). For this, permanent magnets are placed under the sputtering target to enhance the ion density in front of the target. To avoid the Ar ion bombardment on the substrate, a DC bias is widely applied to the substrate assembly during deposition resulting in high-quality DLC coatings. Moreover, the hydrogenated DLC (i.e., a-C:H) coatings can also be deposited via a reactive sputtering process using the combination of Ar plasma and hydrogen or methane. The deposition rate, uniformity, and sp^3 fraction in DLC or hydrogenated DLC coatings can be controlled by the gas pressure and plasma power.

Cathodic arc deposition [30] is another technique widely used for the DLC coatings, and the schematic is illustrated in Figure 6.4d. An arc is used to strike the graphite electrode and to create the energetic plasma with a very high ion density up to 10^{13} cm^3. The major disadvantage of this method is the production of particulates in the film due to high current density and such particulates can be further filtered by using magnetic filters [5]. In the PLD technique, a high energetic short pulses of laser (for example, ArF laser) is used to evaporate the graphite source. The plasma plume carries the evaporated carbon materials and deposited on the substrate as shown in the schematic representation of Figure 6.4e. The deposition rate of the coatings can be controlled through laser fluence. Besides, external plasma source [for example, electron cyclotron resonance (ECR)] assisted PVD techniques are also generally used for the deposition of carbon-based coatings [31].

6.2.2 Chemical Vapor Deposition Techniques

CVD techniques are the most popular deposition methods to prepare highly crystalline and highly hydrogenated carbon-based coatings [32]. The CVD process includes the decomposition of hydrocarbon, such as methane (CH_4) using the external activation technique. The activation process can be done using external plasma and/or high temperature to induce the carbon species from hydrocarbon molecules. Radio frequency or microwave PECVD techniques are the most commonly used techniques for carbon coatings deposition [33, 34]. Owing to the different atomic arrangement of carbon, the crystalline nature of the coatings can be tailored like diamond, DLC, and graphite-like carbon (GLC).

Figure 6.5 shows the schematic representation of a typical PECVD technique. Herein, the plasma discharge is used to induce the carbon from the carrier gases (i.e., CH_4) for the deposition of diamond or other carbon-based

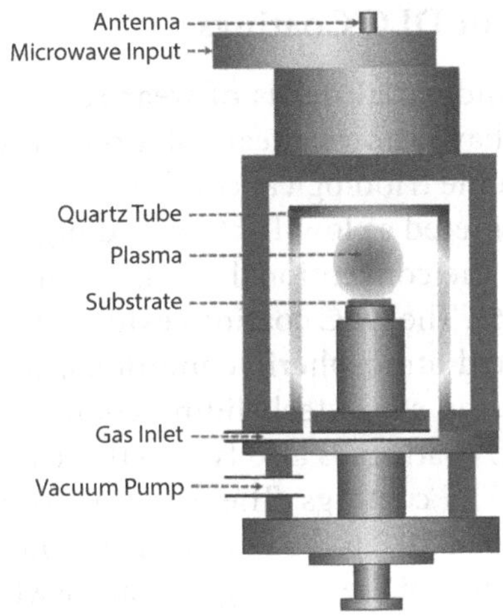

Figure 6.5 Schematic of plasma-enhanced chemical vapor deposition (PECVD) (redrawn based on Hung *et al.* [23]).

coatings. Substrate [typically Si and tungsten carbide (WC) substrates] is placed close to the anode. Diamond powders are ultrasonically seeded over the non-diamond substrates to increase the diamond nucleation to form the diamond coatings [22]. The external plasma aids to convert the carbon atoms to form diamond during the deposition. Higher substrate temperature in the range of 600–950°C is typically required for the deposition of diamond coatings. In general, diamond coatings consist of tetrahedral sp^3 carbons enriched polycrystalline phase with small fractions of sp^2 and sp bonds surrounded over the diamond crystallites. However, the coatings become amorphous due to the change in size and structure of the diamond crystallites with increasing the sp^2 and sp fractions.

6.3 Tribo-Mechanical Characteristics of Carbon-Based Coatings

As described in the previous sections, there are various coating architectures (i.e., DLC and diamond coatings) of carbon-based coatings which have been widely studied by several researchers in the past few decades. In this section, various aspects of carbon-based coatings and selection of coatings for tribological applications are discussed in detail.

6.3.1 H-Free a-C or DLC Coatings

With the emerging industrial needs of wear resistant tribological coatings, DLC coatings have shown great solutions where the components are subjected to extreme tribological conditions. The DLC or H-free a-C coatings can be considered as low-friction and high wear resistant materials as compared to the conventional ceramic tribological coatings (for example, TiN) [35, 36]. The DLC coatings exhibit friction values less than 0.2 under unlubricated atmospheric conditions, which is equivalent to the boundary-lubricated steel-steel sliding contacts. Similarly, the wear resistance of the DLC coatings is also two to three times higher than that of the conventional TiN coatings. The wear rate of the DLC coatings is typically in the range of 10^{-6} to 10^{-10} mm^3/Nm, and it varies depending on the structural features of the coatings. Kasioroswki *et al.* [37] prepared the DLC coatings using different plasma assisted techniques namely, HiPIMS, PECVD, and PIID, and the coatings were composed of different sp^3 hybridization around 30%, 50%, and 20%, respectively. Owing to the presence of enriched sp^3 carbons, the PECVD-deposited coatings exhibited high hardness, low friction (0.1), and enhanced wear resistance (4.79×10^{-8} mm^3/Nm). However, the DLC deposited with other coatings showed wedging spallation failure during the scratch study. Penkov *et al.* [38] have carried out the low-temperature deposition of DLC coatings using the C_{60} ion beam technique with deposition temperature ranging from 100°C to 450°C. Graphitization in DLC coatings increases with increasing temperature beyond 300°C and shows an increased residual stress (>3 GPa) which results in high friction as well as poor wear resistance. Carbon dangling bonds on the surface of the coatings play a key role during sliding interactions between H-free DLC-coated materials. Figure 6.6a illustrates the schematic representation of H-free DLC-coated surfaces interacting under various sliding conditions. The graphitization takes place under dry sliding conditions with increasing humidity and temperature and the chemical reactions occur (i.e., formation of C–O, COO, C–OH) with the available water vapor in atmosphere. Such tribo-chemical layers are reported to reduce the friction coefficient [39]. The most common issues in the DLC coating are lattice mismatch over non-carbon substrates and residual stress of DLC coatings which affect the adhesion and tribological properties of the DLC coatings. This can be rectified by depositing suitable interlayers over the desired substrates to deposit thicker DLC coatings with good adhesion. For example, ceramic interlayers of TiN, CrN, SiC, Si_3N_4, and TiC have been deposited on silicon and tungsten carbide (WC) substrates [12, 40, 41].

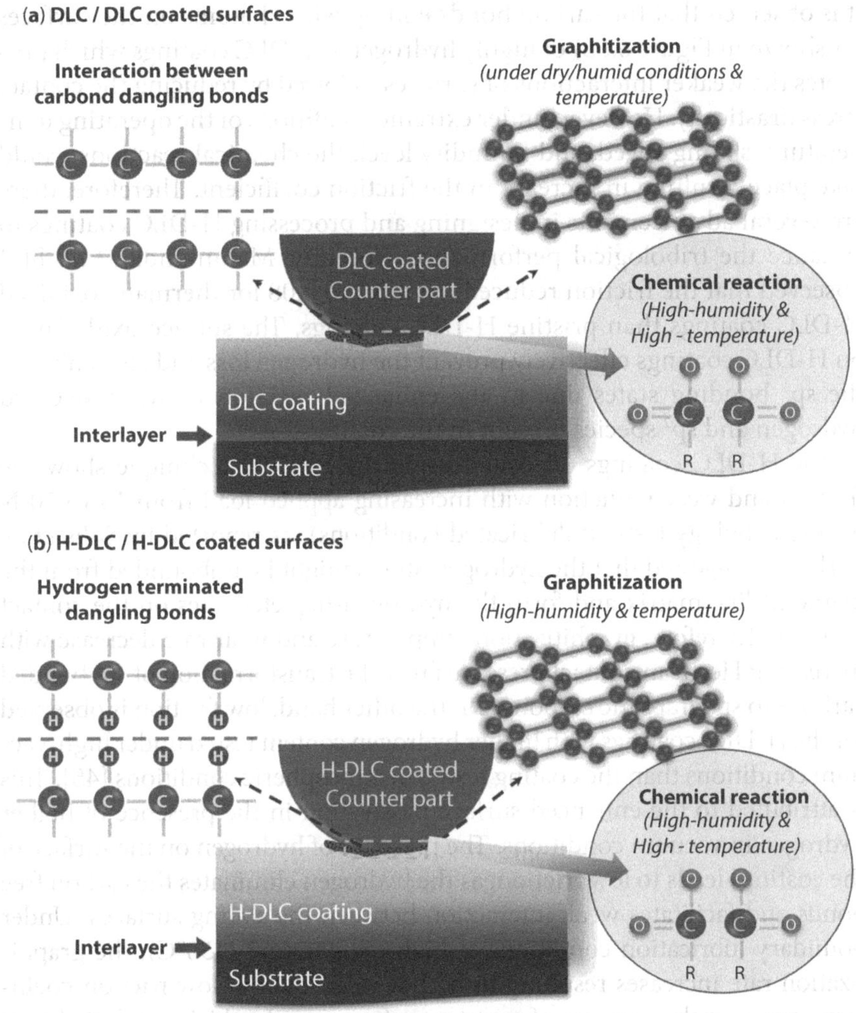

Figure 6.6 Schematics of tribological performance of H-free and H-DLC coatings with respect to change in humidity and temperature conditions. (a) Interactions of H-free DLC/DLC-coated surfaces and (b) interactions of H-DLC/H-DLC–coated surfaces (drawn based on Ronkainen and Holmberg [39]).

6.3.2 Hydrogenated a-C (H-DLC) Coatings

Hydrogen plays a key role in determining the tribochemical interactions between solid surfaces. The hydrogen bonded molecules over the DLC-coated surfaces effectively prevent the C–C bonds at friction interface, resulting in the significant reduction in the friction of DLC coatings [42].

It is observed that the carbon bonds end up with H atoms on the surfaces (as shown in Figure 6.6b) of highly hydrogenated DLC coatings which promotes the weaker interactions of surfaces followed by reducing the contact stress drastically. However, under extreme conditions of the operating temperature, sliding speed, and humidity level, the chemical reactions would take place resulting in increase in the friction coefficient. Therefore, there are several advancements in designing and processing H-DLC coatings to enhance the tribological performance. Recently, Manimunda *et al.* [43] observed that the friction reduced from 0.2 to 0.08 for thermally oxidized H-DLC coatings than pristine H-DLC coatings. The surface oxide layers on H-DLC coatings effectively prevent the hydrogen loss and also enhance the sp^3 bonding states due to the enhanced reaction between released hydrogen and sp^2 species present in the coatings.

The H-DLC coatings deposited using the PECVD technique show the friction and wear reduction with increasing applied load from 10 to 50 N during tribology tests (unlubricated conditions), as reported by Solis *et al.* [44]. It is suggested that the hydrogen atoms might be unbounded from the diamond-like matrix and form the hydrogen-depleted sites on the contact surfaces. Therefore, graphitization temperature and wear rate decrease with increasing Hertzian contact pressure favor the transformation of sp^3 bonded carbons to sp^2 graphitic carbons. On the other hand, low friction is observed for the H-DLC coatings with higher hydrogen content tested under high-vacuum conditions than the coating tested in atmospheric conditions [45]. This is attributed to the enhanced surface passivation in the presence of higher hydrogen in vacuum conditions. The presence of hydrogen on the surface of the coatings leads to low friction, as the hydrogen eliminates the carbon free bonds and facilitates weak interaction between the sliding surfaces. Under boundary lubrication conditions at high temperature (150°C), the graphitization rate increases resulting in higher wear loss but low friction coefficient owing to the easy slip of counter surface over the thick graphitic layers [46]. Cloutier *et al.* [47] studied the long-term stability of H-DLC coatings and their structural as well as tribo-mechanical properties. The structural and mechanical properties remain stable after 3 years of aging, and interestingly, the compressive stress decreases significantly. Owing to the decrease in bonding length rather than the sp^3 conversion, there is an increase in surface oxidation leading to a slight increase of friction coefficient.

6.3.3 ta-C Coatings

The tetrahedral amorphous carbon coatings such as H-free ta-C and ta-C:H coatings show relatively high hardness exceeding 70 GPa which are used in automotive engine parts where higher pressure conditions exist. As seen in Figure 6.2, these coatings are typically enriched with sp^3 hybridized carbon (>80%) and these coatings can be typically deposited using the arc deposition process [48]. The ta-C coatings with 81% of sp^3 hybridized carbon shows a lower friction coefficient due to a large number of dangling bonds reducing the frictional force under lubricated conditions [49]. Under lubricating (glycerol) conditions, ultra-low friction is attained due to the tribofilm formation and weak interactions between both OH-terminated ta-C coatings surface and lubricant [50]. In addition, the ta-C coatings are quite stable with superior tribological properties under low-temperature operations, whereas several cracks are formed under elevated temperatures (>200°C) resulting in the delamination of the coatings [51].

6.3.4 Diamond Coatings: MCD, NCD, and UNCD

As stated in Section 6.1.1.2, the diamond coatings are categorized based on the grain size. The major governing factor which decides the coatings' grain size is the sp^3/sp^2 ratio, which can be altered by various process conditions of the deposition techniques. The MCD coating exhibits larger fraction of sp^3 hybridized carbons which are tightly bound together to form larger grains along with a negligible amount of sp^2 carbons. The MCD coatings are hard and similar to bulk diamond material. Therefore, these coatings are an extraordinary candidate for cutting and drilling performances. The MCD coatings have superior wear resistance and lower friction than that of NCD coatings with a grain size less than 100 nm [52]. The microstructure of the polycrystalline diamond (MCD to NCD) coatings can be tuned by varying the CH_4/H_2 ratio (0.5%–3.0 %) using a hot filament CVD technique and their tribological properties were studied by Ajikumar *et al.* [53]. The grain size of coatings decreased from ~1 µm to ~20 nm with increasing CH_4 concentration. Figure 6.7 illustrates the friction behavior of the polycrystalline diamond coatings with respect to change in CH_4 concentration. The MCD coating shows a low friction coefficient (0.01) due to the mechanism of shear-induced amorphization (sp^2-rich interface), while the NCD coatings undergo shear-induced plastic deformation resulting in slightly increased friction coefficient (0.05) and wear. However, MCD coatings have major limitations concerning several tribological applications not only because of their excessive internal stress but also because of extremely higher hardness and elastic modulus which are not suitable for many base materials.

The UNCD coatings displayed tribo-environment dependent tribological properties as the coatings were tested under different gas (such as

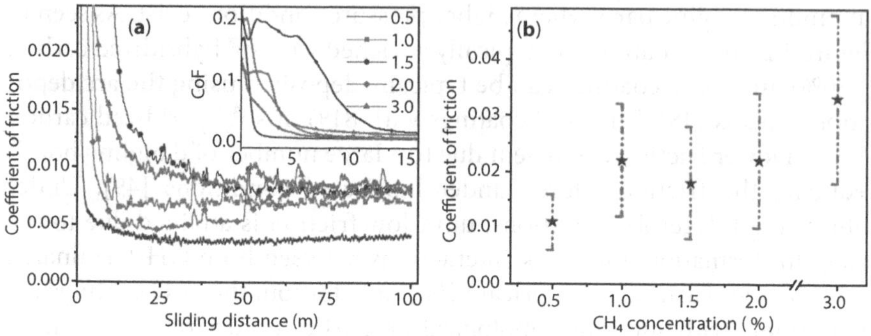

Figure 6.7 Frictional behavior of diamond coatings with varying CH_4 concentration such as 0.5%, 1%, 1.5%, 2%, and 3%. (a) Coefficient of friction versus sliding distance and (b) average friction coefficient versus CH_4 content (%) (adapted with permission from Ajikumar *et al.* [53]).

humid-atmosphere, nitrogen (N_2), Ar, and CH_4) environments [54]. Excess surface oxidation at the sliding interface under humid-atmospheric conditions increases the friction due to the formation of the oxy-functional groups (C–O, COO, and C–OH) which are formed by the water molecules. Besides, nitrogen-terminated carbon bonds under N_2 gas and higher Ar adsorption in Ar environment tend to prevent the carbon covalent bonding resulting in low friction values. Ultra-low friction under the CH_4 environment originates from the formation of higher hydrogen-terminated coating/counterpart interfaces. Moreover, the hardness of the UNCD coatings as well as the nature of counter bodies also has a significant role in determining the tribological properties. The UNCD coating deposited under N_2 plasma shows lower hardness and poor wear resistance because of the formation of sp^2 carbon which easily facilitates shear deformation. While the UNCD coatings deposited under the Ar plasma medium exhibit enhanced hardness with low wear due to the enrichment of sp^3 phases inside the sliding surface, as observed by Rani *et al.* [55]. While the UNCD coatings show high friction in atmospheric conditions than high-vacuum, thermally annealed UNCD coatings show ultra-low friction in ambient conditions. The presence of intrinsic hydrogen in UNCD coatings passivate the dangling bonds under high-vacuum, but the chemical passivation of dangling carbon bonds tend to reduce the friction of the annealed UNCD coatings under atmospheric conditions [56]. Although, higher sp^3 hybridization is present inside the wear track of the annealed UNCD coatings, the absence of chemical passivating medium results in friction.

Concerning the unlubricated tribology applications, significant research works have been recently carried out to design novel architectures of

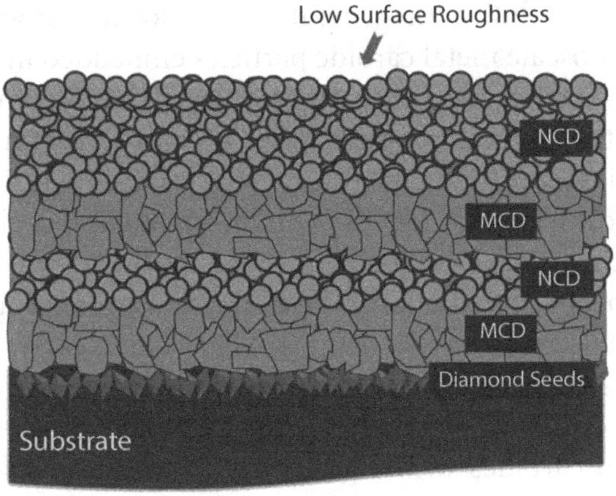

Figure 6.8 Schematic representation of MCD/NCD multilayer coating system with the base MCD layer deposited on Si_3N_4 substrate and smooth NCD top layer with low surface roughness (drawn based on Salgueiredo *et al.* [57]).

diamond coatings. For instance, Salgueiredo *et al.* [57] studied the tribological properties of MCD/NCD multilayer architecture deposited using the hot filament CVD technique. Figure 6.8 illustrates the schematic of MCD/NCD multilayer architecture. In this multilayer structure, the MCD layer was initially deposited on a ceramic substrate to maintain mechanical stability, while the NCD top layers were used to reduce the surface roughness. Such multilayer coating systems can withstand the critical loads during scratch tests greater than 200 N, whereas the bilayer and monolayer coating systems delaminated at ~110 and 100 N, respectively. The NCD coatings have ultra-low friction values of about 0.06 and enhanced wear resistance 2.4×10^{-7} mm^3/Nm. It is observed that the sp^2-rich NCD top layer worn out while the underneath MCD layer reduces the residual stress of the multilayer system. Ultra-smooth (~5nm) and Ar-free UNCD coatings (~2.7-nm grain size) can be prepared with higher deposition rate (1.5 μm/h) under the 1:10 ratio of CH_4:H_2 gases using HFCVD [58]. The coatings show ultra-low friction (~0.06) and a lower wear rate the magnitude of 10^{-8} mm^3/Nm than that of MCD and NCD coatings. This is attributed to the substantial increment of re-nucleation rate under an Ar-free environment as well as the presence of higher amount of energetic carbon precursors.

6.3.5 Metal-Doped Carbon-Based Coatings

In recent years, several research articles have been published on metal- or ceramics-doped carbon-based coatings to develop stress free coatings with high-temperature stability. In general, metal-doped carbon coatings are composed of nanoscale metal carbide particles embedded in the a-C matrix. The structural and tribo-mechanical properties of these coatings are strongly depending on the dopant concentration. Metal to carbon ratios in such Me-DLC coatings is in the range 0.1–0.2. Such hard Me-DLC coatings not only displayed low friction under dry conditions but also decreased the compressive stress than undoped coatings. Since 1980s, several research investigations have been carried out on the evaluation of tribological properties of different metal-doped DLC coatings. Metal (such as W, Ti, Cr, Ni, and Ta)–doped DLC coatings have been widely studied and the detailed survey regarding these coatings has been reported elsewhere [9]. Herein, the nanocrystalline metal carbides with a grain size in the range of 10–20 nm are formed and embedded in the DLC amorphous matrix. The major technical advantage of such duplex coatings is that the ceramic phases (i.e., metal carbides) provide wear resistance and sufficient hardness to the coatings, while the DLC matrix facilitates the self-lubricity. Also, the ceramics like TiC and WC embedded in the a-C matrix have brought much attention to the elevated temperature tribological applications. For instance, WC/a-C coating shows an ultra-low friction coefficient (0.12) at the operating temperature of more than 500°C, and such coatings are currently used to reduce the adhesive wear and tribo-oxidation behavior in many industrial applications [59].

6.4 Factors Influencing Tribological Properties of Carbon-Based Coatings

Besides the hydrogen content, sp^3/sp^2 ratio, dopant concentrations, and mechanical properties, several intrinsic and extrinsic factors also influence the friction and wear characteristics of DLC, H-DLC, and diamond coatings. Detailed review on various governing factors on the tribological properties of carbon-based coatings are reported in the previous literature [39, 60]. Some of the major factors such as internal stress, microstructure, surface properties, and sliding environments which influence the tribological properties are highlighted below:

- ***Internal stress:*** Controlling the internal stress of DLC or other coatings is the primary factor to enhance coatings performance. As mentioned earlier in Section 6.3.1, the carbon coatings generally have high internal stress (>3 GPa), which can be controlled by suitable deposition of interlayers before carbon coatings.

- ***Microstructure:*** Generally, microstructure of the diamond coatings consisting of large grains has low fatigue strength because of higher hydrogen adsorption. High-temperature annealing under vacuum conditions can reduce the grain size through the dehydrogenation process [61].
- ***Surface roughness:*** Surface of the coatings should be smooth since the rough surface can increase the deformation or asperity interlocking of coatings which spontaneously increase friction.
- ***Humidity:*** Humidity has a significant impact on the friction behavior of carbon-based coatings. Figure 6.9 represents the change in friction values concerning humidity level based on the literature [39]. Highly hydrogenated DLC coatings show lower friction in dry sliding conditions due to H-terminated dangling bonds in the presence of water molecules under humid conditions. In the case of H-free DLC, more interactions take place between sliding surfaces resulting in more wear under dry conditions subsequently leading to high friction.
- ***Operating temperature:*** Temperature has a significant effect on the graphitization of carbon coatings, which in one way advantageous to provide low-friction performance but it deteriorates the wear resistance of the coatings.
- ***Atmospheric conditions:*** H-DLC coatings show low friction under dry environments while sliding against DLC-coated

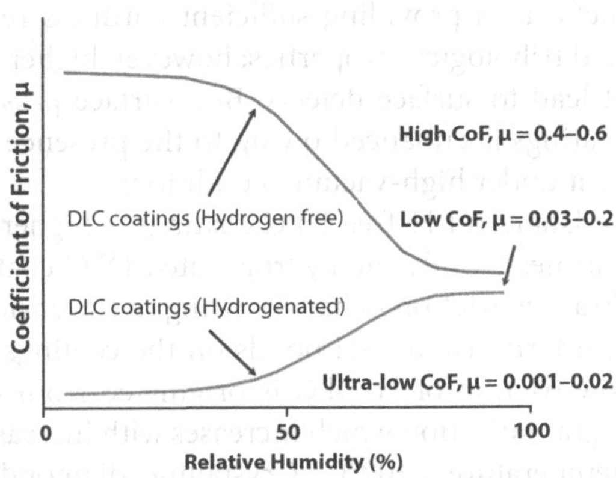

Figure 6.9 Role of humidity level on the frictional characteristics of DLC and H-DLC coatings (redrawn based on Ronkainen and Holmberg [39]).

counter surface because of adhesive friction at molecule level. Under high-vacuum conditions, desorption of hydrogen takes place and consequently increasing wear and friction due to frictional heat.

- ***Tribochemical reactions:*** Under a highly humid and ambient atmosphere, the formation of oxide species considerably influences the tribological properties of carbon coatings. For example, hydrocarbon oxide species (i.e., C=O) are formed, while DLC coatings slide in the humid atmosphere which facilitates further wear.
- ***Third body interactions:*** Development of third-body wear debris from other metallic or ceramic sliding counterfaces would drastically increase the friction and wear.

6.5 Conclusions

The carbon coatings are classified as H-free DLC, hydrogenated DLC, ta-C, and diamond (MCD, NCD, and UNCD) coatings based on the sp^3/sp^2 hybridization ratios, H content, and dopant concentration which can be altered through various process parameters of the PVD/CVD deposition techniques. The main deliverables of this chapter are summarized as follows:

- The enhanced tribo-mechanical properties of the coatings can be achieved by tailoring the sp^3/sp^2 ratio and H content during the deposition process. Higher sp^3 hybridized bonds are beneficial in providing sufficient hardness resulting in enhanced tribological properties; however, higher sp^3 and H content lead to surface defects. But, surface passivation of DLC coatings is enhanced owing to the presence of higher H content under high-vacuum conditions.
- Wear resistance of H-free DLC coatings is higher owing to their hardness, while the hydrogenated DLC coatings provide ultra-low friction behavior owing to the enrichment of hydrogen terminated C-H bonds on the coatings' surfaces. Such low friction of H-DLC is originated from the larger surface graphitization which increases with increasing operating temperature >100°C. Crystalline diamond coatings including MCD, NCD, and UNCD coatings with enriched sp^3 bonds of diamond phases can be deposited by adjusting

the gas precursors of plasma assisted CVD techniques. The MCD coatings possess ultra-hardness nature with higher intrinsic stress. The development of novel architectures like multilayer (i.e., MCD/NCD or UNCD coatings) is beneficial to reduce the internal stress and enhance the coatings performances.
- Various metal (Ti, Ta, Ni, Cr, etc.)-doped DLC/a-C coatings with appropriate dopant concentrations can also increase the hardness, wear resistance, and high-temperature tribological properties. The deposition of suitable metallic/ceramic interlayers can reduce the internal residual stress and durability of the coatings with good adhesion.
- Some of the major factors which include internal stress, microstructure, surface properties, and sliding environments have been found to influence the tribological properties.

References

1. Fujii, Y., Maruyama, M., Cuong, N.T., Okada, S., Pentadiamond: A Hard Carbon Allotrope of a Pentagonal Network of sp2 and sp3 C Atoms. *Phys. Rev. Lett.*, 125, 16001, 2020.
2. Robertson, J., Mechanism of sp3 bond formation in the growth of diamond-like carbon. *Diam. Relat. Mater.*, 14, 942–948, 2005.
3. Efstathiadis, H., Akkerman, Z., Smith, F.W., Atomic bonding in amorphous carbon alloys: A thermodynamic approach. *J. Appl. Phys.*, 79, 2954–2967, 1996.
4. Erdemir, A. and Donnet, C., Tribology of diamond-like carbon films: Recent progress and future prospects. *J. Phys. D. Appl. Phys.*, 39, R311, 2006.
5. Vetter, J., 60years of DLC coatings: Historical highlights and technical review of cathodic arc processes to synthesize various DLC types, and their evolution for industrial applications. *Surf. Coat. Technol.*, 257, 213–240, 2014.
6. Dasgupta, K. and Sathiyamoorthy, D., Disordered carbon - Its preparation, structure, and characterisation. *Mater. Sci. Technol.*, 19, 995–1002, 2003.
7. Meletis, E.I., Erdemir, A., Fenske, G.R., Tribological characteristics of DLC films and duplex plasma nitriding/DLC coating treatments. *Surf. Coat. Technol.*, 73, 39–45, 1995.
8. Gao, G.T., Mikulski, P.T., Harrison, J.A., Molecular-scale tribology of amorphous carbon coatings: Effects of film thickness, adhesion, and long-range interactions. *J. Am. Chem. Soc.*, 124, 7202–7209, 2002.

9. Bewilogua, K. and Hofmann, D., History of diamond-like carbon films - From first experiments to worldwide applications. *Surf. Coat. Technol.*, 242, 214–225, 2014.
10. Charitidis, C., Koumoulos, E., Dragatogiannis, D., Nanotribological Behavior of Carbon Based Thin Films: Friction and Lubricity Mechanisms at the Nanoscale. *Lubricants*, 1, 22–47, 2013.
11. Ray, S.C., Introduction: carbon and carbon nanomaterials, in: *Magnetism and Spintronics in Carbon and Carbon Nanostructured Materials*, pp. 23–45, Elsevier, Amsterdam, Netherlands, 2020.
12. Robertson, J., Diamond-like amorphous carbon. *Mater. Sci. Eng. R Rep.*, 37, 129–281, 2002.
13. Casiraghi, C., Robertson, J., Ferrari, A.C., Diamond-like carbon for data and beer storage. *Mater. Today*, 10, 44–53, 2007.
14. Li, C., Huang, L., Yuan, J., Effect of sp3 Content on Adhesion and Tribological Properties of Non-Hydrogenated DLC Films. *Mater. (Basel)*, 13, 1911, 2020.
15. Tanaka, Y., Miyake, K., Saito, A., Diamond-like carbon film for bearing parts and its mass production technology. *SEI Tech. Rev.*, 88, 141–145, 2019.
16. Jacob, W. and Möller, W., On the structure of thin hydrocarbon films. *Appl. Phys. Lett.*, 63, 1771–1773, 1993.
17. McKenzie, D.R., Tetrahedral bonding in amorphous carbon. *Rep. Prog. Phys.*, 59, 1611–1664, 1996.
18. Hakovirta, M., Tiainen, V.M., Pekko, P., Techniques for filtering graphite macroparticles in the cathodic vacuum arc deposition of tetrahedral amorphous carbon films. *Diam. Relat. Mater.*, 8, 1183–1192, 1999.
19. Tsuda, O., Ishihara, M., Koga, Y., Fujiwara, S., Setsuhara, Y., Sato, N., Characterization of amorphous hydrogenated carbon formed by low-pressure inductively coupled plasma enhanced chemical vapor deposition using multiple low-inductance antenna units. *J. Phys. Chem. B*, 109, 4917–4922, 2005.
20. May, P.W., Ashfold, M.N.R., Mankelevich, Y.A., Microcrystalline, nanocrystalline, and ultrananocrystalline diamond chemical vapor deposition: Experiment and modeling of the factors controlling growth rate, nucleation, and crystal size. *J. Appl. Phys.*, 101, 53115, 2007.
21. Anthony, T.R., Metastable synthesis of diamond. *Vacuum*, 41, 1356–1359, 1990.
22. Narayan, R.J., Boehm, R.D., Sumant, A.V., Medical applications of diamond particles & surfaces. *Mater. Today*, 14, 154–163, 2011.
23. Hung, J.M., Lin, L.H., Shih, Y.H., Liu, C.M., Cheng, H.C., Ou, K.L., Research of microstructural characteristics on nanocrystalline diamond by microwave plasma CVD. *Appl. Surf. Sci.*, 257, 5508–5512, 2011.
24. Wang, H., Song, X., Wang, X., Sun, F., Tribological performance and wear mechanism of smooth ultrananocrystalline diamond films. *J. Mater. Process. Technol.*, 290, 116993, 2021.
25. Kovalchenko, A.M., Elam, J.W., Erdemir, A., Carlisle, J.A., Auciello, O., Libera, J.A., Pellin, M.J., Gruen, D.M., Hryn, J.N., Development of

ultrananocrystalline diamond (UNCD) coatings for multipurpose mechanical pump seals. *Wear*, 270, 325–331, 2011.
26. Gao, G.T., Mikulski, P.T., Chateauneuf, G.M., Harrison, J.A., The Effects of Film Structure and Surface Hydrogen on the Properties of Amorphous Carbon Films. *J. Phys. Chem. B*, 107, 11082–11090, 2003.
27. Aisenberg, S. and Chabot, R., Ion-beam deposition of thin films of diamondlike carbon. *J. Appl. Phys.*, 42, 2953–2958, 1971.
28. Roy, R.K., Ahmed, S.F., Yi, J.W., Moon, M.W., Lee, K.R., Jun, Y., Improvement of adhesion of DLC coating on nitinol substrate by hybrid ion beam deposition technique. *Vacuum*, 83, 1179–1183, 2009.
29. He, X., Li, W., Li, H., Bonding structure and tribological properties of DLC films synthesized by dual ion beam sputtering. *Vacuum*, 45, 977–980, 1994.
30. Takikawa, H., Izumi, K., Miyano, R., Sakakibara, T., DLC thin film preparation by cathodic arc deposition with a super droplet-free system. *Surf. Coat. Technol.*, 163–164, 368–373, 2003.
31. Wu, A.M., Sun, J., Shen, X.K., Xu, N., Ying, Z.F., Dong, Z.B., Wu, J.D., Diamond-like carbon thin films prepared by ECR argon plasma assisted pulsed laser deposition. *Diam. Relat. Mater.*, 15, 1235–1241, 2006.
32. Noda, M. and Umeno, M., Coating of DLC film by pulsed discharge plasma CVD. *Diam. Relat. Mater.*, 14, 1791–1794, 2005.
33. Vercammen, K., Meneve, J., Dekempeneer, E., Smeets, J., Roberts, E.W., Eiden, M.J., Study of RF PACVD diamond-like carbon coatings for space mechanism applications. *Surf. Coat. Technol.*, 120–121, 612–617, 1999.
34. Seker, Z., Ozdamar, H., Esen, M., Esen, R., Kavak, H., The effect of nitrogen incorporation in DLC films deposited by ECR Microwave Plasma CVD. *Appl. Surf. Sci.*, 314, 46–51, 2014.
35. Voevodin, A.A., Donley, M.S., Zabinski, J.S., Pulsed laser deposition of diamond-like carbon wear protective coatings: A review. *Surf. Coat. Technol.*, 92, 42–49, 1997.
36. Ronkainen, H., Varjus, S., Holmberg, K., Friction and wear properties in dry, water- and oil-lubricated DLC against alumina and DLC against steel contacts. *Wear*, 222, 120–128, 1998.
37. Kasiorowski, T., Lin, J., Soares, P., Lepienski, C.M., Neitzke, C.A., de Souza, G.B., Torres, R.D., Microstructural and tribological characterization of DLC coatings deposited by plasma enhanced techniques on steel substrates. *Surf. Coat. Technol.*, 389, 125615, 2020.
38. Penkov, O.V., Pukha, V.E., Zubarev, E.N., Yoo, S.S., Kim, D.E., Tribological properties of nanostructured DLC coatings deposited by C 60 ion beam. *Tribol. Int.*, 60, 127–135, 2013.
39. Ronkainen, H. and Holmberg, K., Environmental and thermal effects on the tribological performance of DLC coatings, in: *Tribol. Diamond-Like Carbon Film. Fundam. Appl.*, pp. 155–200, Springer, Boston, MA, 2008.
40. Weber, M., Bewilogua, K., Thomsen, H., Wittorf, R., Influence of different interlayers and bias voltage on the properties of a-C:H and a-C:H:Me

coatings prepared by reactive d.c. magnetron sputtering. *Surf. Coat. Technol.*, 201, 1576–1582, 2006.

41. Cemin, F., Bim, L.T., Leidens, L.M., Morales, M., Baumvol, I.J.R., Alvarez, F., Figueroa, C.A., Identification of the Chemical Bonding Prompting Adhesion of a-C:H Thin Films on Ferrous Alloy Intermediated by a SiCx:H Buffer Layer. *ACS Appl. Mater. Interfaces*, 7, 15909–15917, 2015.
42. Paulmier, D., Zaidi, H., Nery, H., Le Huu, T., Mathia, T., Tribological behaviour of diamond-like coatings: effect of active gases in atomic and molecular states. *Surf. Coat. Technol.*, 62, 570–576, 1993.
43. Manimunda, P., Al-Azizi, A., Kim, S.H., Chromik, R.R., Shear-Induced Structural Changes and Origin of Ultralow Friction of Hydrogenated Diamond-like Carbon (DLC) in Dry Environment. *ACS Appl. Mater. Interfaces*, 9, 16704–16714, 2017.
44. Solis, J., Zhao, H., Wang, C., Verduzco, J.A., Bueno, A.S., Neville, A., Tribological performance of an H-DLC coating prepared by PECVD. *Appl. Surf. Sci.*, 383, 222–232, 2016.
45. Lugo, D.C., Silva, P.C., Ramirez, M.A., Pillaca, E.J.D.M., Rodrigues, C.L., Fukumasu, N.K., Corat, E.J., Tabacniks, M.H., Trava-Airoldi, V.J., Characterization and tribologic study in high vacuum of hydrogenated DLC films deposited using pulsed DC PECVD system for space applications. *Surf. Coat. Technol.*, 332, 135–141, 2017.
46. Mobarak, H.M., Masjuki, H.H., Mohamad, E.N., Kalam, M.A., Rashedul, H.K., Rashed, M.M., Habibullah, M., Tribological properties of amorphous hydrogenated (a-C:H) and hydrogen-free tetrahedral (ta-C) diamond-like carbon coatings under jatropha biodegradable lubricating oil at different temperatures. *Appl. Surf. Sci.*, 317, 581–592, 2014.
47. Cloutier, M., Harnagea, C., Hale, P., Seddiki, O., Rosei, F., Mantovani, D., Long-term stability of hydrogenated DLC coatings: Effects of aging on the structural, chemical and mechanical properties. *Diam. Relat. Mater.*, 48, 65–72, 2014.
48. Marks, N.A., Thin film deposition of tetrahedral amorphous carbon: A molecular dynamics study. *Diam. Relat. Mater.*, 14, 1223–1231, 2005.
49. Mabuchi, Y., Higuchi, T., Weihnacht, V., Effect of sp2/sp3 bonding ratio and nitrogen content on friction properties of hydrogen-free DLC coatings. *Tribol. Int.*, 62, 130–140, 2013.
50. Matta, C., De Barros Bouchet, M.I., Le-Mogne, T., Vachet, B., Martin, J.M., Sagawa, T., Tribochemistry of tetrahedral hydrogen-free amorphous carbon coatings in the presence of OH-containing lubricants. *Lubr. Sci.*, 20, 137–149, 2008.
51. Deng, X., Kousaka, H., Tokoroyama, T., Umehara, N., Tribological behavior of tetrahedral amorphous carbon (ta-C) coatings at elevated temperatures. *Tribol. Int.*, 75, 98–103, 2014.

52. Lei, X., Shen, B., Chen, S., Wang, L., Sun, F., Tribological behavior between micro- and nano-crystalline diamond films under dry sliding and water lubrication. *Tribol. Int.*, 69, 118–127, 2014.
53. Ajikumar, P.K., Ganesan, K., Kumar, N., Ravindran, T.R., Kalavathi, S., Kamruddin, M., Role of microstructure and structural disorder on tribological properties of polycrystalline diamond films. *Appl. Surf. Sci.*, 469, 10–17, 2019.
54. Rani, R., Panda, K., Kumar, N., Sankaran, K.J., Ganesan, K., Lin, I.N., Tribological Properties of Ultrananocrystalline Diamond Films in Inert and Reactive Tribo-Atmospheres: XPS Depth-Resolved Chemical Analysis. *J. Phys. Chem. C*, 122, 8602–8613, 2018.
55. Rani, R., Kumar, N., Kozakov, A.T., Googlev, K.A., Sankaran, K.J., Das, P.K., Dash, S., Tyagi, A.K., Lin, I.N., Superlubrication properties of ultra-nanocrystalline diamond film sliding against a zirconia ball. *RSC Adv.*, 5, 100663–100673, 2015.
56. Kumar, N., Kozakov, A.T., Sankaran, K.J., Sidashov, A.V., Lin, I.N., Controlled atmosphere dependent tribological properties of thermally annealed ultrananocrystalline diamond films. *Diam. Relat. Mater.*, 97, 107437, 2019.
57. Salgueiredo, E., Abreu, C.S., Amaral, M., Oliveira, F.J., Gomes, J.R., Silva, R.F., Self-mated tribological systems based on multilayer micro/nanocrystalline cvd diamond coatings. *Wear*, 303, 225–234, 2013.
58. Shen, B., Lin, Q., Chen, S., Ji, Z., Huang, Z., Zhang, Z., High-rate synthesis of ultra-nanocrystalline diamond in an argon-free hot filament chemical vapor deposition atmosphere for tribological films. *Surf. Coat. Technol.*, 378, 124999, 2019.
59. Gharam, A.A., Lukitsch, M.J., Balogh, M.P., Irish, N., Alpas, A.T., High temperature tribological behavior of W-DLC against aluminum. *Surf. Coat. Technol.*, 206, 1905–1912, 2011.
60. Zahid, R., Masjuki, H.H., Varman, M., Kalam, M.A., Mufti, R.A., Mohd Zulkifli, N.W.B., Gulzar, M., Nor Azman, S.S.B., Influence of intrinsic and extrinsic conditions on the tribological characteristics of diamond-like carbon coatings: A review. *J. Mater. Res.*, 31, 1814–1836, 2016.
61. Heinrich, G., Grögler, T., Rosiwal, S.M., Singer, R.F., Stöckel, R., Ley, L., The influence of diamond chemical vapour deposition coating parameters on the micro structure and properties of titanium substrates. *Diam. Relat. Mater.*, 5, 304–307, 1996.

7

Performance Evaluation of Journal Bearing Under Thin-Film Lubrication

Sandeep Soni

Tribology Laboratory, Department of Mechanical Engineering, S.V. National Institute of Technology, Surat, Gujarat, India

Abstract

The performance evaluation of hydrodynamic bearing is executed by adopting the viscosity correction approach under thin-film lubrication (TFL). The adsorbed layer thickness is the critically important aspect in TFL, and it also bifurcate regime of thin and thick lubrication. A very thin gap has been seriously recognized between journal surface and lubricant layers. The Reynolds lubrication equation has been modified for evaluating TFL performance parameters that affects the characteristics of hydrodynamic bearing. The finite difference method has been utilized to precisely discretize the clearance space of hydrodynamic journal bearing. The relationship between oil film and adsorbed layer thickness has been established to consider the effectiveness of TFL. A MATLAB code has been developed to calculate the modified pressure distribution and various journal bearing performance parameters including frictional properties. The load capacity, attitude angle, frictional characteristics, flow behavior, frictional torque, and energy loss for the hydrodynamic bearing under the influence of adsorbent thickness have been figured out and the effects on the TFL have been discussed. The comparison has been outlined with and without considering adsorbent thickness in the lubrication regime. This book chapter focuses on the nano-scale TFL which enhances the research understanding and applicability of particular lubrication field to journal bearing application.

***Keywords*:** Thin-film lubrication, finite difference method, journal bearing, adsorbed layer thickness

Email: ssonimed2009@gmail.com

Sarfraj Ahmed and Vinayak S. Dakre (eds.) Tribology and Characterization of Surface Coatings, (139–158) © 2022 Scrivener Publishing LLC

7.1 Introduction

The analytical and practical engineering community have attracted toward hydrodynamic bearing design from many decades. The wider range of application like textile spindles, nuclear reactors, high-speed aircrafts, and precision machine tools need robust hydrodynamic bearing technology to withstand different operating conditions.

On the basis of microscopic view, the formation of lubricant film is crucial factor for successful operation of journal bearing. The advancement in thin-film lubrication (TFL) has catered the need to develop theoretical relationship to divulge practical engineering problems.

In the reports on TFL [1–3], it was realized that the viscosity of the liquid lubricant changes due to reduction in distance between two walls of the surface in the narrow range. The changes in viscosity due to distances are not same for different lubricants.

Tichy [4] proposed a rheological model for assessment of lubrication properties in boundary lubrication (BL) regime. The model was based on material's continuum mechanics considerations. Author has displayed the predictions of load and friction for lubricated contact.

Tichy [5] developed a surface layer model used in TFL. Author has introduced the surface layer viscosity, thickness of the surface layer, and the conventional viscosity parameters in the proposed model.

Wen [6] discussed the distinct phenomena related to EHL and micro-EHL conditions for the lubricant film. They have elaborated analytical and experimental importance of research in TFL.

Luo *et al.* [7] studied the transition mechanism for lubricant film from elastohydrodynamic lubrication (EHL) to TFL in the measurement of the film thickness. They have presented the experimental observation in the variation of film thickness in the EHL and TFL regimes.

Qingwen *et al.* [8] derived Reynolds equation for TFL. They outlined the adsorption and viscous properties of bearing surfaces that affect behavior of the lubricant film.

Qingwen *et al.* [9] described the fluid velocity profiles generated on the bearing surfaces. They described that speed effects played vital role in thin-film flow due to increment in the shear rate.

Qu *et al.* [10] derived an equivalent viscosity model based on theory of adsorption. They applied the model to modify the Reynolds equation and presented the calculation results under thin-film range.

Qingwen *et al.* [11] presented the calculation of adsorbent thickness that characterizes the TFL. They presented the graphical interpretation for frictional properties and load capacity of a bearing operating under thin-film regime.

Zhang [12] summarized the research progress on TFL related to their computational methods, concept, and measurement techniques involved in them.

Liran and Jianbin [13] attempted thorough review of the major advancements in the theory of TFL that bridges the gap in between EHL and BL regimes.

Recently, Ono [14] presented the aftermath of high viscosity surface layers in the simulation of TFL. They considered the viscosity surface layers on the two sides of the bearing surfaces. They compared the one-side and two-side surface layer models for micro-tapered bearing.

The lubricant's viscosity is a decisive parameter in the hydrodynamic lubrication analysis especially for design of journal bearing. Further, for heavy-load and high-speed conditions modifications in viscosity are essentially required. The distribution of viscosity in the lubricant flow through bearing is strongly reliant on the surface adsorbent characteristics and lubricant's velocity.

Hence, it is vital to investigate the steady-state characteristics of hydrodynamic bearing in the TFL regime. This book chapter outlines the performance of short hydrodynamic bearing in the TFL. The distinct performance characteristics at steady state condition have been theoretically calculated by considering the influence of adsorbent thickness (2δ) in the short hydrodynamic bearing.

7.2 Governing Equations

7.2.1 Hydrodynamic Lubrication Modeling

The Reynolds equation, which governs the Newtonian lubricant and incompressible flow in the clearance space of a hydrodynamic lubricated bearing, may be utilized in the following dimensional form and can be written as [16]

$$\frac{\partial}{\partial x}\left(h^3\frac{\partial p}{\partial x}\right)+\frac{\partial}{\partial z}\left(h^3\frac{\partial p}{\partial z}\right)=6U\mu\left(\frac{dh}{dx}\right) \tag{7.1}$$

The short hydrodynamic bearing approximation considers an adequately small in axial direction and disdains the first term of LHS of Equation (7.1), allowing it to be solved numerically or analytically.

$$\frac{\partial}{\partial x}\left(h^3\frac{\partial p}{\partial x}\right)\approx 0 \tag{7.2}$$

The reduced form of Reynolds lubrication equation for proposed bearing is discerned as

$$\frac{\partial}{\partial z}\left(h^3\frac{\partial p}{\partial z}\right)=6U\mu\left(\frac{dh}{dx}\right) \tag{7.3}$$

Actually, by Equation (7.3), the lubricant flow situations in rotational direction do not affect the pressure. Firstly, the governing equation have been converted into dimensionless form using appropriate substitution in Equation (7.3), and it has been further used to enumerate the pressure variation in the clearance space of the short hydrodynamic bearing.

The short hydrodynamic bearing ($L/D = 0.5$) has been solved by using Gumbel's boundary approach. The pressure at the hydrodynamic bearing surface ends is maintained at atmospheric pressure. The following boundary conditions [Equation (7.4)] have been utilized to obtain the pressure variation in the clearance space of the short hydrodynamic bearing.

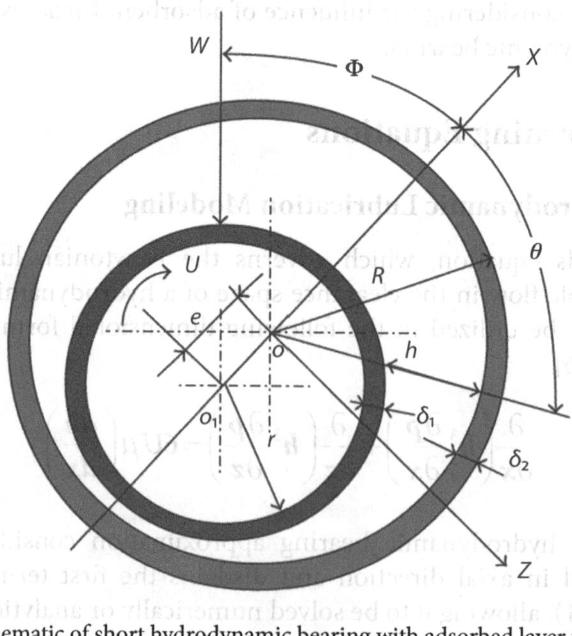

Figure 7.1 Schematic of short hydrodynamic bearing with adsorbed layer thickness.

$$p = 0;\ z = 0\ to\ (\pm L/2)$$
$$p = 0;\ \theta = 0,\ 2\pi \tag{7.4}$$

The lubricant film thickness (Figure 7.1) for the proposed bearing is computed by [16]

$$h = [C + C\varepsilon \cos\theta] \tag{7.5}$$

Figure 7.1 displays the coordinates and geometry of short hydrodynamic bearing with adsorbent thickness (**2*δ***) for the proposed analysis under TFL.

7.2.2 Thin-Film Lubrication Modeling

By taking assumption of similar adsorption properties ($\delta_1 = \delta_2$) at the two bearing surfaces (i.e., at shaft and bearing housing) in the equivalent viscosity correction model for the simulation of hydrodynamic bearing and effect of adsorbent thickness (**2*δ***) of thin-film regime to dynamic viscosity (***μ***) is displayed. Equation (7.6) provides the liaison between lubricant film thickness (***h***) and adsorbent thickness (**2*δ***) for hydrodynamic bearing [10, 15]:

$$\mu = \left(\frac{h}{h - 2\delta}\right)\mu_0 \tag{7.6}$$

The variation in the dynamic viscosity (***μ***) with lubricant film thickness (***h***) and adsorbed layer thickness (**2*δ***) can be achieved.

The influence of TFL in the hydrodynamic bearing has been obtained by solving Equation (7.3) with Equations (7.5) and (7.6) together. The revised form of Reynolds equation utilizing short hydrodynamic bearing approximation including adsorbent thickness (**2*δ***) is calculated through

$$\frac{\partial}{\partial z}\left(h^2\{h - 2\delta\}\frac{\partial p}{\partial z}\right) = \frac{6U\mu_0}{R}\left(\frac{dh}{d\theta}\right) \tag{7.7}$$

Equation (7.7) has been utilized to precisely simulate the lubrication behavior of hydrodynamic bearing under TFL regime and solved with finite difference scheme to achieve the steady-state aspects of the hydrodynamic bearing.

7.3 Solution Strategy

The solution for the oil flow field of the short hydrodynamic bearing assembly under the micro-film lubrication regime with Newtonian lubricant is obtained by finite difference scheme for the applied steady-state load. The Reynolds equation [Equation (3)] is solved together with particular boundary condition [Equation (7.4)], oil film thickness [Equation (7.5)], and viscosity model [Equation (7.6)] simultaneously.

Following dimensionless parameters are introduced in Equation (7.7) to convert it into dimensionless form which can be used to solve it numerically.

$$\begin{aligned} &x=(\theta * R);\ \ h=(C*H);\ \ \mu=(\mu_0\bar{\mu});\\ &\lambda=(D/L);\ \ z=\{(L/2)\bar{z}\};\\ &\delta=(C\bar{\delta});\ \ p=\{(6\mu UR/C^2)\bar{p}\} \end{aligned} \tag{7.8}$$

The Reynolds lubrication equation in dimensionless variables is depicted as

$$\{(\lambda)^2(H^3-2\bar{\delta}H)\}\left(\frac{\partial^2\bar{p}}{\partial\bar{z}^2}\right)=6\left(\frac{dH}{d\theta}\right) \tag{7.9}$$

The unwrapped hydrodynamic bearing film approximated as uniform rectangular grid spacing and central difference formulas are applied to the governing Reynolds equation to obtain the circumferential pressure distribution in the bearing. Numerical solution of Equation (7.9) has been achieved through finite difference scheme. A computer algorithm has been created to simulate the performance parameters of short hydrodynamic bearing without and with considering the adsorbed layer thickness (2δ) in the solution stage.

7.4 Performance Characteristics

Finite difference solution of Equation (7.9) dealing with adsorbed layer thickness (2δ) provides pressure variation in the bearing film through which distinct performance characteristics like load carrying capacity, attitude angle, frictional characteristics, oil flow rate, and torque, as well as power loss have been computed.

7.4.1 Load Carrying Capacity and Attitude Angle

The components of oil-film reactions (W_x and W_y) for short hydrodynamic bearing in the direction of bearing center and normal to it can be computed respectively by

$$\left.\begin{aligned} W_x &= -2\int_0^{\pi}\int_0^{L/2} (p\cos\theta R\, d\theta\, dz) \\ W_y &= 2\int_0^{\pi}\int_0^{L/2} (p\sin\theta\, d\theta\, dz) \end{aligned}\right\} \tag{7.10}$$

The components of oil-film reactions in direction of W_x and W_y are revised by considering the effect of adsorbed layer thickness (2δ) for calculating the total load carrying capacity and attitude angle under TFL of proposed bearing, and they can be expressed as

$$\left.\begin{aligned} W &= (W_x{}^2 + W_y{}^2)^{0.5} \\ \phi &= \left[\tan^{-1}\left(\frac{W_y}{W_x}\right)\right] \end{aligned}\right\} \tag{7.11}$$

7.4.2 Frictional Characteristics

Integration of shear stresses provides the frictional force around the bearing surface. The shear stress in the short hydrodynamic bearing is calculated by

$$\tau = \mu\left(\frac{U}{h}\right)$$

The frictional force F_f is achieved through, over the journal surface area by integrating following relation:

$$F_f = \int_A (\tau dA)$$

By putting $dA = (L * R)d\theta$, the frictional force turns out as

$$F_f = R * L * \int_0^{2\pi} (\tau) d\theta \tag{7.12}$$

Frictional force further improved by the taking the effect of adsorbent thickness (2δ) under TFL.

7.4.3 Oil Flow Rate

The oil flow rate within the bearing is obtained by following:

$$Q_z = \left[2\int_0^{\pi} -\left(\frac{h^3}{12\mu}\right)\left(\frac{dp}{dz}\right)_{z=\frac{L}{2}} R d\theta \right] = (\varepsilon ULC) \tag{7.13}$$

The flow rate calculation is further improved under TFL by considering the adsorbed layer thickness (2δ). The in-flow and out-flow of lubricant through the bearing is also obtained in the TFL regime.

7.4.4 Frictional Torque and Energy Loss

The assessment of frictional torque (T_f) on the bearing surface and energy loss during operation per unit of time (E_f) is figured out by

$$\left.\begin{aligned} T_f &= (F_f * R) \\ E_f &= (F_f * U) \end{aligned}\right\} \tag{7.14}$$

The detrimental effect of adsorbent thickness (2δ) on the frictional drag and energy deficit with time are simulated under thin-film regime.

7.5 Results and Discussion

Based on the solution strategy and steady-state characteristics, a computer code in MATLAB has been generated to simulate the steady-state operational behavior of short hydrodynamic bearing under thin-film regime of lubrication. These steady-state performance parameters have been scrutinized in terms of load carrying capacity, attitude angle, frictional characteristics, oil flow rate, frictional force, and energy loss per unit time with an aspect ratio of 0.50 taking Gumbel's boundary condition in the solution stage. The simulation parameters utilized in the performance assessment of short hydrodynamic bearing under TFL are depicted in Table 7.1.

The two cases are examined in the present simulation work:

(a.) $2\boldsymbol{\delta} = 0.00$ designates short bearing without envisaging ($2\boldsymbol{\delta}$),
(b.) $2\boldsymbol{\delta} = 0.08$ ***to*** 0.80 designates short bearing with envisaging ($2\boldsymbol{\delta}$).

Figures 7.2a and b discern the dimensionless film thickness variations that are achieved by without and with acknowledging the effect of various magnitudes of adsorbent thickness ($2\boldsymbol{\delta}$) at $\boldsymbol{\varepsilon} = 0.15$ and $\boldsymbol{\varepsilon} = 0.75$, respectively. It is observed from Figure 7.2 that when the adsorbent thickness ($2\boldsymbol{\delta}$) enhances from zero to some definite values, dimensionless film thickness decreases sharply at different values of $\boldsymbol{\varepsilon}$.

Table 7.1 Simulation study parameters.

Parameters	Range
Aspect ratio ($\boldsymbol{L/D}$)	0.50
Bearing length ($\boldsymbol{L}$)	$0.05\ \boldsymbol{m}$
Journal diameter ($\boldsymbol{D}$)	$0.10\ \boldsymbol{m}$
Radial clearance ($\boldsymbol{C}$)	$2.50\ \boldsymbol{E} - 05\ \boldsymbol{m}$
Eccentricity ratio ($\boldsymbol{\varepsilon}$)	0.15 & 0.75 0.01 – 0.07
Dynamic viscosity ($\boldsymbol{\mu}$)	$0.05\ \boldsymbol{Pa - s}$
Adsorbed layer thickness ($2\boldsymbol{\delta}$)	0.00 – 0.80

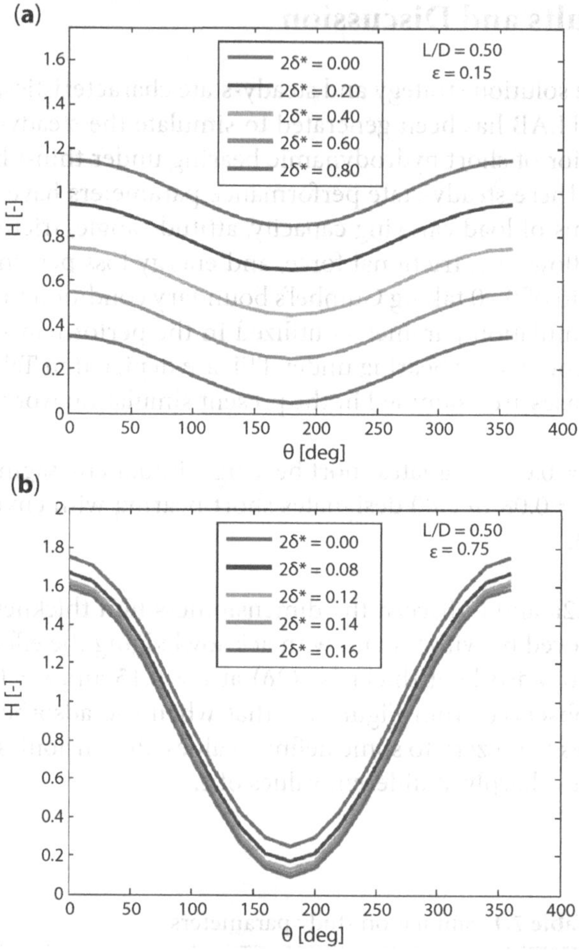

Figure 7.2 (a) Variation of $\boldsymbol{H}$ with $\boldsymbol{\theta}$ at $\boldsymbol{\varepsilon} = 0.15$ for selected values of $2\boldsymbol{\delta}$. (b) Variation of $\boldsymbol{H}$ with $\boldsymbol{\theta}$ at $\boldsymbol{\varepsilon} = 0.75$ for selected values of $2\boldsymbol{\delta}$.

Figures 7.3a and b exhibit the variation of dimensionless pressure distribution with angular coordinate ($\boldsymbol{\theta}$) acquired through proposed lubrication regime of thin-film. As the adsorbent thickness ($2\boldsymbol{\delta}$) increases, dimensionless pressure variation increases significantly toward higher values of $2\boldsymbol{\delta}$. At higher $\boldsymbol{\varepsilon}$, sharp increment is observed in the pressure with the increment in adsorbent layer thickness.

Figures 7.4a and b depict the dimensionless circumferential pressure variation of the short hydrodynamic bearing at distinct eccentricity ratios ($\boldsymbol{\varepsilon} = 0.01$–$0.07$) at adsorbent thickness ($2\boldsymbol{\delta}$) of 0.00 and 0.05, respectively.

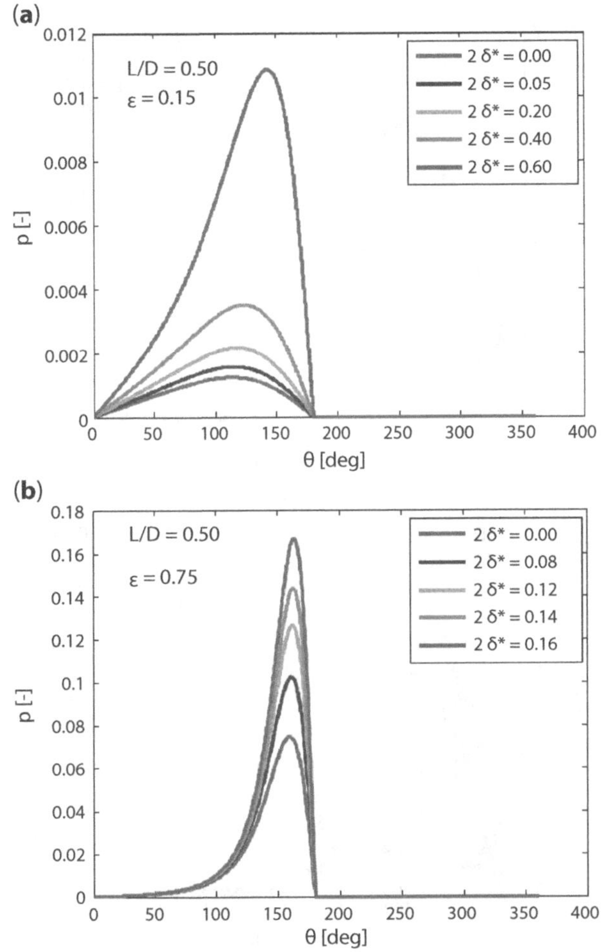

Figure 7.3 (a) Variation of $\boldsymbol{p}$ with $\boldsymbol{\theta}$ at $\boldsymbol{\varepsilon} = 0.15$ for selected values of $2\boldsymbol{\delta}$. (b) Variation of $\boldsymbol{p}$ with $\boldsymbol{\theta}$ at $\boldsymbol{\varepsilon} = 0.75$ for selected values of $2\boldsymbol{\delta}$.

It displays that the pressure $(\boldsymbol{p})$ enhances with increment in the adsorbent thickness $(2\boldsymbol{\delta})$ at smaller magnitudes of eccentricity ratio. The maximum pressure region also varies distinctly with instant change in the magnitude of adsorbent thickness $(2\boldsymbol{\delta})$ from $2\boldsymbol{\delta} = 0.00$ and 0.05 takes place. The lubricant in the TFL regime performs well, when the values of $(2\boldsymbol{\delta} + \boldsymbol{\varepsilon})$ are less than unity. So, judicious selection of the adsorbent thickness $(2\boldsymbol{\delta})$ is required during simulation of bearing performance under proposed lubrication regime.

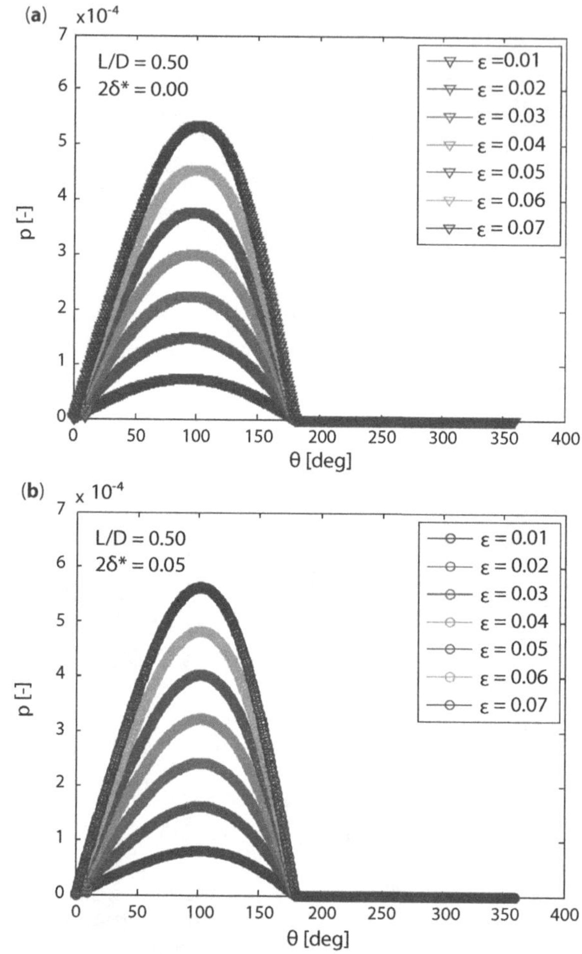

Figure 7.4 (a) Variation of p with θ = at $2\delta = 0.00$ for selected values of ε. (b) Variation of p with θ at $2\delta = 0.05$ for selected values of ε.

The variation of dimensionless load carrying capacity of the short hydrodynamic bearing along with the adsorbent thickness (2δ) is shown in Figures 7.5a and b. At smaller value of 2δ, the variation in the load capacity is minimal while higher load capacity is observed in the zone of higher magnitude of 2δ.

It is clearly observed that when the value of $(2\delta + \varepsilon)$ reaches closer to unity, the dimensionless load capacity of bearing moves toward infinity due to phase transformation of liquid lubricant film. It summarizes that

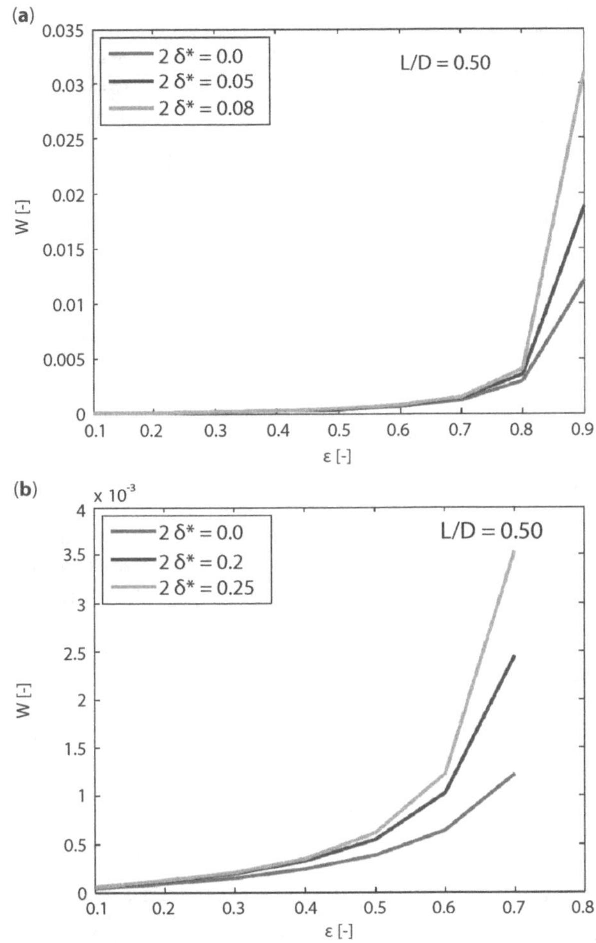

Figure 7.5 (a) Variation of ***W*** with $\boldsymbol{\varepsilon}$ for lower values of $2\boldsymbol{\delta}$. (b) Variation of ***W*** with $\boldsymbol{\varepsilon}$ for higher values of $2\boldsymbol{\delta}$.

the dimensional load capacity under TFL increases with adsorbent layer thickness.

Curves of attitude angle ($\boldsymbol{\phi}$) under thin-film regime for the short hydrodynamic bearing are displayed in Figures 7.6a and b. The distinct combinations of ($\boldsymbol{\varepsilon},\boldsymbol{\phi}$) with adsorbed layer thickness ($2\boldsymbol{\delta}$) play decisive role in achieving the shaft equilibrium for the proposed simulation of bearing. The attitude angle decreases with increment in adsorbed layer thickness ($2\boldsymbol{\delta}$) and these changes are significant toward higher eccentricity ratio ($\boldsymbol{\varepsilon}$) as well as at higher values of adsorbent thickness.

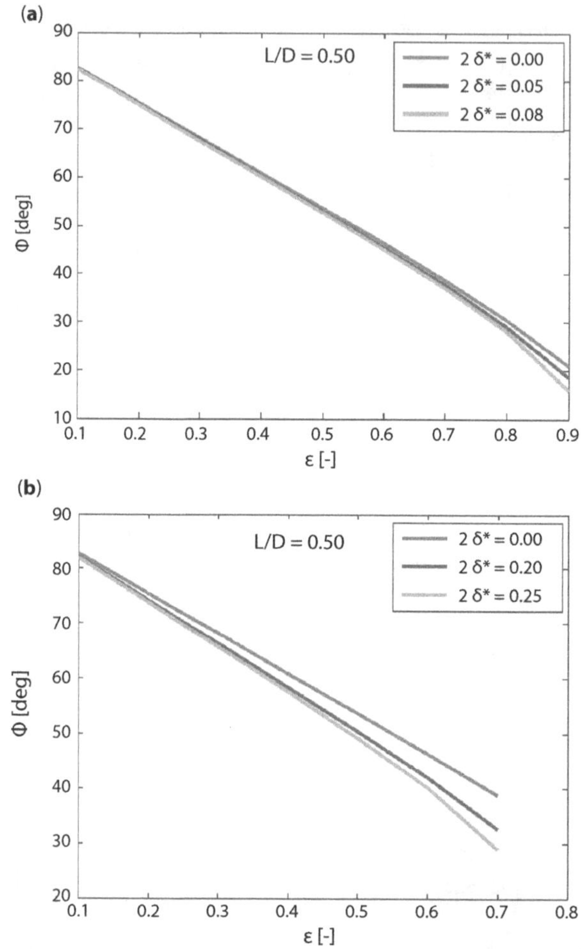

Figure 7.6 (a) Variation of ϕ with $\boldsymbol{\varepsilon}$ for lower values of $2\boldsymbol{\delta}$. (b) Variation of ϕ with $\boldsymbol{\varepsilon}$ for higher values of $2\boldsymbol{\delta}$.

In Figures 7.7a and b, the values of frictional force ($\boldsymbol{F_f}$) are attained for various magnitudes of adsorbed layer thickness ($2\boldsymbol{\delta}$) with respect to eccentricity ratio ($\boldsymbol{\varepsilon}$) for the short hydrodynamic bearing. It is observed that increment in adsorbent thickness increases the frictional force in the bearing assembly. This demonstrates that careful selection of adsorbent thickness is essential for better operational assessment of short hydrodynamic bearing under proposed film regime.

The variation of frictional torque ($\boldsymbol{T_f}$) and energy loss in the bearing assembly per unit of time ($\boldsymbol{E_f}$) with different values of adsorbent thickness

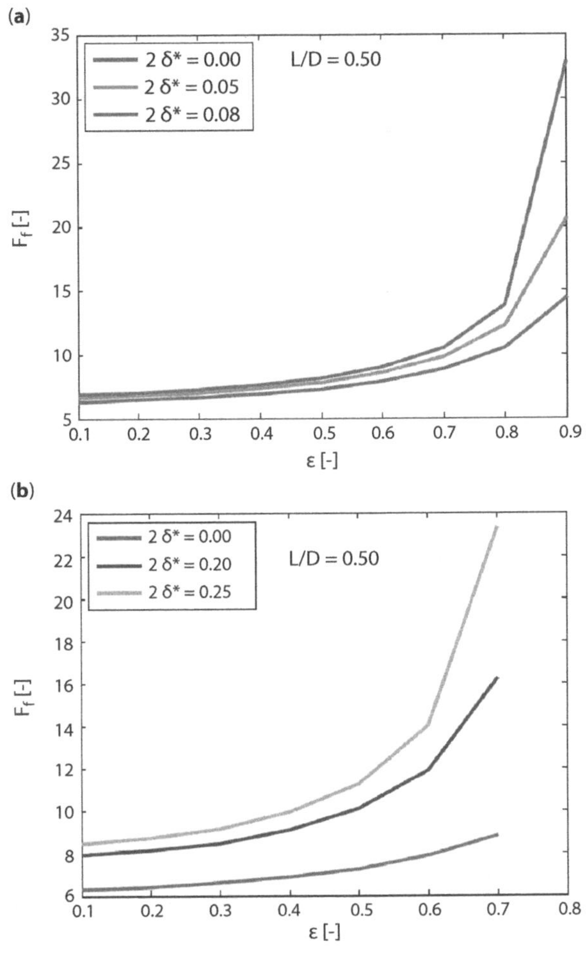

Figure 7.7 (a) Variation of F_f with ε for lower values of 2δ. (b) Variation of F_f with ε for higher values of 2δ.

(2δ) under TFL are plotted against eccentricity ratio (ε) in Figures 7.8a and b and Figures 7.9a and b, respectively. Frictional torque and energy losses in the bearing assembly are showing higher trend in the thin-film regime. Hence, calculations of these parameters are indispensable.

When addition of adsorbent thickness (2δ) and eccentricity ratio (ε) is approaching toward unity, energy losses in the bearing assembly per unit of time (E_f) are increasing abruptly. This reflects that close to phase transformation region, the molecular attractive forces have sharply increasing

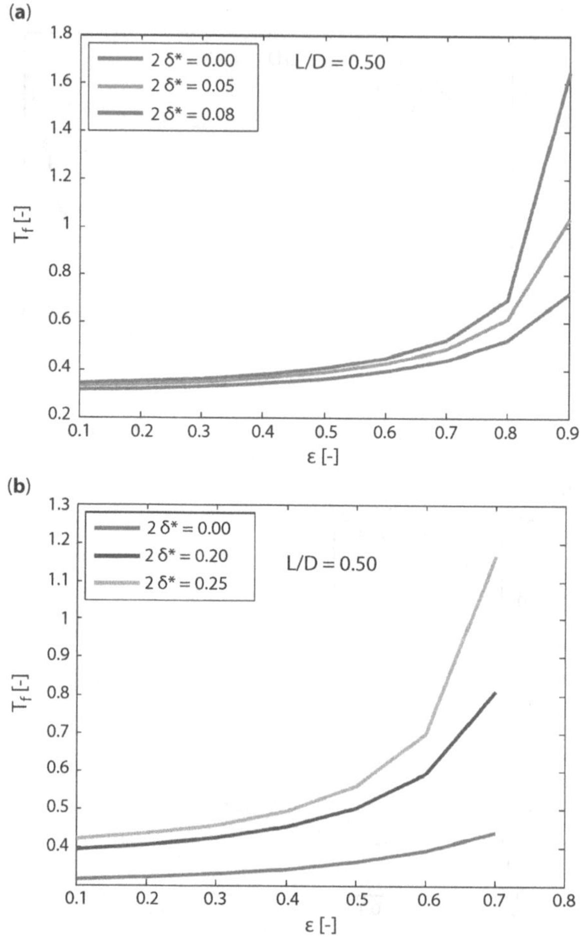

Figure 7.8 (a) Variation of T_f with $\boldsymbol{\varepsilon}$ for lower values of $2\boldsymbol{\delta}$. (b) Variation of T_f with $\boldsymbol{\varepsilon}$ for higher values of $2\boldsymbol{\delta}$.

trend, which authorized that the evaluation of this parameter is essential in TFL.

Figures 7.10a and b describe the interrelationship between the oil flows ($\boldsymbol{Q_{Inlet}}$ and $\boldsymbol{Q_{Outlet}}$) and eccentricity ratio (**ε**) for distinct adsorbed layer thickness (**2δ**). The adsorbent thickness eloquently affects the lubricant flow through the bearing assembly.

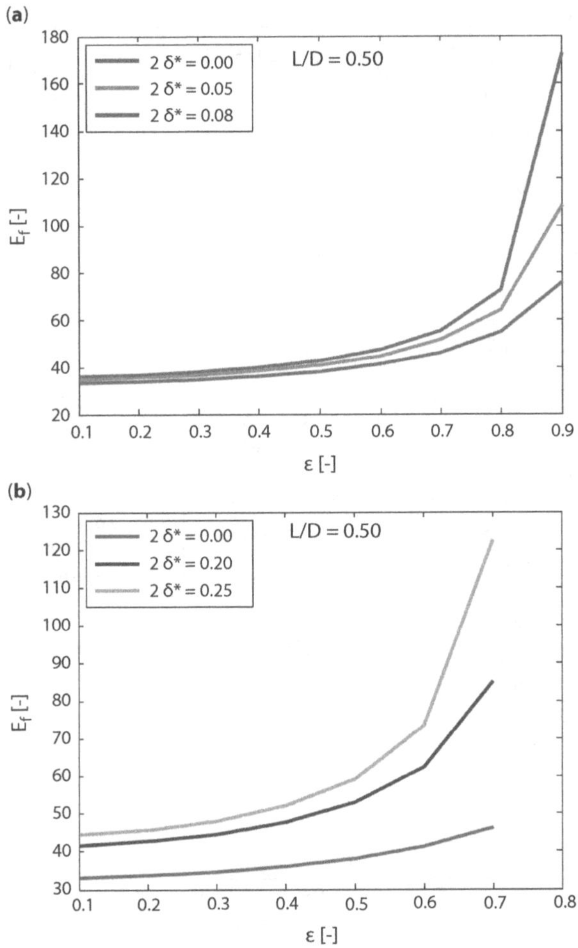

Figure 7.9 (a) Variation of $\boldsymbol{E_f}$ with $\boldsymbol{\varepsilon}$ for lower values of $2\boldsymbol{\delta}$. (b) Variation of $\boldsymbol{E_f}$ with $\boldsymbol{\varepsilon}$ for higher values of $2\boldsymbol{\delta}$.

In short bearing, the increment in the magnitude of adsorbent thickness ($2\boldsymbol{\delta}$) from 0.00 to 0.08, the lubricant's inlet oil flow ($\boldsymbol{Q_{Inlet}}$) decreases in the thin-film regime. Although, the variation of outlet oil flow ($\boldsymbol{Q_{Outlet}}$) changes sharply with increment in adsorbent thickness ($2\boldsymbol{\delta}$) and eccentricity ratio ($\boldsymbol{\varepsilon}$) in the TFL conditions.

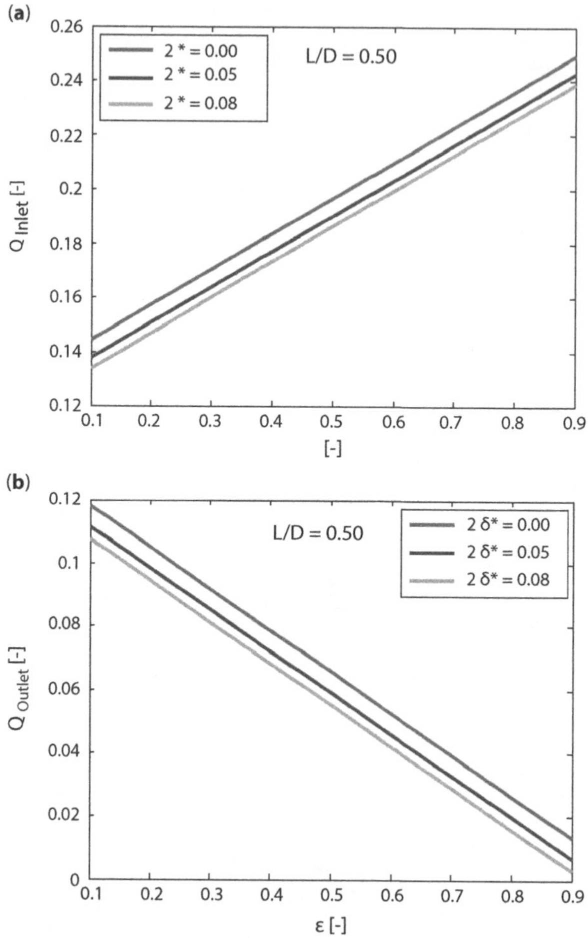

Figure 7.10 (a) Variation of Q_{Inlet} with ε for selected values of 2δ. (b) Variation of Q_{Outlet} with ε for selected values of 2δ.

7.6 Conclusions

A simulation analysis has been achieved on the thin-film regime of short hydrodynamic bearing. Based on the numerical results, the following observations have been drawn:

1. The pressure variation and load carrying ability significantly affected with adsorbent thickness.
2. The calculation of frictional characteristics, torque, and energy losses plays critical role in the analysis of hydrodynamic bearing under thin-film regime.

3. The oil flow behavior changes appreciably due to presence of adsorbent layer.
4. Adsorbent action of surface consideration improves the hydrodynamic performance of bearing assembly.
5. The magnitude of $(2\delta + \varepsilon) < 1.0$ must be established in analysis of TFL.

Nomenclature

C	radial clearance (m)
e	bearing eccentricity (m)
R	radius of journal (m)
L	length of bearing (m)
D	diameter of journal (m)
p	fluid film pressure (N/m^2)
μ	absolute viscosity of lubricant (Pa-s)
μ_0	reference viscosity of lubricant (Pa-s)
U	journal surface velocity (m-s)
2δ	adsorbed layer thickness (m)
h	oil film thickness (m)
W_x	oil film force along the line of centers (N)
W_y	oil film force perpendicular to the line of centers (N)
W	resultant load capacity (N)
ϕ	attitude angle (deg.)
τ	shear stress (N/m^2)
F_f	frictional force (N)
Q_z	total oil flow (m^3/s)
T_f	frictional torque (N-m)
E_f	energy loss per unit of time (N-m/s)

Dimensionless Parameters

$\bar{p}$	$\left[pC^2/(6\mu UR)\right]$
ε	eccentricity ratio (e/C)
θ	circumferential coordinate measured from journal center (x/R)
x	radial coordinate measured from journal center
$\bar{z}$	axial coordinate measured from bearing center $(z/0.50{*}L)$
λ	(D/L)
$\bar{\mu}$	(μ/μ_0)

$\boldsymbol{H}$	$(\boldsymbol{h/C})$
$\bar{\boldsymbol{\delta}}$	$(\boldsymbol{\delta/C})$
$2\boldsymbol{\delta}^*$	dimensionless adsorbed layer thickness

References

1. Israelachivi, J.N., Measurement of the viscosity of liquids in very thin films. *Science*, 110, 1, 263–271, 1986.
2. Johnston, G.J., Wayte, R., Spikes, H.A., The measurement and study of very thin lubrication films in concentrated contact. *Tribol. Trans.*, 34, 187–194, 1991.
3. Wu, S., Empirical determination of effective lubricant rheological parameters. *Tribol. Trans.*, 37, 1, 138–146, 1994.
4. Tichy, J.A., Modeling of thin film lubrication. *Tribol. Trans.*, 38, 1, 108–118, 1995.
5. Tichy, J.A., A surface layer model for thin film lubrication. *Tribol. Trans.*, 38, 3, 577–582, 1995.
6. Wen, S., On thin film lubrication. *Lubr. Sci.*, 8, 3, 275–286, 1996.
7. Luo, J., Wen, S., Huang, P., Thin film lubrication Part I: Study on the transition between EHL and thin film lubrication using a relative optical interference intensity technique. *Wear*, 194, 107–115, 1996.
8. Qingwen, Q., Yahong, H., Jun, Z., An adsorbent layer model for thin film lubrication. *Wear*, 221, 9–14, 1998.
9. Qingwen, Q., Mei, W., Shan, C., Fusheng, Y., Velocity analysis for layered viscosity model under thin film lubrication. *Tribol. Int.*, 34, 517–521, 2001.
10. Qu, Q.W., Meng, L.X., Chai, S., Niu, S.T., A new model for thin-film lubrication. *Part J: J. Eng. Tribol., Proc. Inst. Mech. Eng.*, 215, 103–107, 2001.
11. Qingwen, Q., Mei, W., Lihua, W., Shan, C., A continuous viscosity model for thin film lubrication. *Tribol. Int.*, 35, 459–465, 2002.
12. Zhang, C., Research on thin film lubrication: state of the art. *Tribol. Int.*, 38, 443–448, 2005.
13. Liran, M.A. and Jianbin, L., Thin film lubrication in the past 20 years. *Friction*, 4, 4, 280–302, 2016.
14. Ono, K., Modified Reynolds equations for thin film lubrication analysis with high viscosity surface layers on both solid surfaces and analysis of micro-tapered bearing. *Tribol. Int.*, 151, 106515, 2020.
15. Wen, S. and Huang, P., *Principles of tribology*, 2nd Ed., Tsinghua University Press, John Wiley & Sons Inc., Singapore, 2018.
16. Harnoy, A., *Bearing design in machinery - Engineering, tribology and lubrication*, CRC Press, New York, NY, USA, 2002.

8

Bioactive Coatings for Biomedical Applications

Gobi Saravanan Kaliaraj*, D. Dinesh Kumar and A.M. Kamalan Kirubaharan

Center for Nanoscience and Nanotechnology, Sathyabama Institute of Science and Technology, Chennai, India

Abstract

Artificial implants have gaining interest to repair or replace the defect parts in the human system. Majorly, orthodontic and orthopedic implant materials are playing predominant role in replacing teeth with endosseous parts in oral as well as orthopedic applications such as endossceous implants, fracture fixation, total joint arthroplasty, and ankle fracture replacement. For this, several metallic materials, namely, stainless steel, titanium and its alloy, Ni-Cr, Co-Cr, and Mg-based alloy, have been commercially used. Other non-metallic materials such as ceramics, polymers, and metallic glasses have also explored in implant fixation to improve performance and durability of implants.

Among various materials classifications in biomedical industries, implants used for load bearing applications (femoral heads, knee implant, and ankle prosthesis) have shown significant attention due to dynamic load and more possibilities, and wear debris formation hinders the potential implant functioning. The improvement of surface properties and load bearing ability of implants are mainly related to surface topography, roughness, and its stability in hostile environment. Surface coatings with smooth surface provide better functionality in articulating surfaces, and its surface texture improves the tribological properties which became more popular in biomedical field for the past few decades.

In this chapter, recent advancements in development of bioactive coatings and its role in wear-resistant ism for various articulating surfaces (hip, knee, ankle, and spinal implants) will be discussed by describing root cause of short span of artificial joints using metallic implants. Furthermore, corrosion resistance and biocompatibility of surface modified materials are also discussed in detail.

**Corresponding author*: gsbiotech5@gmail.com

Sarfraj Ahmed and Vinayak S. Dakre (eds.) Tribology and Characterization of Surface Coatings, (159–180) © 2022 Scrivener Publishing LLC

Keywords: Bioactive coatings, bio-tribology, articulating joints, biocompatibility, wear resistance, PEEK

8.1 Introduction

The medical devices which primarily deal with relative rotating or moving parts interact either with native or within the artificial biomaterials along with loading. During motion, the frequent wear and friction of the moving parts affect both functions of implanted devices as well as cause potential adverse effects in surrounding tissues. In general, bio-tribology deals with tribological principles occurred in the biological system, such as wear, lubrication, and friction amidst relatively moving surfaces in the biological systems. This bio-tribology plays a pivotal role in several devices in medical field. The artificial joint implants have received considerable attention followed by cardiovascular, dental, artificial limb, and fracture fixation implant devices. In addition, soft contact lenses (SCLs) are also widely used to reduce the friction on the surface of the eyelid.

Artificial joints are considered as one of the successful implant devices which reveal the achievements in the implant industries. In human body, a total of 206 bones are connected by around 300 joints. Among these, shoulder, hip, knee joints, etc., underly relatively large motions, whereas wrist, elbow, ankles, etc., considered as small joint implants. In addition, temporomandibular joints implant prosthesis and spinal disc replacement is also considered as joint implants. These joints relatively exhibit complex three-dimensional motions along with significant quantity of loading. Probably, there are about one million artificial joins that are implanted worldwide every year. During joint fixations, connection between articulating surfaces, bone fixation, and modular components are important consideration due to its high prevalence to tribological issues. In joint fixations, wear, friction, and lubrication are the key factor for the successful and potential clinical complications.

In articulating joint implants, there are broadly divided into two main categories, namely, hard-on-hard and soft-on-hard implants. The joint implants preferably made by ultra-high molecular weight polytethylene instead of alumina/zirconia or cobalt-based alloys. Titanium alloy is usually used for total disc replacement process in the spine to meet low elastic modulus and wear-resistant properties. Hard-on hard surface is the combinations of ceramic-on-ceramic, metal-on-metal, or metal-on-ceramic [1]. Wear and wear debris are the major challenges currently associated in bearing surfaces of implant joints that can cause hostile tissue reactions and subsequent aseptic loosening of prosthetic components. Improvements in articulate joints are needed owing to implant fracture and wear followed by aseptic loosening. Aseptic loosening is the serious issue which degrades prosthetic joint survival. According to the study, nearly 20% of the implants suffer from implant loosening within 10 years of post-implant surgery made by conventional polyethylene (PE) which leads therapeutic action in revision surgery [2]. The orthopedic materials containing carbon content will tend to form carbide during casting process forming junction between metal and carbon. These much harder particles protrude toward the surface of the alloy which may harmful to articulating counterface. In addition, some of the cobalt-chromium alloy implant materials have small amount of nickel (Ni) element to improve strength of the implant materials. At the same time, Ni element primarily causes metal sensitivity that leads to the pitting corrosion resulting in the formation of hypersensitive reaction in the patient body. The number of revision surgery has been increased drastically worldwide which impact socio-economical and health of the patients.

In addition, post-implant surgery, the metallic materials are prone to degradation due to prolonged interaction of biological fluids that lead to metal ion leaching, and thereby, immunogenic responses are taken place and subsequent implant rejection was occurred. In addition, major drawbacks of implant materials are their poor integration with host cells owing to inappropriate surface chemistry which facilitates post-surgery microbial infection. Most importantly, success of implants not only based on the bulk characteristics of the materials but also the surface chemistry and its property determines its interactions with biological environment. The appropriate surface tailoring facilitates host cellular interaction and modulation of cell-binding peptides, triggering mesenchymal stem cells to express osteogenic growth factors. Recently, surface modified implanted materials enhancing bone growth and bear majority of the load and vanishing subsequently (as degradable material) when a new bone is formed where the load is shared between the implants and healing tissue.

8.2 Surface Coating

Recently, enhancing the wear resistance of bearing surfaces is the main key factor in artificial joints to shield body from metal ions release. For this, several coatings are being developed to enhance the wear resistance of artificial implants such as shoulder, knee, hip, ankle, and elbow implants. Better wear resistance has been accomplished by surface coatings on joint surface for the past few decades. Surface coatings exhibit superior properties by physiochemical properties, topographical design (charge, surface energy, and composition of the coatings), and biochemical adaptations through cellular interactions to exist bulk materials. The surface coatings significantly reduce the generation of wear particles by adverse cellular response. Numerous wear-resistant coatings such as metal nitride, metal oxynitrides, metal carbides, metal carbonitrides, and ceramics are increasingly used in the artificial implant applications. Hence, surface modification engenders success rate and subsequent durability of the implants. Surface coatings enhance physical and mechanical properties like elastic strain, hardness, and wettability of the surface. The wettability nature of surface coatings exhibits low friction and reduces the wear rate, thereby decreasing tribo-corrosion problem. The surface coatings provide physical barrier which improves mechanical behavior and reduce surface roughness, thereby reducing friction and wear rate. In addition, the deposition techniques also play a predominant role in acquiring better quality coatings such as dense with uniform coated surface with improved adhesion to the substrate [3, 4].

By preparing highly wear-resistant coatings onto bulk materials, the accumulation of wear debris could be minimized at joints which solves primary implanted problems and reduces volume loss, thereby enhancing life span and stability of articulating joints. In addition, corrosion resistance and sufficient mechanical property coatings reduce possibilities allergic reactions and bone atrophy, respectively.

In articulating joint surgery, several efforts have been made to reduce wear and friction challenges of implants. Even though metal-on-metal implant materials reduce about 40% linear wear and control the volume of wear about 200 times compare to metal on ultra-high molecular weight polyethylene (UHMWPE), the biocompatibility and toxicity of UHMWPE are limiting their applications due to the generation of UHMWPE debris. Ghosh *et al.* (2016) deposited TiN coating having surface roughness of 169 nm and observed the wear rate of about 6×10^{-4} $mm^3 N^{-1} m^{-1}$ [5]. Similarly, Spriano *et al.* carried out Ta-coated Co-Cr-Mo substrate having two different surface roughness 5–12 nm and 40 nm that exhibited their wear rate of $4–5\times10^{-7}$ $mm^3 N^{-1} m^{-1}$ and $0.755–1.249 \times 10^{-4}$ $mm^3 N^{-1}$, respectively [6]. The results suggest that lower surface roughness or smooth surface is a key role to reduce friction rate. Wang *et al.* stated that CrN (chromium nitride) coating exhibited better corrosion resistant nature in synovial body fluid [7]. Love *et al.* used Cr coating as interlayer in the objective of improving adhesion strength amidst steel substrate and the carbon coating [8]. The intermediate layers such as TiN, CrN, and diamond-like carbon

(DLC) coating acted as intermediate layer to improve surface hardness of the implant material. Moro also observed that CrN coating significantly enhanced adhesion strength amidst substrate and coating material.

8.3 Polyether Ether Ketone

Polyether ether ketone (PEEK) is considered as thermoplastic material that exhibits combinations of toughness and stiffness promotes wide range of applications [9]. PEEK has been mainly used as sliding and bearing material in biomedical applications owing to its better thermal stability and high resistance to wear performance [10, 11]. The better tribological performance of PEEK material facilitates the application of PEEK coating on light weight alloys that usually susceptible to the tribological performances. PEEK and its composite coatings are generally developed onto various substrates using painting and thermal spray techniques [12–14]. Thermally sprayed PEEK coating on engine piston skirts reduced the friction rate of about 3%. PEEK coating was deposited with the thickness of 70–90 μm on Ti-13Nb-13Zr titanium alloy using electrophoretic deposition [15].

Better adhesion strength and uniform PEEK coating was obtained by appropriate pH, applied voltage, and time duration. In addition, the coating was not deformed even at 30 N of load by micro-scratch test. The wear analysis depicted that PEEK coating reduced the wear rate of about 200 times than uncoated Ti-13Nb-13Zr titanium alloy (Figure 8.1). Zhang *et al.* stated that PEEK and PEEK/SiC composite coating on aluminum

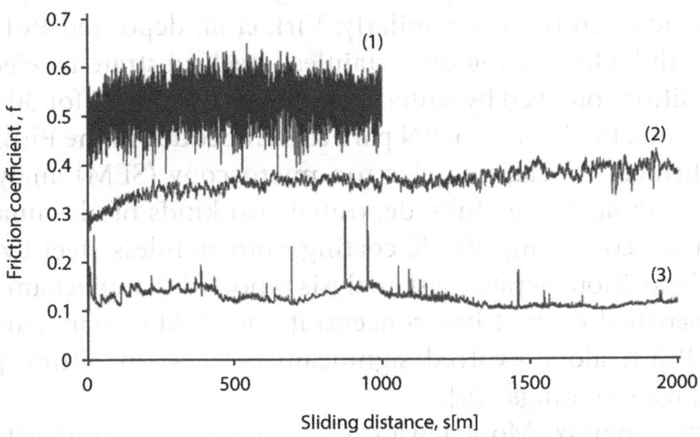

Figure 8.1 Sliding distance vs friction co-efficient. (1) Ti-13Nb-13Zr alloy in dry sliding, (2) PEEK coating in dry sliding, and (3) PEEK coating in in Ringer's solution [15].

significantly reduced the rate of friction than aluminum substrate alone [16]. The detailed protocol was given by Zhang *et al.* to prepare good quality and high dense PEEK coating by thermal spray technique. They suggested that the substrate should be heat treated up to the melting temperature of PEEK, e.g., 400°C. After coating process by thermal spraying, the coating material was subjected to quench in the water to obtain better adhesion and amorphous nature of the coating.

They also stated that thin fillers are required during PEEK coating due to its agglomeration as well as poor fluidity which diminishes quality of the coating. PEEK is considered to be the best alternatives for load bearing applications. However, PEEK was not coated onto titanium alloy till the year of 2015. Jian Song *et al.* is coated PEEK filled by different weight percentage of zirconium oxide (ZrO_2) onto Ti6Al4V for the first time and studied their adhesion strength amidst substrate and coating interfaces. They test the tribological behavior in presence of 25% new born calf serum and optimized that 5 wt % ZrO_2 nanoparticles filled with PEEK coatings performed optimal friction behavior for artificial cervical disc applications [17].

8.4 Nitride Composites Coating

Different wt % of hexagonal phase boron nitride (h-BN) powder were added to PEEK in an ultrasonic condition and deposited by flame spray coating. During tribological performance, 5- to 25-N load was given onto the coated samples, where 25-N load exhibit better performance in the composite coating containing 8% of hexagonal boron nitride [18] due to its self-lubricity properties. Similarly, Virk *et al.* deposited PEEK-based bioglass with h-BN coating onto stainless steel substrate by electrophoretic deposition followed by sintering in furnace at 375°C for 30 minutes. After sintering, the BG and h-BN particles embedded in the PEEK matrix were confirmed by scanning electron microscopy (SEM) analysis [19]. Ilaria Corni *et al.* successfully deposited two kinds of alumina (Al_2O_3) nanoparticles containing PEEK coating onto stainless steel by electrophoretic deposition. Scratch test analysis reported that mechanical property was satisfied even at low concentrations of Al_2O_3 nanoparticles. In contrast, PEEK alone evolved significant increase in pitting potential than composite coatings [20].

Similarly, Tomasz Moskalewicz *et al.* deposited composite Al_2O_3-PEEK coatings on medical grade Ti-13Nb-13Zr substrate by electrophoretic deposition. Micro-scratch test revealed that Al_2O_3/PEEK composite coatings significantly reduced coefficient of friction than PEEK coating in dry conditions (Figure 8.2). However, the scratch test in presence of Ringer's solution increased the friction coefficient Al_2O_3/PEEK rather PEEK-coated alloy [21]. The same group of people investigated Si_3N_4/PEEK nanocomposite coatings by cathodic electrophoretic deposition onto Ti-13Nb-13Zr alloy substrate (Figure 8.3). The suspensions chitosan and polyethylenimine were used during deposition and studied its scratch behavior.

The scratch test divulged that chitosan containing solution for the fabrication of Si_3N_4/PEEK nanocomposite coatings showed better scratch resistance rather polyethylenimine [22]. Multilayer coating exhibited better mechanical property rather single layer coatings. Isabel Claveroa *et al.* deposited ZrN/ZrCN coating on 100Cr6/1.3505 and DIN 17230 steel substrates. The polished and coated rings were tested under real friction torque test procedures. The tests showed that the coating did not encourage an improvement in friction behavior of the bearing [23]. Nanocrystalline WC/a-C with hydrogen-free nanocomposite coating was developed by

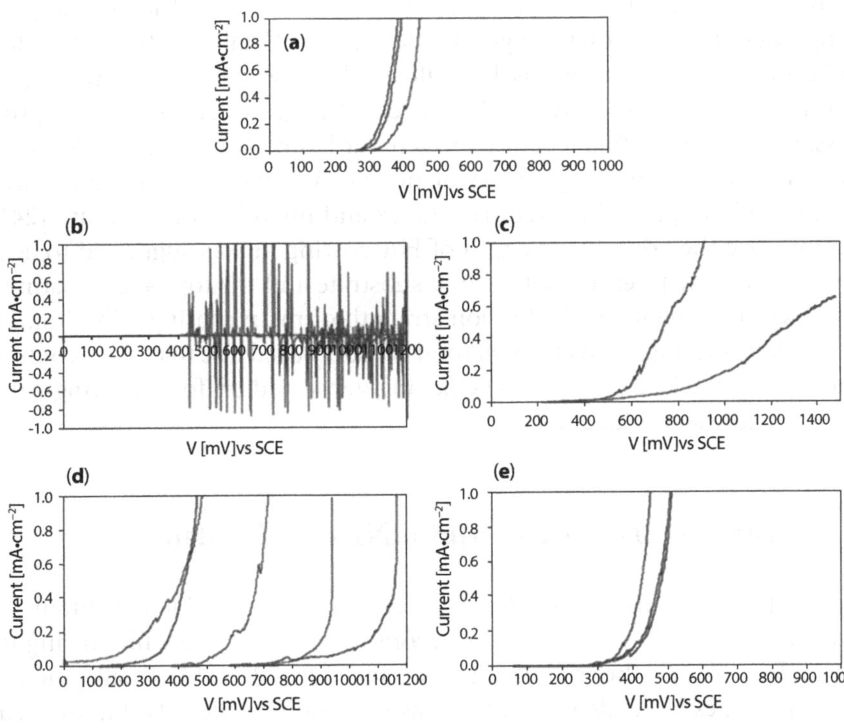

Figure 8.2 Potentiodynamic polarization curves of (a) uncoated 316L SS, (b) PEEK coating, (c–e) Al_2O_3-PEEK films deposited from different suspensions [20].

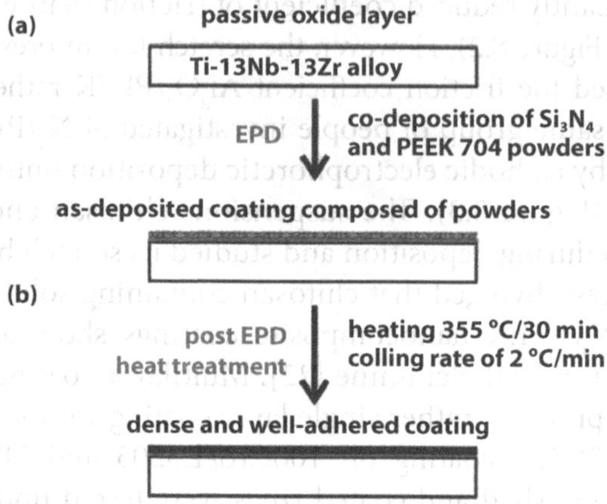

Figure 8.3 Schematic diagram of surface treatment of titanium alloy substrates. (a) Si3N4 and PEEK co-deposited particles. (b) Post-heat treatment process [22].

Moskalewicz *et al.* onto non-harderned (base line) and oxygen-treated Ti-6Al-4V alloy substrate using sputtering technique. The tribological properties divulged that oxygen-treated Ti6Al4V alloy exhibited better adhesion behavior rather baseline alloys which was due to the tuning of surface properties of Ti6Al4V alloy by effective glow discharge (GD) process. GD process influenced the oxygen hardening on Ti6Al4V alloy surface with subsequent deposition of nc-WC/a-C nanocomposite coating significantly improved in wear resistance and microhardness nature [24]. To improve the adhesive strength of HA coating, it was suggested to add an intermediate layer amidst desired substrate and the top layer (coating) enhances the quality and adhesion strength of main coating adhesion. In general, sputtering deposition method is widely used due to its easy control the coating thickness as well as achieving dense and uniform coating with elevated adhesion [25, 26].

8.5 Binary Transition Metal Nitride Coatings

Binary transition metal nitrides, namely, zirconium nitride and titanium nitride, have been used widely for decorative and protective film owing to its golden color aesthetic and wear resistance [27, 28]. Additionally, metallic oxynitrides have also considered as interesting materials due to their unique chemical stability and mechanical behavior [29] along with superior corrosion protection in *in vitro* SBF solution. Zirconium oxynitride

(ZrNxOy) and titanium oxynitride (TiNxOy) coatings are considered as outstanding materials owing to the combination of oxide and nitride of zirconium and titanium [30] which exhibit wide applications such as transparent window [31], memory devices [32], and electrically switchable windows. Gobi Saravanan *et al.* deposited zirconium oxynitride (ZrON) and titanium oxynitride (TiON) coating onto medical grade 316L stainless steel (316L SS) by magnetron sputtering [33]. These films exhibited better nanoindentation and wear properties rather 316L SS substrate. In bacterial adhesion study, both the coatings proved their anti-fouling ability against *Pseudomonas aeruginosa* bacterial adhesion. Further, the coated films showed better corrosion resistance in the presence of hydrogen peroxide (H_2O_2) rather bare 316L SS substrate in ABP solution [33]. Surface modifications by ternary nitride such as $TiAlN_6$, TiZrN, and TiCrN retain its properties in corrosion, hardness, resistance to oxidation, and better biocompatibility by reflecting the constitutent alloy effects which has taken into consideration for the past decades. In this regard, TiZrN film was coated on 316LSS by reactive magnetron co-sputtering method. TiZrN exhibited reduced bacterial adhesion, negligible platelets activation behavior, and increased corrosion protection of TiZrN coating rather uncoated 316L SS [35].

Similarly, Dinesh Kumar *et al.* deposited ZrC and TiC coatings on type 316LSS and studied their bio-corrosion behavior in presence of artificial blood plasma (ABP) for biomedical applications [34]. Biocorrosion analysis by electrochemical test examined in ABP solution containing 1% and 10% H_2O_2. The strong oxide layers formed on carbide-coated 316L SS substrates significantly protected corrosion formation than uncoated SS substrates (Figure 8.4). Further, high degree of protein adsorption and reduction in bacterial attachment lower bacterial adhesion against *P. aeruginosa* was noticed. For cytotoxic test, Alizarin red–based assay was done, and it confirmed the better apatite formation on TiC and ZrC coatings against osteoblast like cells which could be used as implant coating material [34].

In addition, PEEK coating by painting is also effective and economic method to achieve better adhesive and high wear resistance coatings. Zhang developed dense PEEK, PEEK/micron-graphite, and PEEK/micron-SiC coatings by painting. The composite mixture was stirred continuously followed by ultra-sonication for uniform deposition. After homogenization, the composite material in the form of slurry was deposited followed by heat treatment at 400°C for 5 minutes and finally quenched into cold water [16]. The temperature plays an important role in propagation of residual stress in PEEK coating beyond 200°C. The mechanical resistance of the coating was reduced significantly due to breakage of bonding between coating and

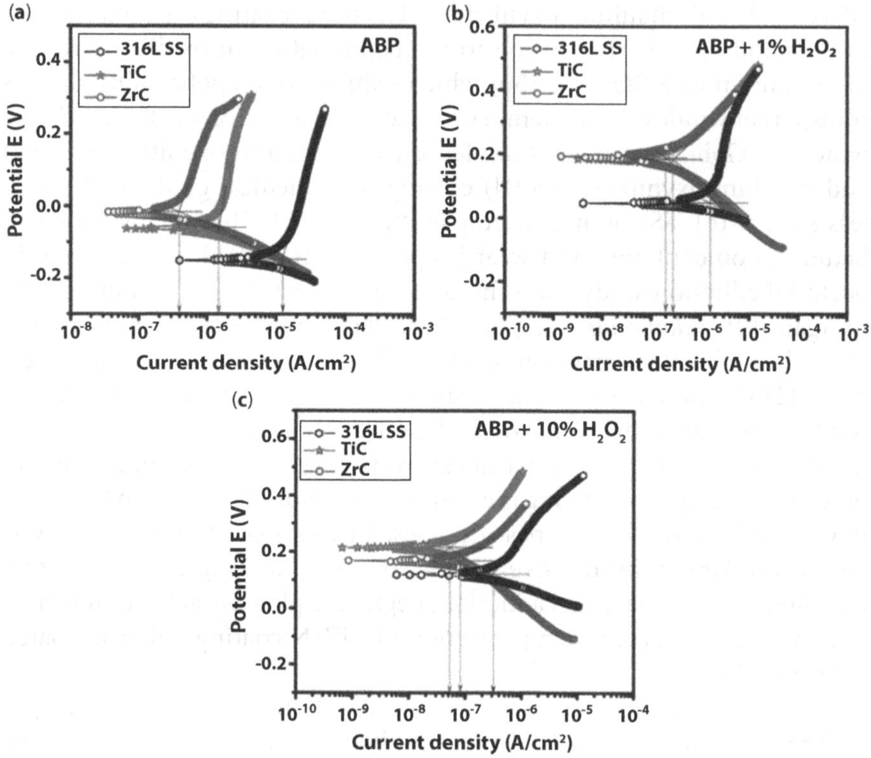

Figure 8.4 Potentiodynamic polarization curves of bare and ZrC- and TiC-coated medical grade 316L SS immersed in ABP solution containing different concentrations of H_2O_2. (a) ABP alone, (b) with 1% H_2O_2, and (c) with 10% H_2O_2 [34].

substrate. Ward *et al.* discussed the crystalline and semi-crystalline nature of PEEK coating and examined their tribological performances. They found that crystalline PEEK coating slightly increase tribological performance; whereas, semi-crystalline PEEK coating significantly increased its wear-resistant behavior due to the presence of amorphous matrix and spherulites [36]. The dispersed spherulites growth in the PEEK coating restricts the slippage and motion of polymer motion. Ajay Kumar Kadiyala *et al.* optimized PEEK-SiC coatings by admixing SiC particles from 3% to 15% of SiC particles that exhibited better wear resistance behavior. Increased scratch hardness and deterioration of PEEK-SiC composite coatings were observed when the percentage of SiC is more than 15% [37].

In implant industries, bacterial invasion is the major challenge for better performance of the implants. During surgery, the bacteria enter into the implant surface through surface of the surgical equipment's or cross-contamination which further form a biofilm. Once the surrounding of implants gets infected, then the recruitment of immune will take place which causes implant loosening. According to the statistical analysis, nearly 3% of implants got affected by primary total joint prosthesis. In order to overcome the above problem, the researchers across the globe are doping antimicrobial agents into the ceramics or other materials, thereby preventing the implant materials from bacterial infection and improve its durability. In this view, zinc-doped hydroxyapatite (HA) composite was prepared by sol-gel process and studied their antibacterial performances. After synthesis, Zn-doped HA annealed into two different temperatures, namely, 500°C and 700°C. The antibacterial study indicated that both ZnHAp500 and ZnHAp700 composited exhibited similar and high antibacterial activity toward *S. aureus* bacteria and *C. albicans* fungal cells [38]. TubaYetim studied TiO_2 coating containing different concentrations of Ag obtained using sol-gel technique on pure titanium (CP-Ti) [39]. The electrochemical corrosion properties of silver-doped TiO_2 coatings were tested with the help of electrochemical impedance spectroscopy (EIS) followed by potentiodynamic polarization (PDP) in simulated body fluid electrolyte medium. From the results, it was seen that Ag-doped TiO_2 coatings enhanced corrosion protection when compared to untreated (Cp-Ti alone) as well as TiO_2-coated samples. Further, highest electrochemical corrosion resistance was observed from Ag-doped samples with 0.05 M Ag concentration [39]. Similarly, silver (Ag)– and strontium (Sr)–doped HA coating was developed onto titanium substrate by plasma spray system. After heat treating, the coated samples were underwent adhesion strength test and it did not exhibit any deleterious effect to the HA coating when doped with bi metallic dopant. Similarly, the coatings constitutes silver drastically reduced the bacterial growth. In addition, the cellular morphology was significantly improved while adding Sr rather Ag-Ha coating alone [40].

8.6 Hydroxyapatite-Based Coating

Carolina H. Navarro *et al.* prepared chitosan/HA composite coatings onto UHMWPE substrate materials and studied their tribological behavior using ball-on-disc tribometer. They stated that the less friction coefficient value was exhibited on chitosan/HA composite coating rather chitosan coating alone [41]. In general, polymers filled with nanoparticles decreased the friction coefficient from one order to 3 order magnitude [42]. Mohammed Hoseini *et al.* deposited new type of micronite and titanium nitride coating onto stainless steel and UHMWPE surfaces and studied their tribological

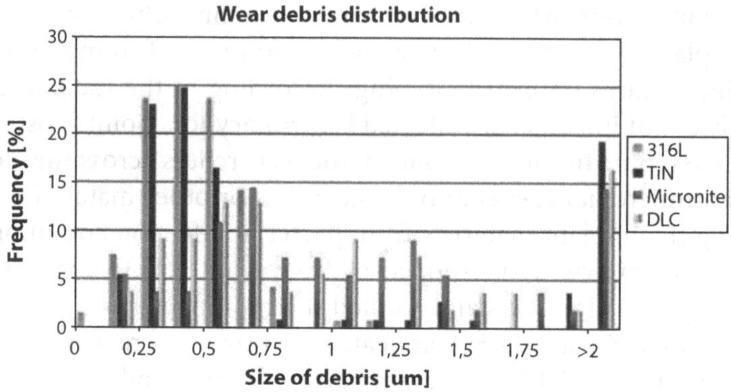

Figure 8.5 Wear debris size distribution of TiN, micronitrile, and DLC coated and uncoated 316L SS [43].

behavior (Figure 8.5). They observed that micronite exhibited higher surface roughness and micronitrie/UHMWPE sliding pair showed better tribological performances for artificial joints [43].

Magnesium (Mg) and their alloys have been used automobile and aerospace industries owing to their low density and high specific strength [44]. Hence, study on electrochemical corrosion of magnesium alloys have taken important consideration especially in biomaterials for the past few decades. For this, zirconium nitride (ZrN) coating deposited on AZ91 magnesium alloy substrate using ion beam–assisted sputtering technique at 100°C with different pressure of ionized nitrogen proportions [F(N_2)] such as 0.3, 0.4, 0.5, and 0.6. From the corrosion study, it was stated that corrosion potential (E_{corr}) of coated samples did not show any significant different than that of uncoated AZ91 alloy. In particular, under the F(N_2) value of 0.6, corrosion current density (I_{corr}) gradually decreased from 14 to 9.7 μA/cm² and drastically improved to 13.5 μA/cm². The study divulged that ZrN coating deposited in presence of F(N_2) value of 0.5 evolved superior corrosion protection and adhesion behavior [45]. Mg alloy also comprises several medical applications because of its superior biocompatibility [46]. However, these materials exhibit less corrosion resistance and degrade very quickly in biological environments [47, 48]. The superior biocompatible properties of Mg-based alloy widen its applications in biomedical industries such as heart stents, screws, and temporary implants. However, drastic release of Mg-ions is the only disadvantage which would be fulfilled by surface modifications. Mihaela Dinu *et al.* deposited HA coating onto Mg-based alloy using RF magnetron sputtering technique [49]. They placed Mg substrates at different locations in substrate holder toward line of sight and studied their tribological behavior analysis. The deposited HA coatings were revealed to divulge an improvement of Ca/P ratio from 1.83 to 1.97, as well as decrease of thickness and (002) texture, as the samples moved toward target erosion zone. By coatings, the roughness (Ra) of Mg-based alloy (AZ31 grade) was diminished (Ra of HA coating = 29 nm and 21 nm; Ra of Mg alloy = 31.3 nm). The Mg-alloy samples were kept at the mid of substrate holder enhanced hardness and Young's modulus ($E = 89 \pm 10$ GPa; $H = 8.3 \pm 0.9$ GPa) rather coating prepared under target erosion zone site ($E = 75 \pm 6$ GPa; $H = 6.9 \pm 1.1$ GPa). The films developed under target erosion zone evolved elevated friction behavior in presence of SBF environment, with friction coefficient of 0.184 (μ), while the Mg-alloy sample kept at center of the substrate holder undergone the friction coefficient value of 0.306 (μ) comparable to Mg-alloy substrate (0.307). Vranceanu *et al.* tried Mg-doped HA coating by new method called galvanostatic electrochemical deposition at pH 5.0 and studied the morphology and degradation rate of the coated samples in SBF solution. From the results, they observed that addition of Mg content controls deposition rate as well as degradation rate [50]. HA coating was developed on calcium-magnesium (Ca-Mg) alloy using RF magnetron sputtering technique. *In vitro* degradation behavior was determined under SBF using electrochemical corrosion methods and it showed that ultrathin film HA significantly protect the deterioration of Mg-Ca alloy by two order of magnitude compare to uncoated Mg-Ca alloy substrate [51].

8.7 Bioactive Glass Coatings

A wide range of bioactive glasses (BGs) can able to bond directly to both soft tissues and bone thereby widely used for soft and hard tissue repair [52, 53]. In general, BGs have gained notable clinical outcomes, because of their in sufficient fracture toughness and intrinsically brittle which limits its wide applications as main load-bearing components. However, numerous researchers are investigating about fracture toughness of glasses across the globe. In order to address this issue, the combination of polymer/BG composites has been prepared to acquire crack resistance and flexibility of polymeric materials [54]. Additionally, an alternate approach is to modify metallic implants with BGs for prominent mechanical strength of the substrate and bioactivity of the coating. Lahiri *et al.* prepared boron nitride nanotube (BNNT) with HA as an alternate for orthopedic applications [54]. During sintering, spark plasma sintering was used for better density and compactness of the BNNT-HA composite materials. From the study, it is illustrated that HA-4 wt % BNNT composite evolved excellent mechanical properties with more than 129% hardness, 120% increase in elastic modulus, and 86% more fracture toughness compared to HA. Enhancement in the fracture toughness and hardness are associated with grain refinement and crack linking by BNNT. BNNT-HA composite showed improvement in the wear resistance up to 75%. In addition, there is no adverse effect was seen in osteoblast proliferation and cell viability against HA-BNNT composite.

Satisfactory wear behavior is essential for *in vivo* fixation for lifetime of an implants [55]. Tribological performance is especially important for bearing surfaces of dental and orthopedic implants, where different glasses have been used. At the same time, the wear resistance behavior of BGs has to be improved as well while bulk BGs are actively involved in orthopedic applications. BGs are increasingly especially in load bearing applications in order to enhance the apatite formation on PEEK surfaces [56]. Atiq Ur Rehman *et al.* investigated the design of experiments (DoE) for optimizing PEEK-based coatings [57]. In the study, a relatively high concentrations of BG (6.6 wt %) and PEEK (2 wt %) was employed for better bioactivity in the presence of SBF. Seuss *et al.* developed BG 45S5 grade embedded matrix of PEEK coating with different concentrations of silver on metallic substrate. The addition of silver into the BG-PEEK matrix helped to improve the antibacterial activity which did not show any negative impact MG-63 osteoblast like cells [58].

8.8 Composite Coatings

Barkallah *et al.* added synthesized β-TCP with titanium dioxide powder (TiO_2) and studied the tribological properties of alumina as nanocomposite coatings. The composite coating materials were made with different concentrations of β-TCP and alumina nanoparticles. The study indicated that Al_2O_3 - 10 wt % TCP - 5 wt % TiO_2 composition coating showed lowest friction coefficient value with better mechanical, micro hardness, and fracture toughness properties [59]. Marsal *et al.* deposited silica or alumina coating by sol-gel technique method on stainless steel 304 substrate. Tribology behavior of the coatings exhibited that silica coating was failed to proof their wear resistance properties; whereas, Al_2O_3 coating showed significant reduction in friction. The results divulged that Al_2O_3 coating reduced the friction coefficient in the order of 1.5 than silica coating [60]. Tribological behavior of different concentrations of ZrO_2 particles reinforced glass ceramic composites were prepared by Santos *et al.* and carried out wear behavior by ball-on-plate methods. The investigation revealed that ZrO_2 particles showed predominant role in the wear and exhibited optimal wear-resistant nature in the ZrO_2 composition of 30 vol %. The better dispersion in reinforcement phase load transfer mechanism in the composite matrix improved the wear-resistant property [61]. ZrO_2 coating onto UHMWPE was improved the wear-resistant properties of UHMWPE for load bearing applications. Berni *et al.* deposited yttria-stabilized zirconia (YSZ) coating by pulsed plasma deposition and studied their tribological behavior in wet and dry condition using ball-on-disc tribometer. The tribological test revealed that 18% of friction reduction was observed in dry condition and 40% of reduction was obtained in presence of fetal bovine serum. The results suggest that ZrO_2 composite scaffold exhibited superior wear resistance performance for biomedical applications [62].

Among the different surface materials, ceramic constitutes several advantages such as corrosion resistance, good mechanical strength, chemical stability, non-allergic, and biocompatibility [63]. Ceramic implant materials have gained insight to surgeons and clinicians owing to its good esthetics, considerable biocompatibility, and wear [64, 65]. Recently, YSZ is used as dental implant [66] due to their better toughness and hardness [67]. Though, YSZ and 3 mol % YSZ failed to remain stable for the period time due to its phase instability. More specifically, nearly 400 femoral heads to be removed shortly post-implantation which results accelerated aging, in 2001 [68]. A mechanism called transformation from tetragonal to monoclinic phase (t-m transformation) is gained important attention for transforming the metastable tetragonal to monoclinic grains of ZrO_2 particles. Gobi Saravanan *et al.* deposited YSZ coating onto titanium (Ti) substrates by electron beam physical vapor deposition (EB-PVD) technique [69]. YSZ coating evolved better hardness and scratch resistance. In addition, YSZ coating is superior in hemocompatibility; whereas activated platelets with pseudopods were observed on Ti substrates. Superior *in vitro* biomineralization behavior with weight gain was observed on YSZ coating. Similarly, different phases of ZrO_2 and its allotrope coatings (*t*-ZrO_2, *m*-ZrO_2, and *c*-ZrO_2) were deposited by EBPVD method. Different phases of ZrO_2 coatings obtained less surface roughness with drastic reduction of pathogenic bacterial invasion. In particular, *t*-ZrO2 coating exhibited superior surface hardness followed by *m*-ZrO_2 and *c*-ZrO_2 allotropes. All the ZrO_2 allotropes showed more blood plasma protein adherence with enhanced corrosion resistance rather uncoated medical grade 316L SS in presence of ABP solution.

8.9 Conclusions

The surface modification is the essential technique to improve performance of metallic implant materials as well as mimicking the surface with desired applications. Surface tailoring is an emerging and ever green trends for interdisciplinary applications ranging from sensors, self- cleaning, biomedical, electronic, and chemical industries. In food packaging industries, surface coatings by antimicrobial and hydrophobic coatings increased significantly to improve quality of the food. In addition, antifouling coatings in urinary catheters restrict the multi-organ damage and protect the bacterial invasion and its colonization. Hydrophobic coating on heart stents avoids early vascularization on heart stents. Further, zirconia and HA coatings mimic the metal implants as self-materials to the immune cells, and thereby, immunogenic response can be avoided. Hence, surface modification is boon to enhance the performance of the materials, thereby increasing the quality and wealth of the human being.

References

1. Fisher, J., Jin, Z., Tipper, J., Stone, M., Ingham, E., Presidential Guest Lecture: Tribology of Alternative Bearings. *Clin. Orthop. Relat. Res.*, 453, 25–34, 2006.
2. Malak, T.T., Broomfield, J.A.J., Palmer, A.J.R., Hopewell, S., Carr, A., Brown, C., Prieto-Alhambra, D., Glyn-Jones, S., Surrogate markers of long-term outcome in primary total hip arthroplasty: A systematic review. *Bone Jt. Res.*, 5, 6, 206–214, 2016.
3. Antunes, R.A. and De Oliveira, M.C.L., Corrosion processes of physical vapor deposition-coated metallic implants. *Crit. Rev. Biomed. Eng.*, 37, 6, 425–460, 2009.
4. Ching, H.A., Choudhury, D., Nine, M.J., Abu Osman, N.A., Effects of surface coating on reducing friction and wear of orthopaedic implants. *Sci. Technol. Adv. Mater.*, 15, 1, 14402, 2014.
5. Ghosh, S., Choudhury, D., Pingguan-Murphy, B., Lubricating ability of albumin and globulin on artificial joint implants: A tribological perspective. *Int. J. Surf. Sci. Eng.*, 10, 2, 193–206, 2016.
6. Spriano, S., Vernè, E., Faga, M.G., Bugliosi, S., Maina, G., Surface treatment on an implant cobalt alloy for high biocompatibility and wear resistance. *Wear*, 259, 7–12, 919–925, 2005.
7. Wang, Q., Zhou, F., Wang, C., Yuen, M.F., Wang, M., Qian, T., Matsumoto, M., Yan, J., Comparison of tribological and electrochemical properties of TiN, CrN, TiAlN and a-C:H coatings in simulated body fluid. *Mater. Chem. Phys.*, 158, 74–81, 2015.

8. Love, C.A., Cook, R.B., Harvey, T.J., Dearnley, P.A., Wood, R.J.K., Diamond like carbon coatings for potential application in biological implants—a review. *Tribol. Int.*, 63, 141–150, 2013.
9. Friedrich, K., Zhang, Z., Schlarb, A.K., Effects of various fillers on the sliding wear of polymer composites. *Compos. Sci. Technol.*, 65, 15–16 SPEC. ISS., 2329–2343, 2005.
10. Lu, Z.P. and Friedrich, K., On sliding friction and wear of PEEK and its composites. *Wear*, 181–183, PART 2, 624–631, 1995.
11. Vande Voort, J. and Bahadur, S., The growth and bonding of transfer film and the role of CuS and PTFE in the tribological behavior of PEEK. *Wear*, 181–183, PART 1, 212–221, 1995.
12. Zhang, G., Leparoux, S., Liao, H., Coddet, C., Microwave sintering of poly-ether-ether-ketone (PEEK) based coatings deposited on metallic substrate. *Scr. Mater.*, 55, 7, 621–624, 2006.
13. Normand, B., Takenouti, H., Keddam, M., Liao, H., Monteil, G., Coddet, C., Electrochemical impedance spectroscopy and dielectric properties of polymer: Application to PEEK thermally sprayed coating. *Electrochim. Acta*, Pergamon, 49, 2981–2986, 2004.
14. Zhang, G., Liao, H., Yu, H., Ji, V., Huang, W., Mhaisalkar, S.G., Coddet, C., Correlation of crystallization behavior and mechanical properties of thermal sprayed PEEK coating. *Surf. Coat. Technol.*, 200, 24, 6690–6695, 2006.
15. Sak, A., Moskalewicz, T., Zimowski, S., Cieniek, Ł., Dubiel, B., Radziszewska, A., Kot, M., Łukaszczyk, A., Influence of polyetheretherketone coatings on the Ti-13Nb-13Zr titanium alloy's bio-tribological properties and corrosion resistance. *Mater. Sci. Eng. C*, 63, 52–61, 2016.
16. Zhang, G., Guessasma, S., Liao, H., Coddet, C., Bordes, J.M., Investigation of friction and wear behaviour of SiC-filled PEEK coating using artificial neural network. *Surf. Coat. Technol.*, 200, 8, 2610–2617, 2006.
17. Song, J., Liu, Y., Liao, Z., Wang, S., Tyagi, R., Liu, W., Wear studies on ZrO2-filled PEEK as coating bearing materials for artificial cervical discs of Ti6Al4V. *Mater. Sci. Eng. C*, 69, 985–994, 2016.
18. Tharajak, J., Palathai, T., Sombatsompop, N., Morphological and physical properties and friction/wear behavior of h-BN filled PEEK composite coatings. *Surf. Coat. Technol.*, 273, 1, 20–29, 2015.
19. Virk, R.S., Ur Rehman, M.A., Boccaccini, A.R., PEEK based biocompatible coatings incorporating h-BN and bioactive glass by electrophoretic deposition. *ECS Trans.*, 82, 1, 89–95, 2018.
20. Corni, I., Neumann, N., Novak, S., König, K., Veronesi, P., Chen, Q., Ryan, M.P., Boccaccini, A.R., Electrophoretic deposition of PEEK-nano alumina composite coatings on stainless steel. *Surf. Coat. Technol.*, 203, 10–11, 1349–1359, 2009.
21. Moskalewicz, T., Zimowski, S., Zych, A., Łukaszczyk, A., Reczyńska, K., Pamuła, E., Electrophoretic deposition, microstructure and selected

properties of composite alumina/polyetheretherketone coatings on the Ti-13Nb-13Zr alloy. *J. Electrochem. Soc.*, 165, 3, D116–D128, 2018.

22. Moskalewicz, T., Zych, A., Kruk, A., Kopia, A., Zimowski, S., Sitarz, M., Cieniek, Ł., Electrophoretic deposition and microstructure development of Si3N4/polyetheretherketone coatings on titanium alloy. *Surf. Coat. Technol.*, 350, 633–647, 2018.
23. Clavería, I., Lostalé, A., Fernández, Á., Castell, P., Elduque, D., Mendoza, G., Zubizarreta, C., Enhancement of tribological behavior of rolling bearings by applying a multilayer ZrN/ZrCN coating. *Coatings*, 9, 7, 434, 2019.
24. Moskalewicz, T., Wendler, B., Zimowski, S., Dubiel, B., Czyrska-Filemonowicz, A., Microstructure, micro-mechanical and tribological properties of the nc-WC/a-C nanocomposite coatings magnetron sputtered on non-hardened and oxygen hardened Ti-6Al-4V alloy. *Surf. Coat. Technol.*, 205, 7, 2668–2677, 2010.
25. Balaceanu, M., Braic, V., Braic, M., Kiss, A., Zoita, C.N., Vladescu, A., Drob, P., Vasilescu, C., Dudu, D., Muresanu, O., Structural, mechanical and corrosion properties of TiOxNy/ZrOxNy multilayer coatings. *Surf. Coat. Technol.*, 202, 11, 2384–2388, 2008.
26. Socol, G., Macovei, A.M., Miroiu, F., Stefan, N., Duta, L., Dorcioman, G., Mihailescu, I.N., Petrescu, S.M., Stan, G.E., Marcov, D.A., Chiriac, A., Poeata, I., Hydroxyapatite thin films synthesized by pulsed laser deposition and magnetron sputtering on PMMA substrates for medical applications. *Mater. Sci. Eng. B Solid-State Mater. Adv. Technol.*, 169, 1–3, 159–168, 2010.
27. Padmavathy, P., Ananthakumar, R., Subramanian, B., Ravidhas, C., Jayachandran, M., Structural and electrochemical impedance spectroscopic studies on reactive magnetron sputtered titanium oxynitride (TiON) thin films. *J. Appl. Electrochem.*, 41, 6, 751–756, 2011.
28. Chen, C.Y. and Chiang, C.L., Preparation of cotton fibers with antibacterial silver nanoparticles. *Mater. Lett.*, 62, 21–22, 3607–3609, 2008.
29. Chappé, J.M., Martin, N., Lintymer, J., Sthal, F., Terwagne, G., Takadoum, J., Titanium oxynitride thin films sputter deposited by the reactive gas pulsing process. *Appl. Surf. Sci.*, 253, 12, 5312–5316, 2007.
30. Chappé, J.M., Martin, N., Terwagne, G., Lintymer, J., Gavoille, J., Takadoum, J., Water as reactive gas to prepare titanium oxynitride thin films by reactive sputtering. *Thin Solid Films*, 440, 1–2, 66–73, 2003.
31. Lazarov, M., Raths, P., Metzger, H., Spirkl, W., Optical constants and film density of TiNxOy solar selective absorbers. *J. Appl. Phys.*, 77, 5, 2133–2137, 1995.
32. Kang, D.H., Ahn, D.H., Kwon, M.H., Kwon, H.S., Kim, K.B., Lee, K.S., Cheong, B.K., Lower voltage operation of a phase change memory device with a highly resistive TiON layer. *Jpn. J. Appl. Phys., Part 1 Regul. Pap. Short Notes Rev. Pap.*, 43, 8 A, 5243–5244, 2004.
33. Kaliaraj, G.S. and Kumar, N., Oxynitrides decorated 316L SS for potential bioimplant application. *Mater. Res. Express*, 5, 3, 36403, 2018.

34. Kumar, D.D., Kaliaraj, G.S., Kirubaharan, A.M.K., Alagarsamy, K., Vishwakarma, V., Baskaran, R., Biocorrosion and biological properties of sputtered ceramic carbide coatings for biomedical applications. *Surf. Coat. Technol.*, 374, 569–578, 2019.
35. Kaliaraj, G.S., Vishwakarma, V., Ramadoss, A., Ramachandran, D., Rabel, A.M., Corrosion, haemocompatibility and bacterial adhesion behavior of TiZrN-coated 316L SS for bioimplants. *Bull. Mater. Sci.*, 38, 4, 951–955, 2015.
36. Ward, I.M. and Sweeney, J., *An Introduction to the Mechanical Properties of Solid Polymers*, 2nd Edition, Wiley, England, 2004.
37. Kadiyala, A.K., Bijwe, J., Kalappa, P., Investigations on influence of nano and micron sized particles of SiC on performance properties of PEEK coatings. *Surf. Coat. Technol.*, 334, 124–133, 2018.
38. Predoi, D., Iconaru, S., Predoi, M., Buton, N., Motelica-Heino, M., Zinc doped hydroxyapatite thin films prepared by sol–gel spin coating procedure. *Coatings*, 9, 3, 156, 2019.
39. Yetim, T., Corrosion behavior of Ag-doped TiO2 coatings on commercially pure titanium in simulated body fluid solution. *J. Bionic Eng.*, 13, 3, 397–405, 2016.
40. Fielding, G.A., Roy, M., Bandyopadhyay, A., Bose, S., Antibacterial and biological characteristics of silver containing and strontium doped plasma sprayed hydroxyapatite coatings. *Acta Biomater.*, 8, 8, 3144–3152, 2012.
41. Navarro, C.H., Moreno, K.J., Arizmendi-Morquecho, A., Chávez-Valdez, A., García-Miranda, S., Preparation and tribological properties of chitosan/hydroxyapatite composite coatings applied on ultra high molecular weight polyethylene substrate. *J. Plast. Film Sheeting*, 28, 4, 279–297, 2012.
42. Sawyer, W.G., Freudenberg, K.D., Bhimaraj, P., Schadler, L.S., A study on the friction and wear behavior of PTFE filled with alumina nanoparticles. *Wear*, 254, 5–6, 573–580, 2003.
43. Hoseini, M., Jedenmalm, A., Boldizar, A., Tribological investigation of coatings for artificial joints. *Wear*, 264, 11–12, 958–966, 2008.
44. Zeng, R.C., Hu, Y., Zhang, F., Huang, Y.D., Wang, Z.L., Li, S.Q., Han, E.H., Corrosion resistance of cerium-doped zinc calcium phosphate chemical conversion coatings on AZ31 magnesium alloy. *Trans. Nonferrous Met. Soc. China (English Ed.)*, 26, 2, 472–483, 2016.
45. Kiahosseini, S.R. and Mojtahedzadeh Larijani, M., Effects of nitrogen gas ratio on the structural and corrosion properties of ZrN thin films grown on biodegradable magnesium alloy by ion-beam sputtering. *Appl. Phys. A Mater. Sci. Process.*, 123, 12, 1–9, 2017.
46. Kiahosseini, S.R., Afshar, A., Mojtahedzadeh Larijani, M., Yousefpour, M., Structural and corrosion characterization of hydroxyapatite/zirconium nitride-coated AZ91 magnesium alloy by ion beam sputtering. *Appl. Surf. Sci.*, 401, 172–180, 2017.

47. Liu, C., Xin, Y., Tian, X., Chu, P.K., Corrosion behavior of AZ91 magnesium alloy treated by plasma immersion ion implantation and deposition in artificial physiological fluids. *Thin Solid Films*, 516, 2–4, 422–427, 2007.
48. Cui, X., Yang, Y., Liu, E., Jin, G., Zhong, J., Li, Q., Corrosion behaviors in physiological solution of cerium conversion coatings on AZ31 magnesium alloy. *Appl. Surf. Sci.*, 257, 23, 9703–9709, 2011.
49. Dinu, M., Ivanova, A.A., Surmeneva, M.A., Braic, M., Tyurin, A.I., Braic, V., Surmenev, R.A., Vladescu, A., Tribological behaviour of RF-magnetron sputter deposited hydroxyapatite coatings in physiological solution. *Ceram. Int.*, 43, 9, 6858–6867, 2017.
50. Vranceanu, D.M., Ionescu, I.C., Ungureanu, E., Cojocaru, M.O., Vladescu, A., Cotrut, C.M., Magnesium doped hydroxyapatite-based coatings obtained by pulsed galvanostatic electrochemical deposition with adjustable electrochemical behavior. *coatings*, 10, 8, 727, 2020.
51. Surmeneva, M.A., Mukhametkaliyev, T.M., Khakbaz, H., Surmenev, R.A., Bobby Kannan, M., Ultrathin film coating of hydroxyapatite (HA) on a magnesium-calcium alloy using RF magnetron sputtering for bioimplant applications. *Mater. Lett.*, 152, 280–282, 2015.
52. Hench, L.L. and Polak, J.M., Third-generation biomedical materials. *Science*, 295, 5557, 1014–1017, 2002.
53. Hench, L.L., Splinter, R.J., Allen, W.C., Greenlee, T.K., Bonding mechanisms at the interface of ceramic prosthetic materials. *J. Biomed. Mater. Res.*, 5, 6, 117–141, 1971.
54. Nganga, S., Zhang, D., Moritz, N., Vallittu, P.K., Hupa, L., Multi-layer porous fiber-reinforced composites for implants: *In vitro* calcium phosphate formation in the presence of bioactive glass. *Dent. Mater.*, 28, 11, 1134–1145, 2012.
55. Lahiri, D., Singh, V., Benaduce, A.P., Seal, S., Kos, L., Agarwal, A., Boron nitride nanotube reinforced hydroxyapatite composite: Mechanical and tribological performance and *in-vitro* biocompatibility to osteoblasts. *J. Mech. Behav. Biomed. Mater.*, 4, 1, 44–56, 2011.
56. Rajeswari, D., Gopi, D., Ramya, S., Kavitha, L., Investigation of anticorrosive, antibacterial and *in vitro* biological properties of a sulphonated poly(etheretherketone)/strontium, cerium co-substituted hydroxyapatite composite coating developed on surface treated surgical grade stainless steel for orthopedic applications. *RSC Adv.*, 4, 106, 61525–61536, 2014.
57. Atiq Ur Rehman, M., Bastan, F.E., Haider, B., Boccaccini, A.R., Electrophoretic deposition of PEEK/bioactive glass composite coatings for orthopedic implants: A design of experiments (DoE) study. *Mater. Des.*, 130, 223–230, 2017.
58. Seuss, S., Heinloth, M., Boccaccini, A.R., Development of bioactive composite coatings based on combination of PEEK, bioactive glass and Ag nanoparticles with antibacterial properties. *Surf. Coat. Technol.*, 301, 100–105, 2016.
59. Barkallah, R., Taktak, R., Guermazi, N., Elleuch, K., Bouaziz, J., Mechanical properties and wear behaviour of alumina/tricalcium phosphate/titania

ceramics as coating for orthopedic implant. *Eng. Fract. Mech.*, 241, 107399, 2021.
60. Marsal, A., Ansart, F., Turq, V., Bonino, J.P., Sobrino, J.M., Chen, Y.M., Garcia, J., Mechanical properties and tribological behavior of a silica or/and alumina coating prepared by sol-gel route on stainless steel. *Surf. Coat. Technol.*, 237, 234–240, 2013.
61. Santos, R.L.P., Buciumeanu, M., Silva, F.S., Souza, J.C.M., Nascimento, R.M., Motta, F.V., Henriques, B., Tribological behavior of zirconia-reinforced glass–ceramic composites in artificial saliva. *Tribol. Int.*, 103, 379–387, 2016.
62. Berni, M., Lopomo, N., Marchiori, G., Gambardella, A., Boi, M., Bianchi, M., Visani, A., Pavan, P., Russo, A., Marcacci, M., Tribological characterization of zirconia coatings deposited on Ti6Al4V components for orthopedic applications. *Mater. Sci. Eng. C*, 62, 643–655, 2016.
63. Kohal, R.J. and Klaus, G., A zirconia implant-crown system: A case report. *Int. J. Periodontics Restorative Dent.*, 24, 2, 147–53, 2004.
64. Jeyachandran, Y.L. and Narayandass, S.K., The effect of thickness of titanium nitride coatings on bacterial adhesion. *Trends Biomater. Artif. Organs*, 24, 2, 90–93, 2010.
65. Gomes, A.L. and Montero, J., Zirconia implant abutments: A review. *Med. Oral. Patol. Oral. Cir. Bucal*, 16, 1, e50–55, 2011.
66. Piconi, C., Burger, W., Richter, H.G., Cittadini, A., Maccauro, G., Covacci, V., Bruzzese, N., Ricci, G.A., Marmo, E., Y-TZP ceramics for artificial joint replacements. *Biomaterials*, Elsevier Sci. Ltd, 19, 1489–1494, 1998.
67. Treccani, L., Yvonne Klein, T., Meder, F., Pardun, K., Rezwan, K., Functionalized ceramics for biomedical, biotechnological and environmental applications. *Acta Biomater.*, 9, 7, 7115–7150, 2013.
68. Chevalier, J., What future for zirconia as a biomaterial? *Biomaterials*, 27, 4, 535–543, 2006.
69. Kaliaraj, G.S., Muthaiah, B., Alagarsamy, K., Vishwakarma, V., Kirubaharan, A.M.K., Role of bovine serum albumin in the degradation of zirconia and its allotropes coated 316L SS for potential bioimplants. *Mater. Chem. Phys.*, 258, 123859, 2021.

9

Study of Tribological Behavior and Wear Mechanism of Nanocomposite Coatings

Lailesh Kumar[1,2], Pankaj Shrivastava[3], Deepankar Panda[3], Arka Ghosh[3] and Syed Nasimul Alam[3]*

[1]*Department of Materials Science Engineering, Seoul National University, 1 Gwanak-ro, Gwanak-gu, Seoul, South Korea*
[2]*Pinetree PosMagnesium Co. Ltd., Suncheon-Si, South Korea*
[3]*Department of Metallurgical & Materials Engineering, NIT Rourkela, India*

Abstract

The requirement of the wear-resistant coatings and materials having better mechanical, endurance, and environmental properties have exceeded compared to the present available lubricating and wear-reducing technologies. There is a huge demand for novel technologies and advanced lubricating materials especially in the field of automobile industries, and to fulfill the requirements, industries are developing environment friendly and effective lubrication systems for present and future vehicles. Nanocomposite coating techniques are gaining importance due to the evolution of highly wear-resistant and very low-friction materials which are traditionally used in specific industrial applications. A single-layered nanocomposite coating comprised of at least two immiscible phases alike as traditional composite materials in which the two phases can be nanocrystalline or one phase is nanocrystalline and the other is an amorphous phase. Nanosized films having thickness upto 100 nm when applied to any surface improves various material performances like corrosion, friction shielding, antifouling, and bacteria resistive properties. The major coating processes by which nanocomposite coatings can be prepared are physical vapour deposition (PVD) and chemical vapour deposition (CVD) techniques, sol-gel and cold spray processes, spin and dip coating techniques, electrodeposition and electroless deposition, *in situ* polymerization, and thermally sprayed coating.

Nanocomposite coatings can be classified into three major group based on the type of nanoreinforcements in it. 0-D coatings comprises of various nanoparticle

**Corresponding author*: syedn@nitrkl.ac.in; nasimulalam@yahoo.com

Sarfraj Ahmed and Vinayak S. Dakre (eds.) Tribology and Characterization of Surface Coatings, (181–210) © 2022 Scrivener Publishing LLC

reinforcements, 1-D coatings comprises of CNTs and whiskers, and 2-D coatings comprises of graphene and its derivatives. These nanocoatings have several impacts on various products like computer hard drives, shaving blades, medical equipments, microelectromechanical systems (MEMS), engine components, and scratch-proof glasses. Nanocomposite coatings provide solutions to the rapidly changing environmental conditions by maintaining exceptional tribological properties through self-adjustment of the surface properties. Various wear mechanisms like fatigue, abrasive, adhesive and cohesive wear can be observed under different tribological environments. Furthermore, nanoscale multilayers coatings shown vast improvement in hardness and wear compared to one or few-layered coating due to high interface density between the nanoscale multilayers which led to the blocking of dislocation motion across the interfaces. This chapter will express some of the major advancements in the field of nanocomposite coatings and also include a descriptive discussion on their properties, characterization, and application. In the last decade, an overwhelming interest in multifunctional nanostructured composite coatings has been seen. It is in demand to use these coatings to enhance the performance, durability, and efficiency of the various tribological systems.

***Keywords*:** Nanocomposite coating, tribological behavior, wear-resistant materials, wear mechanism, testing method for nanocomposite coatings, automobile application

9.1 Introduction

There are many industries and engineering sectors like renewable energy sectors, automotive, locomotive, and aerospace industries that need durable nanocoatings for tribological applications. Generally, the components of different types of equipment fail due to the generation of cracks generated majorly due to the fatigue and abrasion wear and humid corrosion. Due to this, there is a limitation in life of the products and equipment which demands an improvement in the abrasion and wear resistance of the mating surfaces. In general, coatings are defined as employment of a durable material over substrate for its protection from any different environmental conditions which degrade the material [1–3]. The demand of the present requirement for coating characteristic is not fulfilled by the traditionally used coating materials. It was difficult with the traditional coatings to avoid the formation and propagation of cracks which deteriorates the wear resistance of the coating [2–4]. Nanocoatings, on the other hand, suppress cracks, thereby resulting in higher strength and toughness. The grain refinement is the prime reason behind the increase in wear resistance of nanocoatings. When the number of grain boundaries increases due to the increase in the number of nucleation sites, it also leads to a simultaneous decrease in grain size. Due to the grain size refinement, the grain boundaries in the material increase which obstructs the formation of crack and its further propagations [3–5]. A nanocoating can be defined as either having a thickness of the coating in nanoscale or dispersion of nanosized grains in the coatings. The composition of the nanocoating can be custom-tailored according to the wear and friction requirements of a specific application. In a nanocomposite coating, a particular phase added to the nanocomposite can enhance the wear behavior of the nanocoating. Conventional materials do not show better properties compared to the nanomaterials due to the absence of nano-effects in them. The wear resistance and toughness of the coating can be improved utilizing the nano-effect of the nanomaterial coatings. Nanoadditives in nanocoatings can solve complex tribological problems with respect to the design considerations regarding reliability and durability. Nanocoatings are found to have good toughness. Nanoparticles can be used to increase the abrasion, wear, and scratch resistances by utilizing the boosted tribological properties [4, 5].

Comparative to the traditionally used coatings, the nanocoatings provide better material and functional properties. There is an ever-growing interest in nanocoatings due to their wide range of applications and fascinating surface protective properties. In past few years, advancement in the materials has been made attributed to the nanomaterials technology. When the thickness of the coatings or the reinforcing phase in the coating is in the nanosized range of less than 100 nm, the coatings are considered to be nanocoating. Nanocoatings are being widely used in different fields of corrosion protection and wear resistance [3–6]. Nanocoatings are a vast area of research and a lot of highly effective nanocoatings have already been developed. Nanocoatings are referred to the nanosized lubricating films that are employed over wear prone materials to improvise the functionalities such as corrosion protection, wear resistance, protection from water or ice, antibacterial, antifouling and antifriction properties, self-cleaning and radiation, and thermal resistance. Nanotechnology application in coatings includes the use of nanoparticles as additives in coatings, synthesis of *in situ* nanostructures in nanocoatings, coatings of the deposited nanofilms. Nanocomposite coatings could have a thickness of a few hundred microns. Nanocoatings are currently commercially available and are widely applied in various surface coating applications. Nanocomposite coatings have enhanced mechanical strength, thermal properties, and wear and tear resistance for an enhanced lifecycle performance. They have unique properties that are not seen in the phases that make the nanocomposite when used individually [6, 7]. The nanocomposite surface coatings are engineered in a way that they could provide upgraded properties along with lesser costs. These coatings successfully provide wear and environmental oxidation resistance, higher surface hardness and toughness, and many others wear-reducing properties. The properties like wear resistance of the nanocoatings are governed by different aspects considering grain reduction, phase evolution, and transformation, strengthening mechanisms and nano-effects. The improved properties of nanocomposite coating can be ascribed to the morphological improvement with a nanosized phase-separated domain. The nanocomposite properties like hardness are directly connected to its nanostructures. Due to the absence in the activity of dislocations, the hardness generally increases. As nanocoatings are very thin they are also transparent. As nanocoatings are usually transparent, they are useful in applications where we need a transparent coating. Coatings can make a window resistant to heat and UV rays. Nanocoatings are also used in currency to prevent forgery. A wide range of materials can be coated with the nanocomposite coatings and chemical bonding can be established with the non-porous material surfaces. These coatings can be applied to almost all the types of materials including plastics, ceramics, polymers, and metals and their corresponding composites. In addition to it, they can be applied to protect the other nanocoatings also. A nanocomposite coating is consisting of two materials immiscible to each other separated from one another by their distinct interphases. Also, one of the dimensions of the filler added should be in nanoscale and a uniform dispersion of it is maintained within the matrix [8–10]. Broadly, the nanocomposites are categorized under organic and inorganic forms of materials. Nanocoatings have applications in aerospace mainly in the landing gears and engine, defense, medical, marine, hydraulic shafts, turbines, and oil industries to prevent wear and corrosion. Figures 9.1 and 9.2 show coating deposited on a substrate to make it hydrophobic. Figure 9.3 shows different applications of the nanocoatings.

Nanocoatings are hydrophobic if the contact angle is more than 90° and super hydrophobic if the contact angle is more than 150° (Figure 9.2). Surface coating technologies available today are capable of functionalising numerous substrate types to be hydrophobic. Hydrophobic coatings and segments are highly water repellent offering numerous performance benefits. In addition, these hydrophobic nanocoatings are ultrathin and nanoscale in thickness, optically clear and durable. Hydrophobic treatments act as an anti-wetting coating solution.

Nanocomposite coating processes provide an optimizing property ratio between hardness and elastic modulus to provide better wear resistance

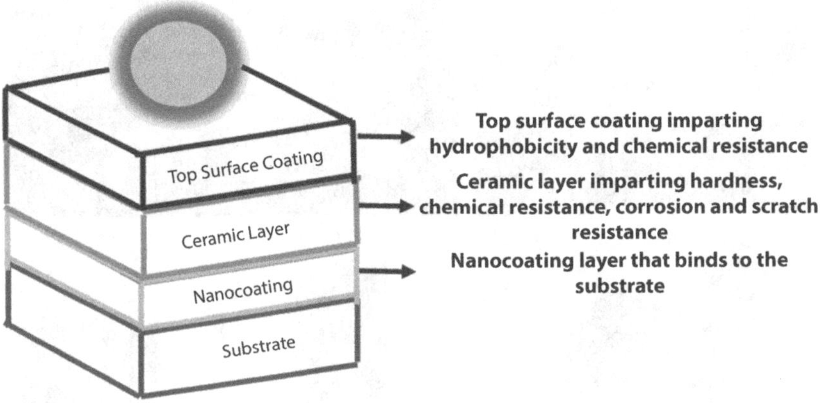

Figure 9.1 A schematic figure showing the different coatings on the substrate.

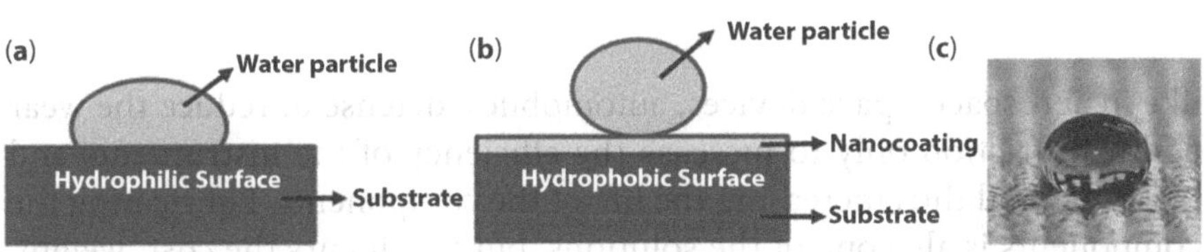

Figure 9.2 Nanocoating to make a surface hydrophobic. (a) Hydrophilic surface, (b) hydrophobic nanocoating on surface, and (c) hydrophobic material.

and better mechanical strength [11]. The increasing trend in the field of nanotechnology application in coatings is due to an increase in the availability of nanosize particles and advancement in the field of coatings at the nanoscale. The requirement for high hardness and thermally stable nanocomposite coating does not only depend on the grain sizes but is strongly affected by the structural interfacial states, i.e., grain boundaries under grain and interfacial disorientation, the value of the free volumes, and defect concentration at the grain boundaries. The incorporation of the nanofillers within the matrix increases the coating's barrier resistance, by providing a zigzag diffusion path and decreasing the porosity. Recently, a group of different researching bodies developed a process of size-selective deposition of nanocomposite coating intending to improve the physical, chemical, and physicochemical properties of the newly coated materials. There is a need in several advances and broad fields

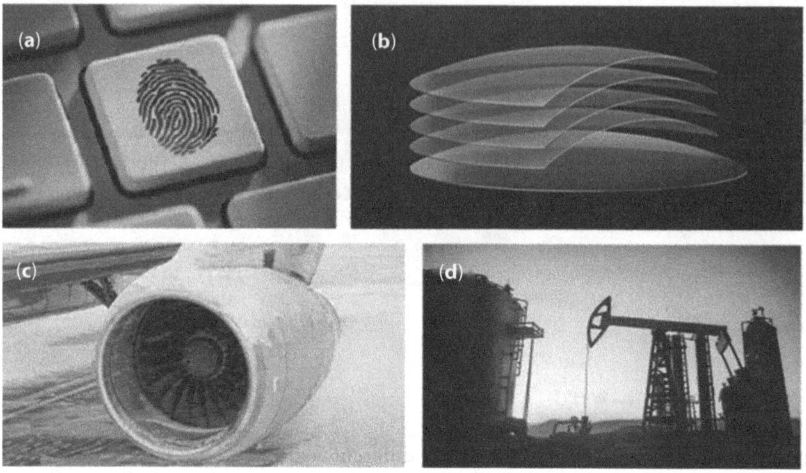

Figure 9.3 (a) Antifigrerprint, (b) antireflective, (c) antiicing, and (d) antifouling nanocoatings.

like in aerospace, space devices, automobiles, defense to reduce the wear and tear, friction only to increase the efficiency of the instruments and machines and thus increasing the life of the components. Lubricating the components is also one of the solutions, but this incurs the cost, weight, and complexity of these systems. So, it is better to coat such devices such that it could help in reducing friction as well as achieving the lubrication applications [12, 13].

In a metal, ceramic, or polymer-based matrix composite coating, the second phase of material are most commonly considered in the form of plates, nanowires, particles or whiskers, depending on the application of the coatings. A nanocomposite coating can be understood in a way that it is a layer provided on to the other materials to provide the other materials from any foreign phenomenon which could be corrosion, abrasion, wear and tear, inclusion, etc. The layer provided is made from a material composed of two similar or dissimilar phases separated from one another through their in-between interfacial region. It is supposed to have very high interfacial contact between the materials. For this, it is desired and recommended to disperse the nanoparticles to increase the interfacial contact. There is an increase of about 22% in the interfacial material content when the size of the nanoparticles has been decreased from 300 to 10 nm. There are, in general, two types of matrices are found for the production of nanocomposite coatings, i.e., one is organic which is polymer-based and second is inorganic which is metal or alloy-based. Depending on the type of application, the matrix selection has been done. Now, for the inorganic type of matrices, i.e., for alloy and metals, the nanoreinforcements are carbides, oxides, metallic particles, carbon isotopes, carbides, etc., and for the organic type of matrices, i.e., polymer-based matrices, the reinforcing materials are cellulose, PTFE, PEO, etc. [14–16].

With the metal matrix ceramic reinforced nanocomposite coatings, in general, the second phases are TiO_2, Al_2O_3, Cr_2O_3, SiC, WC, talc, etc., and are used in the application of engineering processes like surface finishing applications. Coatings were found to be the best cost-effective and practical solution to enhance the corrosion resistance and mechanical properties of metallic components. It has been seen that even stainless steel is also get corroded in an extreme environment, and thus, the most non-susceptible material also needs to be protected from the environment. To lubricate the surfaces with other elements increases in price and weight, and thus, a cost-effective way to protect these non-prone to the environment, as well as other environment prone materials, is coating [17, 18].

This chapter will also elaborate on the different processes to develop nanocomposite coatings, for instance, PVD, CVD, thermal spraying, electroplating or electrochemical codeposition, brush plating, dip coating, and electrophoretic deposition. Today, nanocoatings are used to solve several common problems and challenges. They have extensive areas of applications. This chapter will also thoroughly elaborate the nanocoatings and highlights the recent activities and achievements in this area. Various processing techniques for their development have also been discussed. Conventional coating techniques like CVD and PVD along with the recently developed different surface coating techniques like thermal spraying, sol-gel technique, and laser cladding have been reviewed in this chapter in detail. The application prospect of nanocoatings has also been discussed in detail. The present chapter will help in understanding how the nanocoatings are deposited on the substrate, their properties, and their industrial applications.

9.1.1 Techniques Employed for Depositing Nanocoatings

There are several techniques available for depositing nanocomposite coatings. Few laboratory and industrially accepted techniques are mentioned below.

9.1.1.1 Electrodeposition Coating

Electrodeposition is a very well-known technique to develop *in situ* metallic coatings by the application of current onto a conducting material immersed in metal powder solution which is to be deposited. In electrochemical codeposition or electrodeposition, the nanoreinforcement is required to be added with the metal mixture. It is needed for the process to be effective and while the electrochemical process is in function, the suspended nanoreinforcement should not precipitate at the bottom and should be properly dispersed in the electrolyte. Maintaining this increases the reliability of the coating to prevent corrosion. In addition to this, it is also needed to be taken care of that the suspended particles should not be agglomerated in the electrolyte solution [19]. Cathode generally remains the working electrode that is to be coated in the electrodeposition process during electroplating. In this technique, the process of electrolysis or electroplating takes place in between the interface of the electrode (conducting material) and electrolyte, i.e., conductive solution. The general thickness of the coating remains in range within 40 μm and always free from pores which makes it an effective technique compared to PVD and CVD processes. Also, this process is less expensive and no special arrangement or environmental conditions are required for the experimentation. Commonly, Zn, Cr, Ni, Cd, Cu, and Ag metal coatings are produced from this coating technique [20, 21]. When required to coat on conductive material, electrodeposition is a preferable method which could deposits coatings of different metal and their alloy like Ni, Cu, Ni-Cu, Zn, Cr, Co, Ni-Co, and Cu-Zn. It was recently found that the matrix in the coating tends to increase the corrosion and wear resistances of the coating and the nanoreinforcement added tends to greatly enhance the mechanical and tribological properties of the coatings. With compared to pure metal coating, the nanocomposite coating exhibit an enhancement synergistically to the microhardness, anti-corrosion, wear resistance, modulus of strength, and other tribological properties. The final result of the coating depends on many factors like pH of the solution, suspension mixing speed, concentration, and size of the reinforcements added in the electrolyte, solution temperature, etc. [22, 23].

9.1.1.2 Plasma Detonation Technology

The plasma detonation coating is also called detonation spray coating (DSC), and it is a process similar to the thermal spraying where powder for coating is heated and expedited by gaseous products which were formed from controlled burning of the measured mixture of acetylene and oxygen. This is done to provide a uniform and desired layer of coating. The thickness of coating in this technique is received by providing layer by layer of the plasma spray such that the protecting coating could reach a thickness in the range of 100–500 μm. The bond between the material to be coated and spray coating is maintained mechanically [24, 25].

9.1.1.3 Vacuum Arc Deposition

Vacuum arc deposition (VAD) is generally employed either as anodic arc deposition or cathodic arc deposition. In either way, they provide high MP metals in evaporating form at low pressures such as Fe and Ti. Initially, an arc melts the metal in a crucible by providing a 50-V current, and then, it ionizes in vapor in the anodic source and is provided on to the substrate. This technique provides a vacuum arc at very low voltage. The arc is initiated when the cathode or anode touches the striker electrode to produce plasma which provides a huge amount of carbon particles. The range of particles depends on the application of the coating. Industrial used VAD DC sources with pulse modifications. This technique developed a few decades ago as one of the competitive ion deposition techniques. When a dense and adherent coating is required with a wide range of compositions, VAD is a suitable technique for it. Due to applications and better results, nowadays, this technique has applications in the fields of electronics and optics too [26, 27].

9.1.1.4 Sol-Gel Technique

In the sol-gel process, the thickness of the coating is having its limitation up to the micron level. Producing a nano-thin coating is not possible with the sol-gel process. Also, during drying, thermal stresses generate, and due to this, cracks initiate whenever the thickness of the film is more than the critical value. To obtain advanced results, the sol-gel process can be used with other processes like with the electrodeposition process for the addition of organic and inorganic nanosized reinforcements into inorganic or organic matrices [28].

9.1.1.5 Cold Spray Method

The cold spray method works at the temperatures well below the sprayed material's melting point. This makes this technique different from the traditional thermal spraying method. Due to working at lower temperatures, thus oxidation of the materials does not take place due to which deterioration and decomposition of the materials do not happen. In general, this technique is used to provide nanocomposite coating of metallic or alloy matrices with inorganic nanofillers like CNT, graphene, and nitrite [29, 30].

9.1.1.6 Chemical Vapor Deposition

The chemical vapor deposition (CVD) technique is, in general, employed for the preparation of inorganic matrices reinforced by inorganic nanofillers. This method produces very high-quality films of coating and has very good reproducibility. The CVD synthesized nanocoatings are widespread in various industrial and commercial applications. They are considered as eco-friendly non-hazardous thin coatings. In this process, a volatile precursor is reacted by injecting it under a vacuum chamber. Later, the vacuum chamber is heated till the precursor gas starts to react and break down and get dispersed over the substrate surface [31]. Figure 9.4 is the schematic diagram of the CVD furnace.

9.1.1.7 Physical Vapor Deposition

Like the CVD method, the physical vapor deposition (PVD) technique is also generally employed for the preparation of inorganic matrices reinforce with inorganic nanofillers. The PVD technique involves few methods like thermal evaporation and laser-based deposition for coating deposition on the substrate [32]. Figure 9.5 is the schematic diagram illustrating the PVD technique.

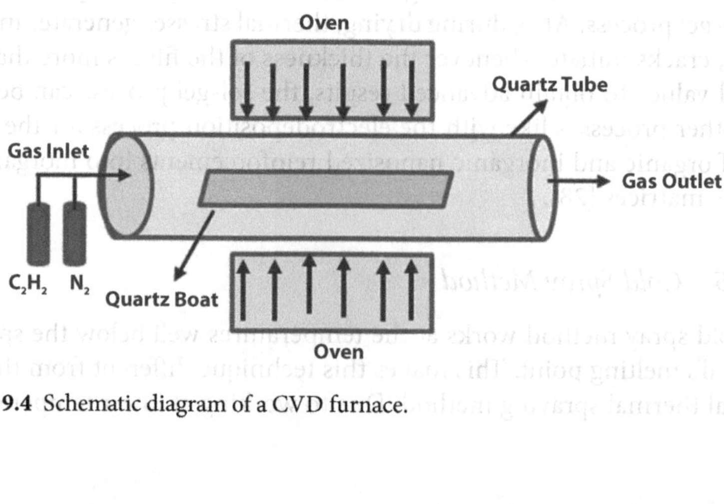

Figure 9.4 Schematic diagram of a CVD furnace.

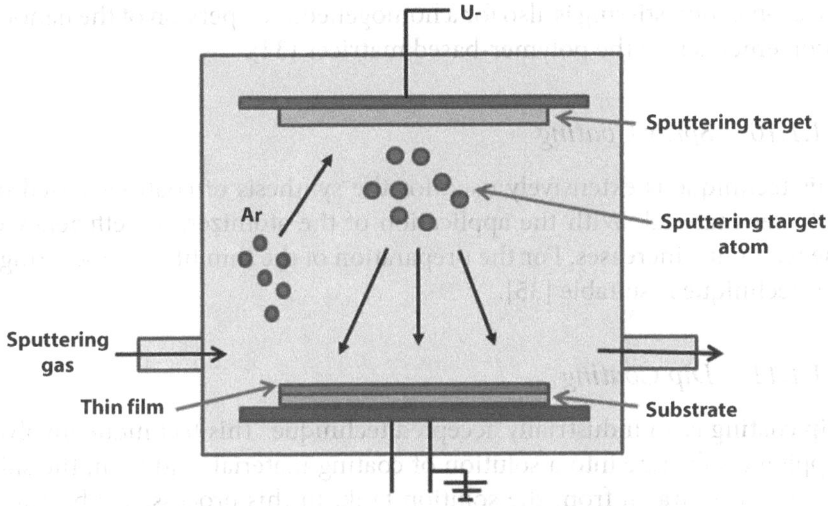

Figure 9.5 Schematic diagram of the physical vapor deposition (PVD).

9.1.1.8 Thermal Spray Method

The thermal spraying method protects the surface of the material from the wear and corrosion contaminants. This method also favors to coat a wide variety of materials. In this process, a consumable is made to be spread over the substrate in form of fine semi-molten drops. Layer coating of different category of materials including metals, polymers, and ceramics can be done with maintaining a coating thickness in the range of 0.1–10 mm. The material that can be melted or having plasticity in it can be deposited by this technique. On the substrate, the deposited particles from the spraying technique can form platelets which can form an interlink with each other to give the coating. These coatings are suitable to exhibit surface protective properties like provide resistance to corrosion, wear and tear, erosion and abrasion, and cavitation and heat. The material that is sprayed is produced by ball milling and the coating is provided with a matrix of metal or alloy, i.e., inorganic matrix. When the dispersion is associated with suspension, the method is called suspension plasma thermal spraying. The plasma-assisted spray method employs a DC current arc to form a high-temperature flow of ionized plasma gas [33].

9.1.1.9 Solution Dispersion

For polymer-based matrices, i.e., for organic matrices, this method is most appropriate. In this method, along with traditional mechanical stirring, ultrasonication stirring is also for a homogeneous dispersion of the nanoreinforcements into the polymer-based matrices [34].

9.1.1.10 Spray Coating

This technique is extensively used for the synthesis of coatings based on polymer material. With the application of the atomizer, the efficiency of the technique increases. For the preparation of the thin film nanocoatings, this technique is suitable [35].

9.1.1.11 Dip Coating

Dip coating is an industrially accepted technique. This technique involves dipping a substrate into a solution of coating material, and then, the substrate is withdrawn from the solution tank. In this process, a substrate is soaked into a tank and then allowed to drain. For a nanocomposite coating, the tank containing a nanocomposite into which the substrate is dipped and later pulled out at a controlled speed. The substrate is lifted vertically at a certain speed. A wet film of thickness λ is dragged from the liquid in the tank upward along with the substrate. The substrate along with the wet coating is then allowed to drain. Due to the controlled pulling out rate, the coating thickness can be controlled. Later, baking or pressure-drying is done to cool the coated substrate. Dip coating is a very economical technique for depositing aqueous-based liquid phase coating solutions over any material that includes fibers and polymers along with metal and ceramics [36]. The various steps of the dip coating process have been illustrated in the schematic diagram in Figure 9.6.

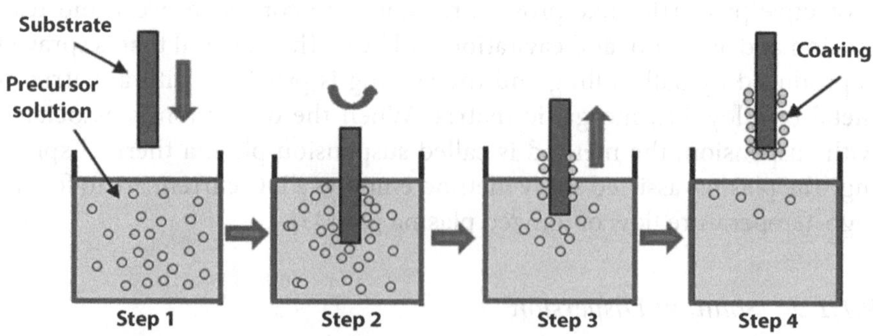

Figure 9.6 Dip coating technique.

9.2 Applications of the Different Nanocomposite Coating Techniques

Several techniques have their applications, and on basis of their ease of use and product affinity, let us understand the applications of the different nanocomposite coating techniques. It is well known that the investigation in the field of applications of nanocomposite coating is a promisingly progressing field of advanced material science. To obtain a hard nanocomposite coating having a considerable toughness, the coating should have low plastic deformation, enhanced cracking resistance, and a low value of engineering modulus. The application of nanocoating requires specialized equipment, complex procedures, and extreme heat. The successful development of nanocoatings is the basis for the application of nanocoatings. These techniques can be categorized into two separate groups, namely, vapor-phase and liquid-phase techniques. Vapor phase techniques include various coating methods like CVD, vapor condensation, laser ablation, and plasma arc processes. On the other hand, in liquid-phase technique processes are electrolysis, hydrothermal, sol-gel, and precipitation. New developments in coating technology are making nanocoating much accessible [37, 38]. There are wide variation and diversity in the fields of application of the coating.

Few of the examples includes house and building interior paints, various category of vehicles like earth movers, airplanes, and spacecraft, different structures like bridges and dams, and glass coating of double to multi-story buildings and also include varieties of industrial maintenance sections. Whereas, on a small scale, there is a separate wide variety of applications are there starting from microelectronic products to biomedical products. Before the development of the wide variety of inorganic and organic nanoparticles for reinforcements, silica was the only available choice due to its similar refractive index (RI) (1.46) to common organic binders to provide clear coatings for eyeglass lenses, polycarbonate made scratch-resistant sheets, automobile top coatings, etc. But, nowadays, due to advancement in the technology, the materials with higher RI like alumina (1.76), by reducing the particle size, they are being used in clear coating without haze problems. Many products today in the market include polyamide and PP composites which provide barrier films for packaging applications. For UV, IR, etc., kind of radiation barrier, different varieties of organic and inorganic pigments are used depending on the need. Continue exposure to the UV or IR exposures lead to mortifying the coated films and special coatings needed to resist these prolonged exposures. Here, the use of a combination of both organic and inorganics is suitably found to be the best approach. TiO_2, SiO_2, CeO_2, etc., are the most effective pigment for providing radiation barriers to many surfaces [39, 40].

The deterioration of the surfaces of the materials is a serious problem under the dominance of electrochemical corrosion facing today's world. Its primary and enduring prevention is an application of the coating on the surfaces. In today's world, it is important to understand the deterioration mechanism and reactive behavior of the surfaces to the aggressive dynamic outer environment. The best for it, it is important to cite the use of coatings and, in particular, nanocomposite coatings. This chapter will deal with the types of it, their applications, and mechanical advantages of different types of nanocomposite coatings. Nanocomposite coatings can be considered under functional coating as other than its usual corrosion protection properties, these coatings additionally provide wear and abrasion resistance, mechanical strengthening, as well as thermal and electrical conduction/isolation. Depending on the need of the application, several different functional coatings could be produced based on ceramic, metallic, polymeric, or most popularly nanocomposites [41, 42]. Nanocomposites have so many fields of applications, few include corrosion and lubrication protection of aircraft electrical and sliding mechanical systems, car accessories and drilling tools, nuclear apparatus, mining high-pressure valves, engineering and marine precision devices, medical equipment, and nuclear apparatus, which are some of the major fields which have adverse nanocomposite applications [43].

9.2.1 The Future of Nanocoatings

The nanocomposites today not only protect the materials, but their role is also very vast due to various organic/inorganic combination of the matrices and reinforcements. Today, they are incorporated in the material strength too. There is the coating of the future which shows self-cleaning property that could be used on the building structures or in the monuments so that efforts and cost could be saved in their maintenance. Nanocoating is a master engineering technique that provides a cost-efficient substrate coating for wide variety of materials. Nanocoatings are going to be the most promising candidates for the next-generation protection system. For a further improvisation regarding protection of the substrate surfaces, research in the area of nanocoatings should be focused. It is absolute that in the coming future, the nanocoating will get associated with the advanced 3D technologies. The 3D technology will lead to lower cost of the nanocoating materials, promote technology updated of nanocoating materials, and will reduce the production cycle of nanomaterials. In the future, nanocoatings are going to be multifunctional, whereas a single nanocoating would be self-cleaning, self-healing, antibacterial, antifouling, UV-resistant, and thermal resistant. The demand of the nanocoatings in the field of automotive and medical industry is speeding and will thrust the demand in the near future. Since the traditional coatings are based on solvent-based formulations, there is a risk of the emissions related to volatile organic compounds (VOCs). Different researchers and manufacturers are working in the field of water-based formulation-based nanocoatings which will fulfill the industrial needs and will exhibit high market demand globally in the coming years. Finally, new nanocoatings systems also need to be developed.

9.3 Effect of Different Coating Techniques on the Mechanical Properties of the Materials

It is well known that the mechanical properties of the nanocomposite coated products depend on their microstructures, grain sizes, coating parameters, reinforcements, the orientation of deposit, etc. In the case of MMnC coatings, when the size of the reinforcements is in the range of microns, this results in the premature failure of the composite coatings due to the formation of the defects. Few researchers have found that the enhanced hardness of the nanocomposite coating could be two times the hardness of its harder component due to the following mechanisms responsible, (i) the in-between atomic cohesion, (ii) the plastic deformation induced through dislocations, and (iii) the nanostructured feature of the materials. When grain size is smaller than 10 nm, then dislocations dependent hardness enhancement is dominating, whereas when the grain size is higher than 10-nm size, then the nanofeature of the materials dominates the hardness mechanism [22].

Thus, it is required for the higher strength requirement, and the size of the reinforcements is desired to be in the nanorange. It was also proven that the smallest the particle size used in the reinforcements, the highest was the nanohardness of the composite coatings. Due to smaller particle size and finer grain size, the obstruction is provided to the movements of the dislocations, due to which the hardness of the coating increases [44, 45]. Carbonaceous materials like graphene and MWCNT which are having very high surface areas are found to improve the particle dispersion within the matrix and improve the mechanical properties and thermal stability. It has been also seen that hybrid reinforcement coating also provides better resistance and property enhancement than mono reinforced nanocomposites [46]. Nanofillers have several effects on the mechanical and tribological properties of the nanocoatings. The impact resistance of the epoxy-based (polymers) coating can be increased by incorporating the nanoparticles in the matrix. The mechanism behind increasing the stiffness and toughness of the coating is that the nanoreinforcements get settled in the voids and pores of the coatings which make them impermeable due to which the film gets interconnected throughout the surface; this increases the cross-linking density of the coating which ultimately enhances the mechanical properties of the coating. Sometimes, the addition of nanoreinforcement in the polymer matrix increases the property several times. The property enhancement simultaneously depends on the integrity and internal properties of coating surfaces [47]. Literature survey shows the use of several nanoreinforcements and various processing techniques for developing nanocomposite coatings. Table 9.1 gives a list of nanomaterials and the properties they enhance of the nanocomposites [11, 15, 28, 36, 43, 48, 49]. Table 9.2 gives the list of different processing techniques used for different nanomaterial reinforced nanocoatings [14, 21–23, 28, 29, 32, 44, 50].

For the calculation of the hardness of the coatings, it is advised to measure through AFM-based nanoindentation (<50-nm tip size) in the place of evaluation through microhardness, so that engineering stress can also be calculated from the nanoindentation reading. To obtain super hardness in a coating, i.e., a strength of >40 GPa, the deformation executed through

Table 9.1 Nanomaterials reinforced property enhancement.

Nanomaterials	**Enhanced properties**
CuO, TiO_2, ZnO	Anti-microbial
Nanoclays	Gas barrier
Nanoclays, boehmite, graphene, CeO_2	Corrosion
Nanoclays, TiB_2, lanthanum zirconate	Fire and high-temperature resistance
Fe_2O_3	Magnetic property
Al_2O_3, SiO_2, ZrO_2	Mechanical/scratch resistance
Graphene, nickel,	Hardness
Graphene, CNT, non-metallic wires, ceramics	Tensile strength and Young's modulus
SiC, SiO_2, Al_2O_3	Wear resistance

Table 9.2 Nanomaterial reinforced property enhancement from different techniques.

Coating reinforcing material	Processing technique	Enhanced properties/ application	Comment
Organic/ inorganic hybrid	Sol-gel	Corrosion resistance	Replacement for aircraft Al alloy primers
polycarbonate and plastics	Sol-gel	Scratch resistance	Aircraft photostability Improvement
Ag, Au sandwiched between SiO_2, TiO_2, or CeO_2 layers	PVD	IR resistance	IR blocking coatings for window glasses
Indium tin oxide (ITO),	PVD, CVD, Sol-gel	Electrical and magnetic conductivity, UV-absorption	Displays, solar cells, optoelectronic devices, gas sensors
CNT, boride	Cold spray	High strength, strong adhesion	Al and Cu matrix
Organic nanofillers such as PEO, PTFE	Electrodeposition	Wear resistance and corrosion properties	Pulse current method
Nano-SiO_2 up to 10 wt.%	Spray coating	Fracture toughness	Increase in critical crack length
Nanosilica or carbon nanotube	Sol-gel, CVD	Hydrophobicity	By surface roughness improvement

plastic flow should be prohibited. High strength grain boundaries could result in the higher hardness of the coatings, whereas metallic matrices could guarantee to enhance their toughness [48]. There could be several applications of the nanocomposite coatings, such as anti-corrosion, anti-wear, high hardness, superhydrophobic, superhydrophilic, antifouling, and antibacterial, antibiofouling, and self-cleaning. Most applications are from the combination of the organic matrix and inorganic nanoreinforcement. The nanostructured hydrophobic and super-hydrophobic primer coatings act as a barrier to environmental humid and avoid water absorption to the surface and thus enhance the service life of the alloys and metals. For enhancement of the hardness up to >40 GPa, thermal stability up to 1,200°C, and oxidation resistance >800°C, a super hard nanocomposite coating is provided onto the substrate [49].

Universal challenges still exist in the field of coating which is uniform dispersion of the reinforcement, coat incurred in materials and techniques of coatings, deagglomeration of the coating constituents, coating materials having variations in the property enhancement, i.e., if mechanical strength is enhanced, then it has been seen that the corrosion resistance has been decreased and vice versa, and environmental hazardous issues, i.e., coatings, should not harm the environment. High viscosity due to high surface area caused by the nanoparticle dispersion is also a problem needed to be dealt with as due to it, a limitation is provided in the addition of the amount of the nanoparticles that can be incorporated [49]. It is well understood that the hardness of the nanoparticles is a very important factor to consider, enormous efforts have been put and several attempts have been made in research to improve scratch resistance, thermal properties, and bulk composite mechanical properties. Research is needed in these areas and success is yet to achieve in it.

9.4 Corrosion Behavior in Nanocoatings

There are various forms of corrosion and has \$3 trillion global investment per year. Currently, the market size for anticorrosion coatings is over \$12 billion. The annual market growth is estimated to be around 4.5% as demands from energy and mining resources has increased. Growth in nanocoatings will be above the compound annual growth rate (CAGR) as nanocoatings can meet the organizational requirements by providing exceptional wear and chemical resistance. A large fraction of the current products in anti-corrosion coatings led toward balancing the stiff environmental standards dealing with VOC emissions particularly in European countries and North America. The major growth markets are in water treatment and filtration, oil and gas pipeline coatings, and wind turbine coatings. Today, there is a growing demand for a cost-effective non-epoxy–based anticorrosion coating that can operate under temperature swings and in strong acid, water, and road salts. Nanoparticles comprise of very high specific surface areas due to which, when the surface is chemically functionalized, it can extensively help in inhibiting the organic corrosion. The functionalized tailored nanoparticles are considered as the ideal carriers for the delivery of appropriate levels of active corrosion inhibitors. As of November 2011, there are about 250 nanocoating companies worldwide and 65 of them are targeting medical applications for their coatings. In 2012, the demand of nanocoatings in the global market reached upto 255 kilotons with an estimated revenue of 1,495 million US$. In the year 2019, the estimated revenue of the global nanocoating market was 5.9 billion US$. The surge in the extent of research and development in the area of nanocoating is mainly due to the preference for nanocoating in the area of medical and automotive industries. Looking at the present global scenario, there is an increase in the demand of the self-cleaning and anti-microbial types of coatings in the fields of mechanical as well as aerospace applications to increase the life span of the rubbing parts. This demand is going to get increase in the upcoming near future very rapidly. There is a growing need for advanced materials in the medical and automotive industry, and this has resulted in a surge in research and development in nanotechnology, which is expected to boost nanocoatings market growth. The nanocoating market includes coatings developed using nanotechnology, which possesses superior properties including abrasion resistance, ductility, hardness, lubricity, and transparency as compared to other conventional coatings. Shifting demand toward nanocoatings instead of polymer coatings is mostly due to superior properties and low VOC emissions. These are expected to be major the driving forces for the nanocoatings market. Increasing anti-microbial coatings awareness and self-cleaning coatings in mechanical and aerospace applications for the prolonged life span of parts is expected to augment market demand. The global nanocomposites market is widespread majorly in the Asia Pacific, North America and Latin America, Europe, Africa, and middle Eastern countries as shown in the chart in Figure 9.7 [51, 52].

The key players in the nanocoatings industry are BASF, Buhler, NanoTech, NanoCoating, Nanogate, AdMat Innovations, Nanofilm, Tesla NanoCoating, Nanomech, Integram, Cleancorp, Eikos, Nanovere, etc. The improvement in technology and increase of funding in R and D sectors of the nanocomposite development has enabled the development of nanocoatings at a much lower cost and has also provided the growth of the global nanocoatings market. Depending on the type of nanocoatings, the division of market in different areas like anti-fingerprint, self-cleaning, anti-microbial, anti-fouling, and easy-to-clean type depending on the final use the market is divided into automotive, food packaging, electrical and electronics, construction, and

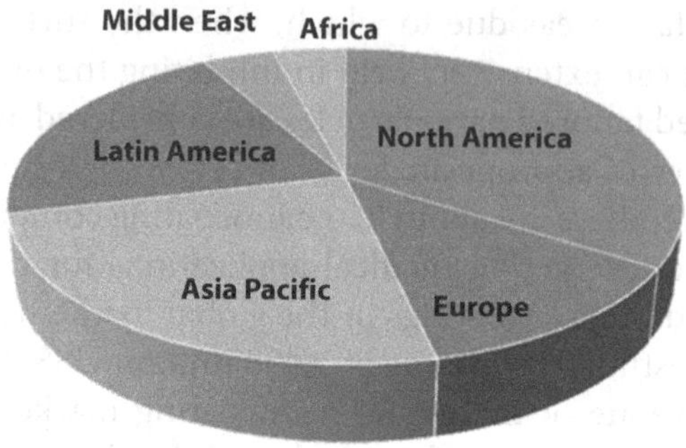

Figure 9.7 Global nanocoating market size by geography.

several others. The increasing demand for various unique type of coatings like UV and light protection coatings, depolluting coatings, anti-corrosion coatings, insulative nanocoatings, water beading coatings, antifouling coatings, anti-graffiti coatings, self-cleaning textiles, and electronics with waterproof properties are fueling innovations in nanocoatings [53, 54].

9.5 Wear Mechanism in Nanocoatings

In the past decade, nanocomposites coatings have been gaining much more attention regarding increment in the wear and tear resistance of the substrate materials. Various coatings have been developed, such as sol-gel method, cold spray method, PVD and CVD methods, thermal spray method, *in situ* polymerization method, electroless deposition, electrodeposition method, solution dispersion, magnetron sputtering, and dip coating.

Recently, Wasekar *et al.* [55] have investigated regarding wear behavior of Ni-W/SiC nanocomposite synthesized by the pulsed electrodeposition method. In their research work, they have used nano-SiC particles for the synthesis of nanocomposite coatings over Ni-W matrix phase. Both volume fraction and interparticle spacing of SiC nanoparticles in the composite coating could be attributed to the enhanced wear resistance and reduction in groove dimension on the wear surface. However, similar results have been reported by the same author while SiC used for coating having a particle size in the range of microns [56]. Furthermore, direct current mode (DC) is also applied for carrying out the electrodeposition method for coating. Zhao *et al.* [57] proposed the effect of ZrO_2 coating over the wear behavior of Ni-W nanocomposite. The authors have shown that virgin Ni-W alloy had abrasive groves over its worn surface; however, coating of ZrO_2 nanoparticle considerably decreases the amount of grooves as well as the depth of wear track of Ni-W/ZrO_2 nanocomposite coating. With the increase in the amount of ZrO_2 nanoparticles, the wear behavior of nanocomposite coating was prominently enhanced. This could be attributed to the dispersion and particle strengthening of ZrO_2 nanoparticles. Moreover, during the application of load, the ZrO_2 particles were positioned in such a manner, so that load can be easily transferred from the matrix phase to the nanoparticles. Thus, it not only improves the surface properties but also serves as a better solid lubricant.

Among a wide variety of production methods, recent developments in PVD techniques for the fabrication of nanolayered coatings have become popular. In recent times, many researchers have shown about the fabrication of nanocomposite coatings via sputtering and electron beam (EB) evaporation (two types of PVD techniques) process and the importance of hardness (H), as well as elastic modulus (E) of materials in deciding their tribology behavior, have reported [58–60]. They investigated that the elastic strain to failure ratio, i.e., H/E plays an important role in deciding the performance of coatings, i.e., the high value of H/E ratio will be the indicator of better wear resistive materials. This suggests that higher hardness and lower elastic modulus value are the two vital parameters that influence the wear behavior of materials.

Nowadays, synthesis of nanocomposite coatings via plasma spray process is becoming popular. Yang *et al.* [61] have investigated the fabrication of $FeAl_2O_4$-based nanocomposite coating through plasma spray of Fe and Al_2O_3 powders. In his research work, he showed a far better wear characteristic of the composite coating compared to a single Al_2O_3 coating. As we know the size, shape, amount, and spreading of pores as well as defects are the deciding factors of the mechanical and tribological properties of materials. However, during thermal spray of Al_2O_3 coating, cracks were easily formed and directed for the propagation. This was attributed to the high melting temperature of Al_2O_3 which resulted in presence of a large number of pores. Therefore, easy formation and propagation of crack occurred. However, in composite coatings, the exothermic reaction occurred among composite powders; subsequently, more amount of heat was generated, and thus, the melting of powder mixtures as well as products formed during reactions occurred. It was seen that the least cracks were present in the composite coating because of the increase in density. This suggested an enhancement in the wear resistance of the nanocomposite coatings. The uniform distribution phases in the composite led to the enhancement in grain boundary strength between matrix and reinforcement/second phase particles. Another reason for the improvement in wear resistance of composite coatings can be concluded as the presence of melting and spreading of droplets over the surface of the substrate during the thermal spray process. Further, the bonding strength between particles increased all over the coatings. This resulted in better wear behavior of nanocomposite coatings as compare to individual coatings.

Tribology study of carbon-based coating is found to be very difficult, since it is influenced by two important parameters, such as internal assets, i.e., properties of a material, and external assets, i.e., applied load and atmospheric condition. Usually, different wear mechanisms like adhesive, abrasive, and corrosive are seen in carbon-based coating materials. Pardo *et al.* [62] have investigated the wear properties of hydrogenated amorphous carbon (a-C:H)–based nanocomposite coating fabricated via the CVD process. In this work, nano-Ag and Cr particles were deposited in the matrix phase via ECR-CVD process for the synthesis of nanocomposite coatings (Me-C:H). Incorporation of these metallic particles reduced the coefficient of friction of the coatings and, because of their high reactivity with oxygen or humid air present nearby, restricted the contact between carbon matrix and the surrounding environment during the wear test. Incorporated nanoparticles behave as a safeguard for substrate material and this could be confirmed from the smaller size and depth of the sliding track. Further, the enhanced bonding nature of the metallic particles with carbon network film was continuously providing appropriate support to minimize the surface degradation during the wear test of nanocomposite coatings.

The superior coating is the one that can repair by adding organic and inorganic inhibitors to coatings. These inhibitors behave like a barrier against corrosion and decrease the influence of defects like pores, holes, and cracks to increase the efficiency of coating in the course of mechanical and tribological study. Therefore, researchers have implemented a simple and easy method for the development of nanocomposite coatings, i.e., sol-gel method. Generally, this method is carried at a low temperature followed by control of process parameters and results in better adherence ability and uniform deposition over the substrate surface. Shanaghi *et al.* [63] in their investigation, sol was ready by mixing of isopropoxide zirconium, aluminum tri-sec-butylate, isopropanol, ethylacetoacetate (EAA), and nitric acid. Whereas Benzotriazole was added as an inhibitor and trimethoxy-silane ($C_9H_{20}O_5Si$) acted as both encapsulation agent and stabilizer. After gone carrying out of different heat treatment process of Al alloy, it was used as the substrate over which deposition was carried out by dip coating. In the course of the low-temperature sol-gel process, the presence of defects in coatings was minimized and showed better corrosion resistance of the amorphous phase. Benzotriazole, an inhibitor, was reacted with the other chemical species used during preparation of sol mixture and delayed the formation as well as propagation of defects in the composite coating during wear test. This inhibitor initially reacts with the solution for the prevention of defect formation and is also absorbed in the metal-sol mixture interface for the development of a thick layer. Another reason for enhanced wear resistance could be ascribed to the good interfacial surface to substrate bonding of coating at their interface and also having outstanding adhesion ability and elasticity. Further, the incorporation of nanoparticles in nanocomposite coatings could be acted as preservers for showing good agreement between the organic inhibitors and coating surface for the enhancement in wear resistance.

Usually, a nanocomposite coating consists of hard nanoparticles and accommodates a matrix for the formation of dense phases at the interface that acts as a barrier for crack propagation and subsequently cracks growth. Voevodin *et al.* [64] have proposed that nanocomposite coatings provide a large number of interface boundaries that restrict the crack growth and assists for the crack deflection as well as termination of crack propagation. Pei *et al.* [65] have investigated the study of wear behavior of TiC/a-C:H nanocomposite coatings synthesized via magnetron sputtering process. In their work, they revealed that induction of transfer film over the surface of the ball, self-lubrication of coating surface occurred during the ball-on-disc wear test. This brings a decrement in the friction coefficient value significantly. Further, with the rise in the thickness of transfer film on the ball surface, the COF value gradually decreasing. Once the ball surface was fully covered with transfer films, the wear rate of the coating surface got reduced and the volume of wear debris was found to be less. Therefore, there was a gradual reduction of the thickness of film occurred until completely disrupted. Subsequently, the COF values suddenly raised and forming more wear debris. This offers new materials for the formation of the fresh and new film on the surface of the ball, and so, this cycle was occurring periodically. Further, nano-TiC particles were distributed in the amorphous hydrogenated carbon matrix phase; thus, it helps in tolerating the load provided during the wear test. Due to the amorphous nature of the matrix phase, nucleation and propagation of crack are significantly prohibited. Because the amorphous phase has a better toughness value as compared to the crystalline phase, a combination of amorphous and crystalline phases results in improved wear resistance of nanocomposite coatings.

Sometimes, these nanocoatings could combine the higher hardness with low coefficient of friction and thereafter exhibited considerably lower wear when compared to conventional coating. A newly nanocomposite category exhibiting good wear resistance has been transpired which shows higher hardness emerged from the microstructural engineering of the individual phases. To enhance the protective ability of the conventional coatings, nanoparticles addition could result in this regard. However, the amount of nanoparticles should be appropriate; otherwise, it would act adversely and reduce the protective properties of the coatings. Nanoparticles in nanocomposites can provide a lubricating effect and wear characteristic of the material could be augmented. In the case of traditional coatings, there is an epitaxial to non-epitaxial growth in the interfacial grains when nano-Al_2O_3 particles are added. The added nano-Al_2O_3 particles get dispersed near the grain boundaries and cell substructures and eliminate the crack formation at the matrix interface. It not only hinders the new phase formation but also helps in preventing the diffusion of any alloying elements. In improving the microstructure of the coatings, the nano-effects of the dispersed nano-Al_2O_3 particles plays a very important role. The lack of activity in the dislocation, when coated with nanocoatings, is the prime reason behind the hardness enhancement of the nanocoatings. Due to the presence of the large interfaces provided by the nanoparticles, the atomic arrangement of the interface is very chaotic and it is very easy for the atoms to migrate under the application of an external force. Due to this, the nanocoatings exhibit excellent toughness and reasonable ductility. Nanosized particles play a very important role in the development of different types of coatings, and recently, with the introduction of innovative nanomaterials, such as graphene, carbon nanotubes, and cellulose nanofibers in the coating, the demand for nanocoatings has increased manifold [66–68].

Another mechanism by which the wear resistance is enhanced by the nanocoatings is the phase transformation toughening mechanism.

In this mechanism, the enhancement in wear properties of the compo-coating is ascribed to the phase evolution toughening related with tetragonal zirconia (ZrO_2). A highly ductile tribofilm is produced while conducting the wear test to take part in the pores of the coating. Pure ZrO_2 can exist in three crystal forms, and it can be reversed back to single crystal form due to the presence of phase transition process. Since ZrO_2 is a polymorphic material, it exists in three different temperature-dependent forms. From room temperature to ~1,200°C called as monoclinic, from ~1,200°C to ~2,400°C called as tetragonal, and from ~2,400°C to 2,700°C called as cubic form which is the only stable phase among all three. The ZrO_2 toughening mechanism is shown in the schematic diagram given below in Figure 9.8. A tetra-directional monoclinic transformation from tetragonal to monoclinic occurs when stress is applied such that the generated stress field around the neck region could obstruct the crack propagation [69, 70].

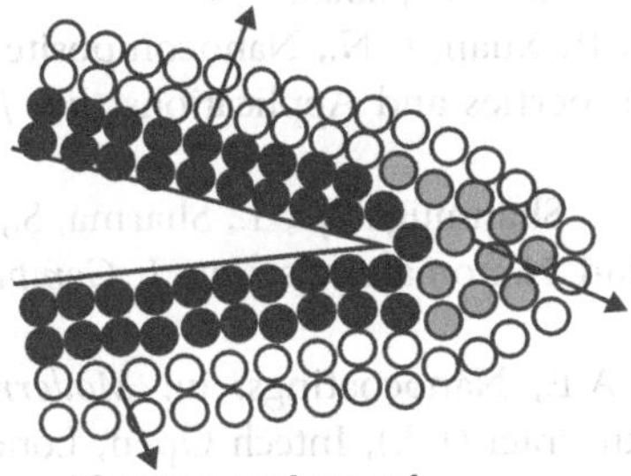

Figure 9.8 Phase transformation toughening mechanism.

The effect of the formation of the third body layer and tribofilms is one of the main reasons for the increased lubricating behavior when nanoparticles are added. Nanocoatings can also play a key role in preventing tribocorrosion. When a material degrades under different types of interactions like mechanical, chemical, or electrochemical between two surfaces that are in relative motion, it results in an irreversible transformation of the material property. This phenomenon is called tribocorrosion [68, 71, 72].

References

1. Gu, Y., Xia, K., Wu, D., Mou, J., Zheng, S., Technical Characterization and Wear-Resistance Mechanism of Nano Coatings: A Review. *Coatings*, 10, 3, 23, 2020.
2. Hingdon, C., Cook, B., Harringa, J., Russell, A., Goldsmith, J., Qu, J., Blau, P., Friction and Wear Mechanism in $AlMgB_{14}$-TiB_2 Nanocoatings. *Wear*, 271, 9–10, 2111–2115, 2011.
3. Bajwa, R., Khan, Z., Nazir, H., Chacko, V., Saeed, A., Wear and Friction Properties of Electrodeposited Ni-Based Coatings Subject to Nano-enhanced Lubricant and Composite Coating. *Acta Metall. Sin.*, 29, 10, 902–910, 2016.
4. Bugnicourt, E., Kehoe, T., Larorre, M., Serrano, C., Phillipe, S., Schmid, M., Recent Prospects in the Inline Monitoring of Nanocomposites and Nanocoating by Optical Technologies. *Nanomaterials*, 6, 8, 150, 2016.
5. Bao, W., Deng, Z., Zhang, S., Ji, Z., Zhang, H., Next-Generation Composite Coating System: Nanocoating. *Front. Mater.*, 6, 6–11, 2019.
6. Singh, A., Mittal, S., Mudgal, D., Gupta, P., Design, Development and Application of Nanocoatings, in: *Nanmaterials and their Applications*, Khan, Z.H. (Ed.), Chapter 7, pp. 191–207, 2018.

7. Patscheider, J., Nanocomposite Hard Coatings for Wear Protection. Superhard Coating Materials. *MRS Bull.*, 28, 3, 180–183, 2003.
8. Tri, P.N., Nguyen, T.A., Carriere, P., Xuan, C.N., Nanocomposite Coating: Preparation, Characterization, Properties and Applications. *Int. J. Corros.*, Article ID 4749501, 1–19, 2018.
9. Chinglenthoiba, C., Ramkumar, K., Shanmugaraja, T., Sharma, S., Study of Nanotechnology, Nanocoating and Nanomaterials. *Int. J. Comput. Aided Eng. Technol.*, 3, 1, 17–25, 2017.
10. Abdel-Karim, R. and Waheed, A.F., Nanocoatings, in: *Modern Surface Engineering Treatments*, M. Alifkhazraei (Ed.), Intech Open, London, UK, 2013.
11. Hau, T.V., Trinh, P.V., Nam, N.P.H., Tu, N.V., Lam, V.D., Phuong, D.D., Minh, P.N., Thang, B.H., Electrodeposited nickel–graphene nanocomposite coating: effect of graphene nanoplatelet size on its microstructure and hardness. *RCS Adv.*: Royal Society of Chemistry, 10, 22080–22090, 2020.
12. Pogrebnjak, A.D. and Beresnev, V.M., Hard nanocomposite coatings, their structure and properties, in: *Nanocomposites, New Trends and Developments*, pp. 123–161, Intech open, London, UK, 2012.
13. Bertolini, L., Elsener, B., Pedeferri, P., Redaelli, E., Polder, R.B., *Corrosion of Steel in Concrete: Prevention, Diagnosis, Repair*, 2nd Edition, pp. 1–434, Wiley-VCH, Weinheim, Germany, 2004.
14. Diebold, M.P., Bettler, C.R., Mukoda, D.M., Mechanism of TiO2/ZnO instability in latex paints. *J. Coat. Technol.*, 75, 29–36, 2003.
15. He, S., Lu, C., Zhang, S., Facile and efficient route to polyimide-TiO2 nanocomposite coating onto carbon fiber. *ACS Appl. Mater. Interfaces*, 3, 12, 4744–4750, 2011.
16. Kaboorani, A., Auclair, N., Riedl, B., Landry, V., Mechanical properties of UV-cured cellulose nanocrystal (CNC) nanocomposite coating for wood furniture. *Prog. Org. Coat.*, 104, 91–96, 2017.
17. Meng, Q., Frenkel, G.S., Colijn, H.O., Goss, S.H., Stainless-steel corrosion and MnS inclusions. *Nature*, 424, 1–389, 2003.
18. Abbott, S. and Holmes, N., Nanocoatings: Principles and practise from research to production. DEStech Publications, Inc., Lancaster, PA, USA, 2013.
19. Rezende, T.G.L., Cesar, D.V., do Lago, D.C.B., Senna, L.F., A review of corrosion resistance nanocomposite coatings: Nanocomposites, in: *New Trends and Developments*, pp. 147–207, Intech open, London, UK, 2012.
20. Borkar, T. and Harimkar, S.P., Effect of electro deposition conditions and reinforcement content on microstructure and tribological properties of nickel composite coatings. *Surf. Coat. Technol.*, 205, 17–18, 4124–4134, 2011.
21. Shi, L., Sun, C., Gao, P., Zhou, F., Liu, W., Mechanical properties and wear and corrosion resistance of electrodeposited Ni-Co/SiC nanocomposite coating. *Appl. Surf. Sci.*, 252, 10, 3591–3599, 2006.

22. Sen, R., Das, S., Das, K., Effect of stirring rate on the microstructure and microhardness of Ni-CeO2 nanocomposite coating and investigation of the corrosion property. *Surf. Coat. Technol.*, 205, 13–14, 3847–3855, 2011.
23. Sun, K.N., Hu, X.N., Zhang, J.H., Wang, J.R., Electrodeposited Cr-Al2O3 composite coating for wear resistance. *Wear*, 196, 1–2, 295–297, 1996.
24. Cashon, E.P., Wear resistant coatings applied by the detonation gun. *Tribol. Int.*, 8, 111–115, 1975.
25. Sundararajan, G., Dameracharla, S.R., Govindarajan, S., Govindarajan, S., Joshi, S.V., *Detonation Spray Coatings, Encyclopedia of Tribology*, Q. J. Wang and Y. W. Chung (Eds.), pp. 699–812, Springer, Boston, MA, 2013.
26. Ma, W., Ruys, A.J., Zreiqat, H., 16 - Diamond-like carbon (DLC) as a biocompatible coating in orthopedic and cardiac medicine, in: *Cellular Response to Biomaterials*, L. Di Silvio (Ed.), pp. 391–426, Woodhead Publishing Series in Biomaterials, Cambridge, UK, 2009.
27. Brown, I.G., Cathodic arc deposition of films. *Annu. Rev. Mater. Sci.*, 28, 243–269, 1998.
28. Radhamani, A.V., Lau, H.C., Ramakrishna, S., Nanocomposite coatings on steel for enhancing the corrosion resistance: A review. *J. Compos. Mat.*, 54, 5, 681–701, 2019.
29. Kim, J.S., Kwon, Y.S., Lomovsky, O.I., Dudina, D.V., Kosarev, V.F., Klinkov, S.V., Kwon, D.H., Smurov, I., Cold spraying of *in situ* produced TiB2-Cu nanocomposite powders. *Compos. Sci. Technol.*, 67, 11–12, 2292–2296, 2007.
30. Neerinck, D., Persoone, P., Sercua, M., Goel, A., Kester, D., Bray, D., Diamond-like nanocomposite coatings (a-C:H/a-Si:O) for Tribological applications. *Diamond Relat. Mater.*, 7, 2–5, 468–471, 1998.
31. Mattox, D.M., *The Foundations of Vacuum Coating Technology*, Springer, Heidelberg, Germany, 2003.
32. Wang, C., Wang, Y., Wang, L., Hao, G., Sun, X., Shan, F., Zou, Z., Nanocomposite Lanthanum Zirconate Thermal Barrier Coating Deposited by Suspension Plasma Spray Process. *J. Therm. Spray Technol.*, 23, 7, 1030–1036, 2014.
33. Shi, X., Nguyen, T.A., Suo, Z., Liu, Y., Avci, R., Effect of nanoparticles on the anticorrosion and mechanical properties of epoxy coating. *Surf. Coat. Technol.*, 204, 3, 237–245, 2009.
34. Steele, A., Bayer, I., Loth, E., Inherently superoleophobic nanocomposite coatings by Spray Atomization. *Nano Lett.*, 9, 1, 501–505, 2009.
35. Grosse, P., Hertling, R., Muggenberg, T., Design of low emissivity systems based on a three-layer coating. *J. Non-Cryst. Solids*, 218, 38–43, 1997.
36. Tang, X. and Yan, X., Dip coating for fibrous materials: mechanism, methods and applications. *J. Sol-Gel Sci. Technol.*, 81, 378–404, 2017.
37. Abu-Thabit, N.Y. and Makhlouf, A.S.H., Recent Advances in Nanocomposite Coatings for Corrosion Protection Applications, in: *Handbook of Nanoscience and Nanocomposite Coatings and Materials*, Makhlouf, A.S.H. and Scharnweber, D. (Eds.), Chapter 24, pp. 515–549, 2015.

38. Saji, V.S., The impact of nanotechnology on reducing corrosion cost, in: *Corrosion Protection and Control using Nanomaterials*, V.S. Saji and R. Cook (Eds.), pp. 3–15, Woodhead Publishing Series in Metals and Surface Engineering, Cambridge, UK, 2012.
39. Hwang, D.K., Moon, J.H., Shul, Y.G., Jung, K.T., Kim, D.H., Lee, D.W., Scratch Resistant and Transparent UV-Protective Coating on Polycarbonate. *J. Sol-Gel Sci. Technol.*, 26, 783–787, 2003.
40. Oniciu, L. and Muresan, L., Some fundamental aspects of leveling and brightening in metal electrodeposition. *J. Appl. Electrochem.*, 21, 7, 565–574, 1991.
41. Musiani, M., Electrodeposition of composites: An expanding subject in electrochemical materials science. *Electrochim. Acta*, 45, 20, 3397–3402, 2000.
42. Stankovic, V.D. and Gojo, M., Electrodeposited composite coatings of copper with inert, semi conductive and conductive particles. *Surf. Coat. Technol.*, 81, 2–3, 225–232, 1996.
43. Moghadam, A.D., Schultz, B.F., Ferguson, J.B., Omrani, E., Rohatgi, P.K., Gupta, N., Functional metal matrix composites: self-lubricating, self-healing, and nanocomposites- an outlook. *J. Miner. Met. Mater.*, 66, 6, 872–881, 2014.
44. Jagtap, S.B., Patil, V.D., Suresh, K., Ram, F., Mohan, M.S., Rajput, S.S., Patil, S., Shukla, P.G., Shanmuganathan, K., Functionalized carbon nanotube reinforced polymer nanocomposite microcapsules with enhanced stiffness. *Colloids Surf. A: Physicochem. Eng. Asp.*, 550, 82–89, 2018.
45. Nguyen, T.A., Nguyen, H., Nguyen, T.V., Thai, H., Shi, X., Effect of nanoparticles on the thermal and mechanical properties of epoxy coatings. *J. Nanosci. Nanotechnol.*, 16, 9, 9874–9881, 2016.
46. Ragosta, G., Abbate, M., Musto, P., Scarinzi, G., Mascia, L., Epoxy-silica particulate nanocomposites: Chemical interactions, reinforcement and fracture toughness. *Polym. J.*, 46, 23, 10506–10516, 2005.
47. Gryaznov, V.G., Solov'ev, V.A., Trusov, L.I., The peculiarities of initial stages of deformation in nanocrystalline materials (NCMs). *Scr. Mater.*, 24, 8, 1529–1534, 1990.
48. Mackay, M.E., Tuteja, A., Duxbury, P.M., Hawker, C.J., Horn, B.V., Guan, Z., Chen, G., Krishnan, R.S., General strategies for nanoparticle dispersion. *Science*, 311, 5768, 1740–1743, 2006.
49. Sung, L.P., Comer, J., Forster, A.M., Hu, H., Floryancic, B., Brickweg, L., Fernando, R.H., Scratch behavior of nano-alumina/polyurethane coatings. *J. Coat. Technol. Res.*, 5, 419–430, 2008.
50. Fernando, R.H., Nanotechnology Applications in Coatings: Recent Advancements, in: *ACS Symposium Series American Chemical Society*, Washington, DC, 2009.
51. Elvin, G., Building Green with Nanotechnology. *Technical Proceedings of the 2007 NSTI Nanotechnology Conference and Trade Show*, vol. 4, Tech Connect Briefs, pp. 683–686, 2007.

52. *Nanocoatings-Global Market Trajectory & Analytics*, Global Industry Analytics, Inc, ID-4721438, San Jose, USA, 2020.
53. Tator, K., Nanotechnology: The Future of Coatings-Part 1. *Mater. Perform.*, 53, 3, 34–36, 2014.
54. *Insights into What's Next in Nanocoatigs-R&D Analysis*, Global News wire, Dublin, 2020.
55. Wasekar, N.P., Bathini, L., Ramakrishna, L., Rao, D.S., Padmanabham., G., Pulsed electrodeposition, mechanical properties and wear mechanism in Ni-W/SiC nanocomposite coatings used for automotive applications. *Appl. Surf. Sci.*, 527, 146896, 2020.
56. Wasekar, N.P., Bathini, L., Sundararajan, G., Tribological Behavior of Pulsed Electrodeposited Ni-W/SiC Nanocomposites. *J. Mater. Eng. Perform.*, 27, 5236–5245, 2018.
57. Zhao, C., Yao, Y., He, L., Electrodeposition and characterization of Ni–W/ZrO_2 nanocomposite coatings. *Bull. Mater. Sci.*, 37, 1053–1058, 2014.
58. Oberle, T.L., Properties influencing wear of metals. *JOM*, 3, 438–439, 1951.
59. Richardson, R.C.D., The wear of metals by relatively soft abrasives. *Wear*, 11, 245–275, 1968.
60. Rebholz, C., Leyland, A., Schneider, J.M., Voevodin, A.A., Matthews., A., Structure, hardness and mechanical properties of magnetron-sputtered titanium–aluminium boride films. *Surf. Coat. Technol.*, 120–121, 412–417, 1999.
61. Yang, Y., Chen, X.G., Wang, L., Chu, Z.H., Liu, Z., Dong, Y.C., Yan, D.R., Zhang, J.X., He, J.N., Wang, L., Sliding wear behavior of *in-situ* $FeAl_2O_4$ matrix nanocomposite coating fabricated by plasma spraying. *Tribol. Int.*, 81, 97–104, 2015.
62. Pardo, A., Aleixandre, C.G., Celis, J.P., Buijnsters, J.G., Friction and wear behavior of plasma assisted chemical vapor deposited nanocomposites made of metal nanoparticles embedded in a hydrogenated amorphous carbon matrix. *Surf. Coat. Technol.*, 206, 3116–3124, 2012.
63. Shanaghi, A., Souri, A.R., Rafie, M., Chub, P.K., Nano-mechanical properties of zirconia-alumina-benzotriazolenano-composite coating deposited on Al_20_{24} by the sol-gel method. *Thin Solid Films*, 689, 137417, 2019.
64. Voevodin, A.A. and Zabinski, J.S., Supertough wear-resistant coatings with 'chameleon' surface adaptation. *Thin Solid Films*, 370, 223–231, 2000.
65. Pei, Y.T., Galvan, D., De Hosson, J. Th. M., Nanostructure and properties of TiC/a-C:H composite coatings. *Acta Mater.*, 53, 4505–4521, 2005.
66. Branagan, D.J., Swank, W.D., Haggard, D.C., Fincke, J.R., Wear Resistant Amorphous and Nanocomposite Steel Coatings. *Metall. Mater. Trans. A*, 32, 2615–2621, 2001.
67. Stepien, M., Chinga-Carrasco, G., Saarinen, J.J., Teisala, H., Tuominen, M., Aromaa, M., Haapanen, J., Kuusipalo, J., Makela, J.M., Toivakka, M., Wear Resisiatnce of Nanoparticle Coating on Paperboard. *Wear*, 307, 1–2, 112–118, 2013.

68. Sankaranarayanan, T.S.N., Nanocoatings to improve the tribocorrosion performance of materials, in: *Corrosion Protection and Control using Nanomaterials*, 1st Edition, V.S. Saji and R. Cook (Eds.), Woodhead Publishing Series in Metals and Surface Engineering, Cambridge, UK, 2012.
69. Bocanegra-Bernal, M.H. and de la Torre, D., Phase Transisition in Zirconium Dioxide and Related Materials for High Performance Engineering Ceramics. *J. Mater. Sci.*, 37, 4947–4971, 2002.
70. Fathy, A., Elkady, O.A., Abu-Oqail, A., Microstructure, mechanical and wear properties of Cu-ZrO_2 nanocomposites. *Mater. Sci. Technol.*, 33, 17, 2138–2146, 2017.
71. Achanta, S., Drees, D., Celis, J.P., Nanocoatings for tribological applications, in: *Nanocoatings and Ultra-Thin Films*, pp. 355–396, Woodhead Publishing Series in Metals and Surface Engineering, Cambridge, UK, 2011.
72. Grandin, M. and Wiklund, U., Friction, Wear and Tribofilm Formation in Electrical Contact Materials in Reciprocating Sliding Against Silver-Graphite. *Wear*, 302, 1–2, 1481–1491, 2013.

10

Investigation of Automotive Disc Brake's Material Based on Tribological Parameters by Using Computational Structural Analysis

Arul Prakash R.[1], Vijayanandh R.[1*], Ramesh G.[1], Hariaran S.[1], Janardhanan Y.[1], Senthil Kumar M.[1] and Jagadeeshwaran P.[2]

[1]*Aeronautical Engineering, Kumaraguru College of Technology, Coimbatore, Tamil Nadu, India*

[2]*Department of Humanities and Sciences, Rajalakshmi Institute of Technology, Chennai, Tamil Nadu, India*

Abstract

Shear effect (tribology) is one of the prime factors, which has drastically affected both solid and fluid properties. In the perspective of solid, the shear stress effects are lot in the failure occurrence of a specimen. Especially, area proximity–based regions are severely affected from shear forces. The ultimate aim of this work is to numerically analyze the shear effects on the two-wheeler disc brake for various relevant materials. Generally, shear effects are non-linear in nature, which means these react effectively at peak of the component's design. So, the numerical study about shear stress and its side effects on various materials is mandatory for the successful execution of vehicle's mission. In this work, innovative disc brake is constructed and thus conceptual design of the disc brake is modeled in CATIA, in which runners and brake pad are separated purposively with a tiny interval for different material allocation. The structural simulations on brake disc are executed by using ANSYS Structural Tool for various materials because of the flexibility of tool. Finally, the comparative analyses are executed on shear stress basis, and thereby, the suitable material is optimized. The short-listed materials for this comparative investigation are stainless steel, tungsten carbide, aluminum alloy, titanium alloy, carbon fiber reinforced polymer (CFRP), glass fiber reinforced polymer (GFRP),

**Corresponding author*: vijayanandh.raja@gmail.com

Sarfraj Ahmed and Vinayak S. Dakre (eds.) Tribology and Characterization of Surface Coatings, (211–238) © 2022 Scrivener Publishing LLC

Kevlar fiber reinforced polymer (KFRP), sandwich composites, and nanocomposites. To increase the trustworthiness of the material optimization outcome, two computational sensitivities are organized, which are mesh convergence investigation and analytical formula-based validations. The comparative error is calculated, which is less than 10% so the same computational procedures verified, and thereby, the computational optimizations are strongly recommended for real time applications.

***Keywords*:** Automotive vehicle, composite material, disc brake, nanocomposites, optimization, tribology

10.1 Disc Brake

10.1.1 Introduction

When a vehicle is in motion, the only system which assists in stopping or retarding the motion is the braking system. As the automotive technology developed, they came across drum brake [1]. The working of drum brake is quite simple, the effective function which is the brake shoe pressure or the force applied against the drum, it stops the rotating moment of the drum. The next leap in the brake system was the invention of disc brake [2]. The disc brake is in contact between the stationary parts of the brake pad situated inside the brake caliper. In this paper, we are presenting the study of shear stress analysis over a disc brake [3]. To achieve an optimal design, it should implement all available economic technologies to solve the technical problems, thus observing the experimental studies [4]. In the aerospace and automotive industry, many parts are subjected to simultaneous thermal and mechanical loads, due to these loads the stresses cause deformation and may even damage the system [5]. The principle objective of this paper is to numerically analyze the shear effect on disc for relevant materials. The braking system is designed in CATIA software and the simulation is done using ANSYS structural tool 16.2 [6]. The comparative analyses for different materials are mentioned below: stainless steel, titanium alloy, tungsten carbide, glass fiber reinforced polymer (GFRP), and carbon fiber reinforced polymer (CFRP).

10.1.2 Materials Used in Disc Brakes

Selections of brake pad material is done on the basis of friction consistency and wear properties under varying conditions of load, velocity, temperature, environment, and high durability. There are many parameters to be taken

Table 10.1 The useful mechanical properties of short-listed materials.

Sl.no.	Material property	Thermal conductivity [W/(m·K)]	Density (g/cm^3)	Specific heat (J/kg k)	Thermal expansion (μ/K)	Elastic modulus (GPa)	Coefficient of eriction (μ)
1	Stainless steel	30	7.8	460	11	210	0.4
2	Titanium alloy	21.6	4.70	525	7.6–9.8	80–125	0.4–0.5
3	Tungsten carbide	84.02–110	15.63	292	7.1	530–700	0.20–0.25
4	GFRP	1.2	1.8	800	4.9	72	0.10
5	CFRP	2.01	1.6	800–1,800	6	70	0.15

in consideration while selecting materials in which the primary factor is atmospheric conditions such as corrosion and thermal expansion [7]. The material must also be selected with respect to cost. At last, the comparative analyses of entire mechanical properties for all the materials are listed in Table 10.1, in which most of the data are extracted from literature survey.

10.2 Literature Survey—Problem Description and Solution Technique

10.2.1 Problem Descriptions

The ultimate aim of this work is to provide a good rotating platform for a moving vehicle by generation of low shear force and thereby increase the lifetime of a short-listed and specified disc brake. In general, the heat generations due to frictional contact between the components in the disc brake are quite high because of its nature of the work [8]. As a result of this heat increment, the performance of a disc brake affected drastically, and thus, the reduction in lifespan also taken place. Naturally, the lifetime of the vehicle and its user are predominantly depended on good performance of a disc brake, and therefore, its life affecting problem needs to be sorted out in an efficient manner [9]. Hence, the ultimate aim of this work is perfectly coincided with existing lifetime problem of a disc brake, which confirmed that the objective, and thereby, the methodologies used are good. In this work, the advanced composite materials' implementation-based proposal has been suggested in the disc brake [10]. The primary reason behind this accomplishment of these advanced composite materials is that it has the potential to resist the tough environments with the help of its integrative effect. Basically, the composite materials are good in strength to weight ratio, but in this work, the short-listed composite materials are mainly chosen based on its high thermal conductibility and shear force rather than stiffness to weight ratio [11]. From the standard literature studies, it is strongly believed that the composite materials are more suitable to tackle these critical environments such as huge heat generation and shear force resisting property in the disc brake [12]. In order to confirm the assumption, the computational structural analyses are executed on the short-listed advanced composite materials. The short-listed materials are Epoxy S Glass Wet, Epoxy Carbon UD 230 GP–Prepreg, and Epoxy Carbon Woven 230 Wet. Moreover, the existing materials also included and analyzed for the purpose of comparative analyses.

10.2.2 Methodology Used—Computational Structural Analysis

The previous and existing real-time problems have been solved effectively with the help of standard engineering approaches such as theoretical calculation, experimental testing, and computational simulations [13]. The standard theoretical formula-based solution on real-time problems is called as theoretical approach, in which normal and medium levels of problems are solved in an efficient manner, whereas high-level problems are solved in an ineffective manner [14]. The second approach is experimental testing, which have the competence to solve all kinds of real-time problems in the well-organized method. But the issues accomplished with experimental testing are nonflexible in test procedure, high possible of human error, high cost, and time-consuming process, etc. Finally, the advanced approach is computational simulation, which is fully relayed on computer-based solution technique [15]. In all the perspectives, the computational simulations are great to undertake this targeted problem so, in this paper, computational approach is used as a primary solver to analyze the disc brake in various perceptions [16]. Moreover, in this work, shear stress and deformation of disc brake play an imperative role in the selection process of better material. Also, the flexible test procedures of analyses of computational simulations provided the confident to extend the shear stress scrutiny for advanced composite materials [17]. In the computational analyses, the standard and fine conditions are employed, which provided the reliable output and thus the validational work for counter check the outputs are eliminated. For all class of computational structural analyses, the FEA (Finite Element Analysis) tool, i.e., ANSYS Workbench 16.2, and its facilities are used for further optimization.

10.3 Structural Analysis

10.3.1 Conceptual Design

The modeling of an automotive disc and brake pad are modeled in CATIA, in which the reference dimensions are taken from the success bike model, that was manufactured from one of the leading up-growing companies in industry [18]. The dimensions of the disc brake and brake pad are reveled in Figures 10.1 and 10.2, in which all the dimensions are in millimeters.

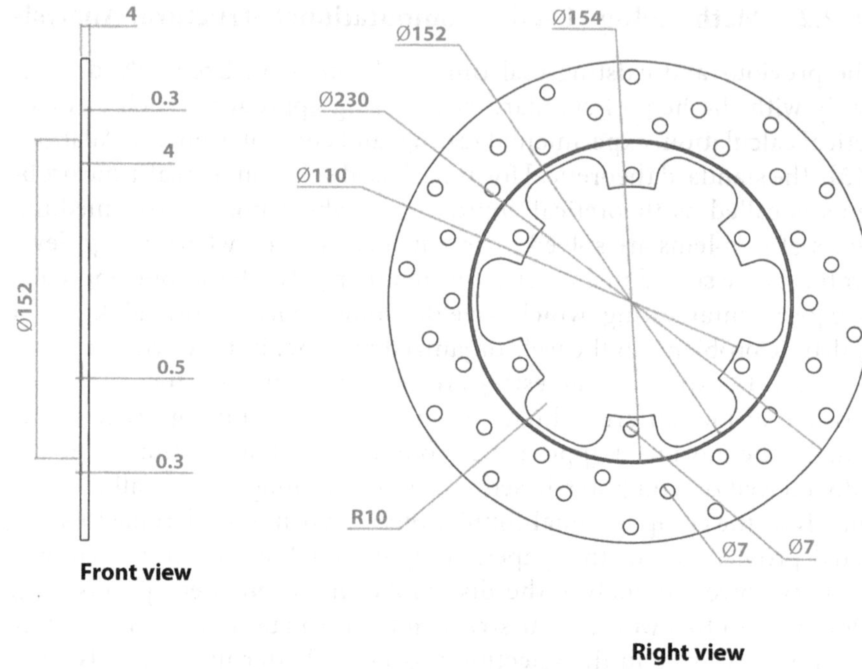

Figure 10.1 Shear stress of Epoxy Carbon Woven 230 Wet.

10.3.2 Pre-Processing

In computational structural analysis, three primary stages are available to provide the clarity in the structural analysis, and thereby to complete the simulation process in perfect manner. In the prep-processing, the following processes are needs to be carried out: provision of computational model, execution of discretization, application of supports, implementation of external and internal loading conditions, and solver selection. The first stage is import or creates geometry (assembly of disc and pad), which has been imported to ANSYS 16.2 static structural tool in the format of IGS. After that, the discretization process of brake disc and pad, that is more complicated shape so the fine-tuned operations are executed through ANSYS Mesh tool 16.2 and also explained in separate topic. The boundary conditions are as follows: speed of the vehicle is 615 RPM or 50 km/hour, coefficient of friction is given as per material behavior, and support of inner disc is given as fixed. The entire boundary conditional details are revealed in Figure 10.3. The resultshave been analyzed in order to understand how the applied loads affected the model. The change in cross-sectional area and the total deformation of the material is observed, and

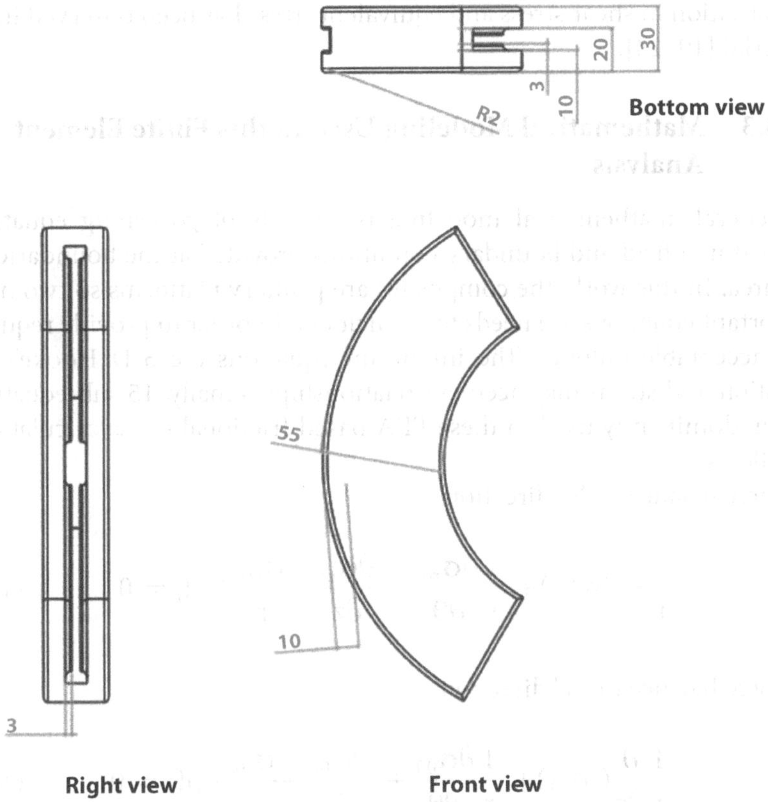

Figure 10.2 Shear stress of Epoxy Carbon Woven 230 Wet.

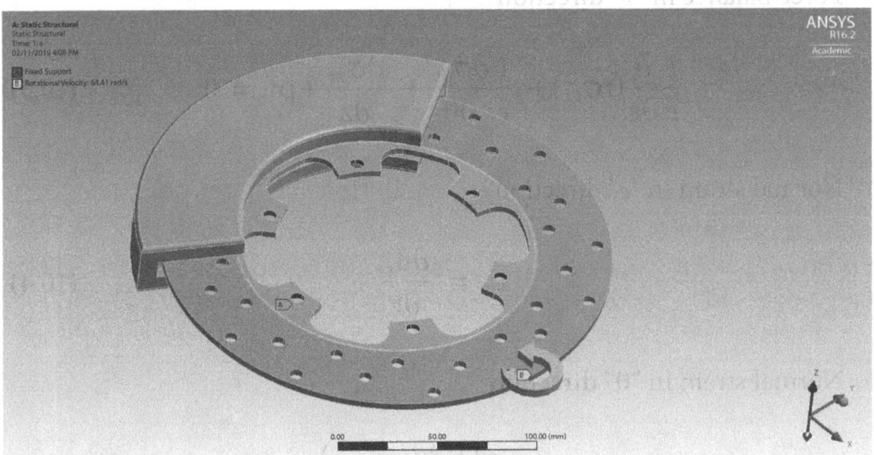

Figure 10.3 Details of boundary conditional parameters applied at brake disc and pad.

the variation in shear stress and equivalent stress has been observed in the material [19–24].

10.3.3 Mathematical Modeling Used in this Finite Element Analysis

In general, mathematical modeling is consists of governing equations defined in a field and boundary conditions provided at the boundaries of the area. In this work, the composites are primary platforms so two more important equations are needs to be included in order to provide required and acceptable outputs. The important equations are 3-D Hooke's law equation and strain-displacement relationships. Finally, 15 sub-equations are predominantly used in these FEA-based frictional stress calculations, as follows:

Force balance in "r" direction

$$\frac{1}{r}\frac{\partial}{\partial r}(r\sigma_{rr})+\frac{1}{r}\frac{\partial\sigma_{r\theta}}{\partial\theta}+\frac{\partial\sigma_{rz}}{\partial z}-\frac{\sigma_{\theta\theta}}{r}+\rho f_r=0 \tag{10.1}$$

Force balance in "θ" direction

$$\frac{1}{r}\frac{\partial}{\partial r}(r\sigma_{r\theta})+\frac{1}{r}\frac{\partial\sigma_{\theta\theta}}{\partial\theta}+\frac{\partial\sigma_{\theta z}}{\partial z}-\frac{\sigma_{r\theta}}{r}+\rho f_\theta=0 \tag{10.2}$$

Force balance in "z" direction

$$\frac{1}{r}\frac{\partial}{\partial r}(r\sigma_{rz})+\frac{1}{r}\frac{\partial\sigma_{\theta z}}{\partial\theta}+\frac{\partial\sigma_{zz}}{\partial z}+\rho f_z=0 \tag{10.3}$$

Normal strain in "r" direction

$$\varepsilon_{rr}=\frac{\partial u_r}{\partial r} \tag{10.4}$$

Normal strain in "θ" direction

$$\varepsilon_{\theta\theta}=\frac{1}{r}\left(\frac{\partial u_\theta}{\partial\theta}+u_r\right) \tag{10.5}$$

Normal strain in "z" direction

$$\varepsilon_{zz} = \frac{\partial u_z}{\partial z} \tag{10.6}$$

Shear strain on "r" plane

$$\Upsilon_{r\theta} = \frac{1}{2}\left(\frac{1}{r}\frac{\partial u_r}{\partial \theta} + \frac{\partial u_\theta}{\partial r} - \frac{u_\theta}{r}\right) \tag{10.7}$$

Shear Strain on "θ" plane

$$\Upsilon_{\theta z} = \frac{1}{2}\left(\frac{1}{r}\frac{\partial u_z}{\partial \theta} + \frac{\partial u_\theta}{\partial z}\right) \tag{10.8}$$

Shear strain on "z" plane

$$\Upsilon_{zr} = \frac{1}{2}\left(\frac{\partial u_r}{\partial z} + \frac{\partial u_z}{\partial r}\right) \tag{10.9}$$

Normal stress in "r" direction

$$\sigma_r = \left[\frac{1-\upsilon_{23}\upsilon_{32}}{E_2E_3\Delta}\right]\varepsilon_{rr} + \left[\frac{\upsilon_{21}+\upsilon_{23}\upsilon_{31}}{E_2E_3\Delta}\right]\varepsilon_{\theta\theta} + \left[\frac{\upsilon_{31}+\upsilon_{21}\upsilon_{32}}{E_2E_3\Delta}\right]\varepsilon_{zz} \tag{10.10}$$

Normal stress in "θ" direction

$$\sigma_\theta = \left[\frac{\upsilon_{21}+\upsilon_{23}\upsilon_{31}}{E_2E_3\Delta}\right]\varepsilon_{rr} + \left[\frac{1-\upsilon_{13}\upsilon_{31}}{E_1E_3\Delta}\right]\varepsilon_{\theta\theta} + \left[\frac{\upsilon_{32}+\upsilon_{12}\upsilon_{31}}{E_1E_3\Delta}\right]\varepsilon_{zz} \tag{10.11}$$

Normal stress in "z" direction

$$\sigma_z = \left[\frac{\upsilon_{31}+\upsilon_{21}\upsilon_{32}}{E_2E_3\Delta}\right]\varepsilon_{rr} + \left[\frac{\upsilon_{32}+\upsilon_{12}\upsilon_{31}}{E_1E_3\Delta}\right]\varepsilon_{\theta\theta} + \left[\frac{1-\upsilon_{12}\upsilon_{21}}{E_1E_2\Delta}\right]\varepsilon_{zz} \tag{10.12}$$

where

$$\Delta = \frac{(1-\upsilon_{12}\upsilon_{21}-\upsilon_{23}\upsilon_{32}-\upsilon_{13}\upsilon_{31}-\ 2\upsilon_{21}\upsilon_{32}\upsilon_{13})}{E_1E_2E_3}$$

Shear stress on "θ" plane

$$\tau_{r\theta} = [G_{12}]\gamma_{r\theta} \tag{10.13}$$

Shear stress in "z" direction

$$\tau_{\theta z} = [G_{13}]\gamma_{\theta z} \tag{10.14}$$

Shear stress in "r" direction

$$\tau_{zr} = [G_{31}]\gamma_{zr} \tag{10.15}$$

The primary governing equations are involved in the FEA analysis, which are mentioned in Equations (10.1) to (10.15). These complicated equations are need to be transformed to so call weak form using Galerkin's method, and thereby, the element formulations are constructed. For static analysis, this equation is

$$[K]\{u\} = \{F\} \tag{10.16}$$

In general, static conditions are very low to represent the real-time problems, and thus, the dynamic conditions are more preferable to solve all kind of structural problems. Equation (10.17) provides the dynamic analysis formulation:

$$[M]\{\ddot{u}\} + [C]\{\dot{u}\} + [K]\{u\} = \{F\} \tag{10.17}$$

where [M] is the mass matrix, $[\ddot{U}]$ is the second time derivative of the displacement [U] (i.e., the acceleration), $[\dot{U}]$ is the velocity, [C] is a damping matrix, [K] is the stiffness matrix, and [F] is the force vector. The computation tool first computes the displacements and then strains, and from strains, it calculates stresses using constitutive equation [1, 9].

10.3.4 Generation of Grid and Its Convergence Study

Meshing is an integral part of engineering simulation process where the complex geometries are divided into simple elements that can be used as discrete local approximations of the larger domain. The mesh influences the accuracy, convergence, and speed of the simulation results; the better and more automated the meshing tools, the faster and more accurate the solution. In this case, the disc and pad must be divided into number of small pieces known as finite elements, which provides the platform for the calculation of numerical integration. The shape of the element chosen for meshing the model is tetrahedral three-dimensional configuration. Structurally, the elements are medially influences the sensitivity of computational simulation than fluid dynamic simulation. Even though this medium has influence, in this work, the grid convergence study is executed because of the non-linear working conditions of two-wheeler disc brake. The mesh-based sensitivity test conducted on computational structural analysis on Epoxy Carbon Woven 230 Wet–based disc brake, in which four different cases are implemented for these convergence investigations. The four cases are coarse, medium, fine, and fine with face mesh setup used, which are classified as Case-1, Case-2, Case-3, and Case-4, respectively. The different classifications are based on elemental counts and qualities, which are revealed in Figures 10.4 to 10.7 in the low to high order manner. The elemental counts with respect to its mesh cases are clearly listed in Table 10.2 for the purpose of future reference. The comprehensive structural analyses are carried out with the aforesaid material and boundary conditions, and the entire results are revealed in the Figures 10.8 and 10.9,

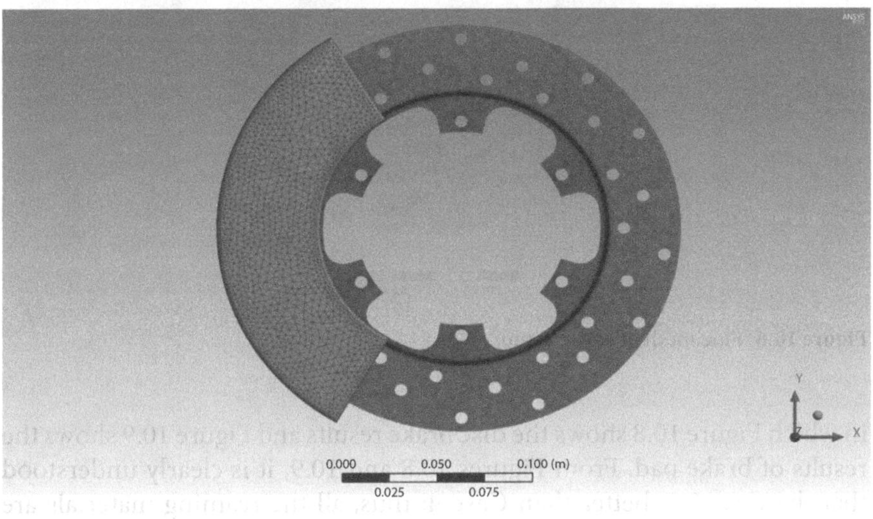

Figure 10.4 Coarse mesh of entire setup.

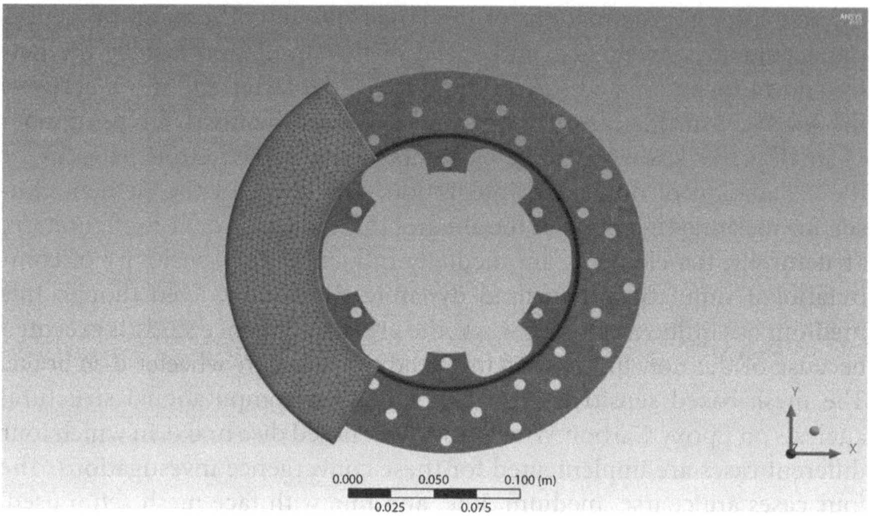

Figure 10.5 Medium mesh of entire setup.

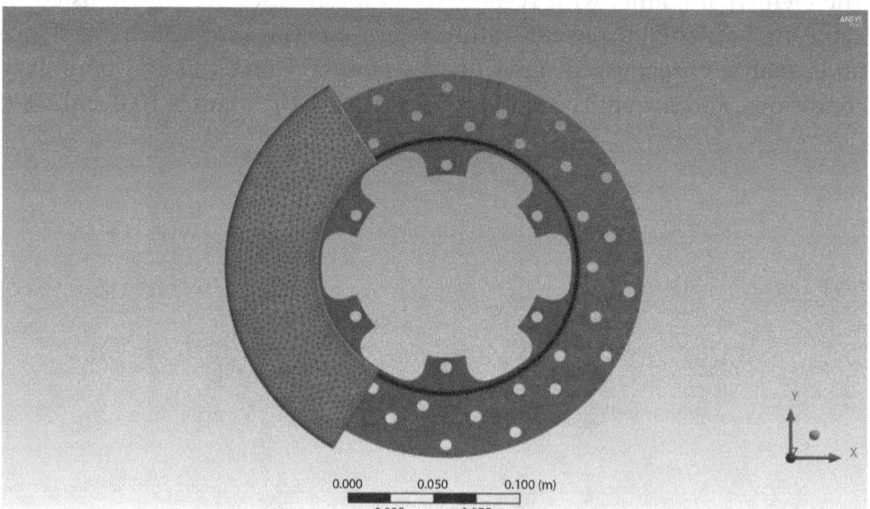

Figure 10.6 Fine mesh of entire setup.

in which Figure 10.8 shows the disc brake results and Figure 10.9 shows the results of brake pad. From Figures 10.8 and 10.9, it is clearly understood that the Case-3 is better than Case-4; thus, all the reaming materials are analyzed with fine level of meshing [20–27].

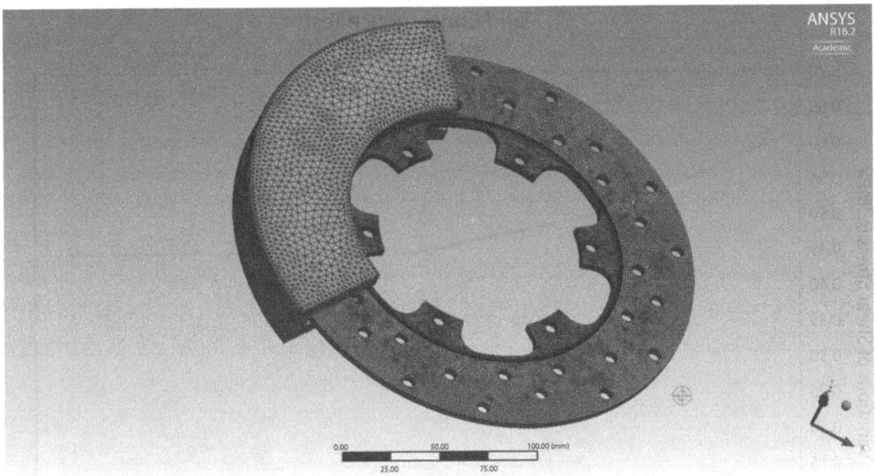

Figure 10.7 Fine with individual setup mesh of entire setup.

Table 10.2 The comparative data of various meshes and its elements on entire setup.

Sl. no.	Description	Types	No. of nodes	No. of elements
1	Coarse Mesh	Case-1	1,400,422	2,155,056
2	Medium Mesh	Case-2	1,692,925	2,601,007
3	Fine Mesh	Case-3	1,983,610	3,033,962
4	Fine with Face Mesh	Case-4	2,387,684	3,956,248
5	Fine with Inflation Mesh	Case-5	2859463	4727256

10.3.5 Numerical Results on Disc

The numerical simulations have been carried out with the help of aforementioned boundary conditions for various advanced materials. In that, the three different material groups are involved, which are existing materials used in disc brake, specialized materials to handle frictional stress, and advanced composite materials. In the existing materials group, stainless steel was implemented, and it is used in this comparative analysis, and while in the case of composites, two different perfect materials such as GFRP and CFRP are employed in this work. Finally, the titanium alloy

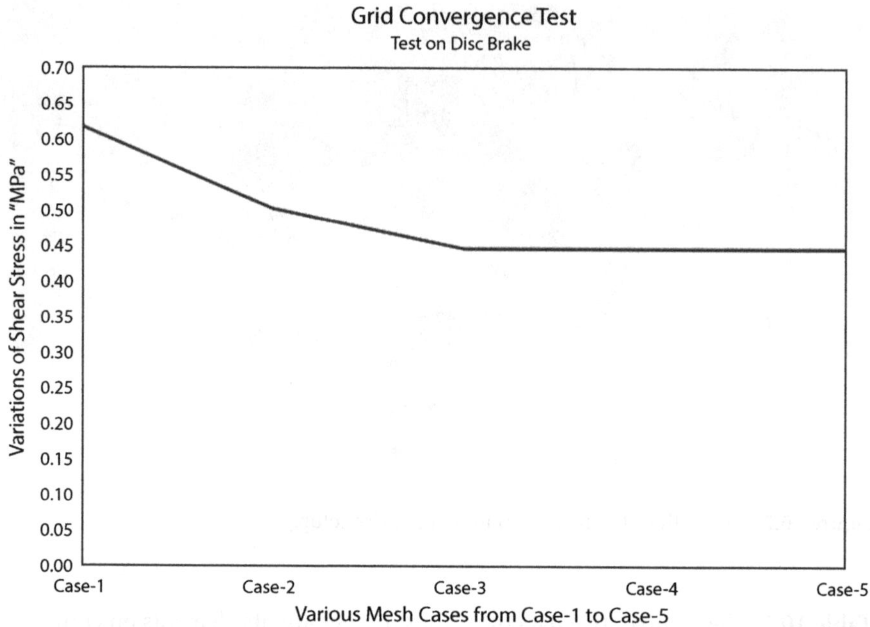

Figure 10.8 Grid convergence test for CFRP-based disc brake.

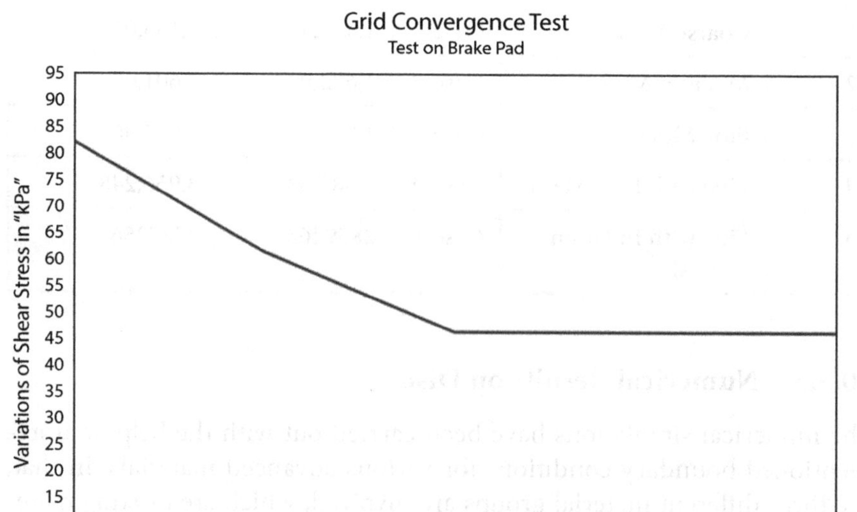

Figure 10.9 Grid convergence test for CFRP-based brake pad.

and tungsten carbide are selected in the specialized material category, which have the capability to tackle thermal stress due to friction with the help of its chemical nature. The total deformations, equivalent stress, and shear stress variations on disc brake, which is assigned with Epoxy Carbon Woven 230 Wet property are revealed in Figures 10.10, 10.11, and 10.12, respectively.

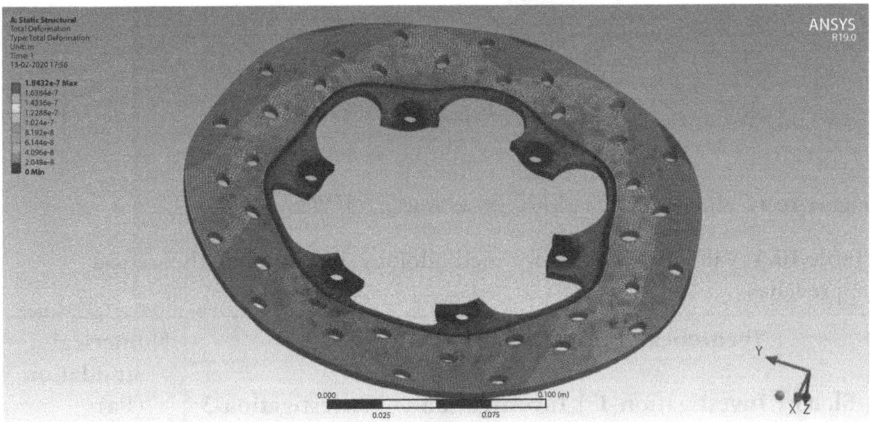

Figure 10.10 Total deformation of Epoxy Carbon Woven 230 Wet.

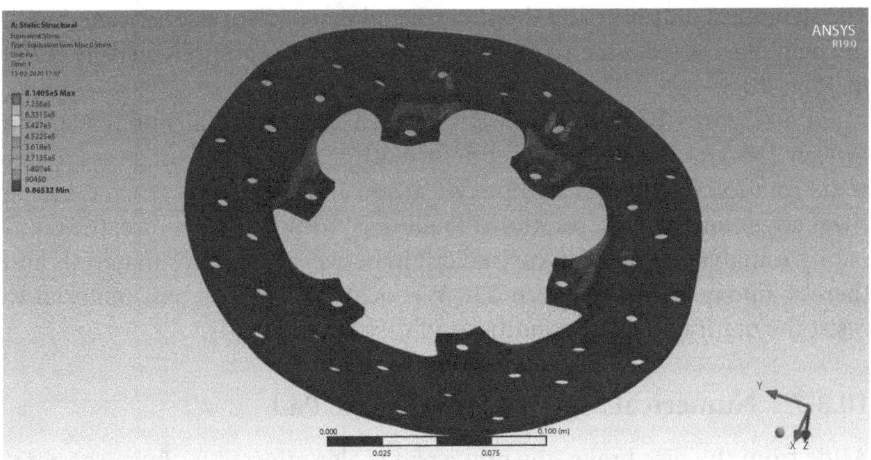

Figure 10.11 Equivalent stress of Epoxy Carbon Woven 230 Wet.

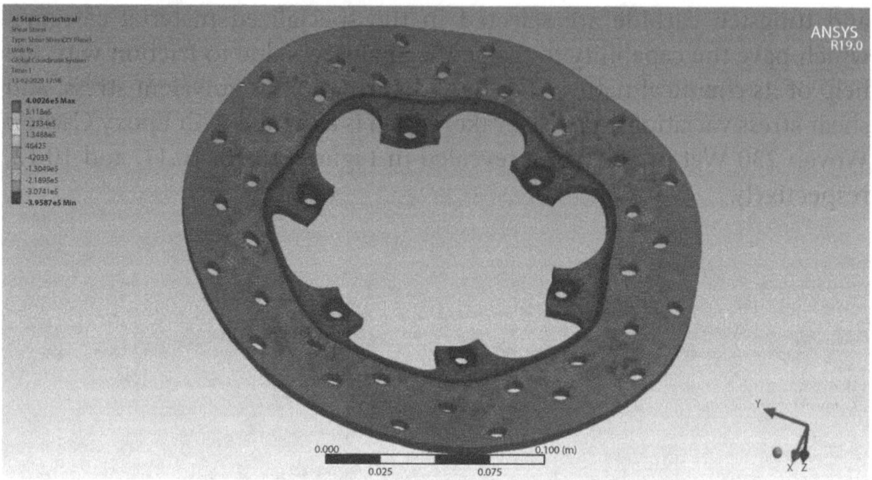

Figure 10.12 Shear stress of Epoxy Carbon Woven 230 Wet.

Table 10.3 Validational study for methodology used through theoretical approaches.

Sl. no	Theoretical investigation (Pa)			Numerical simulation (Pa)
	Investigation-1	Investigation-2	Investigation-3	
1	963.822919	1376.85367	1286.008	1206.515376

10.3.6 Comparative Analysis of Disc

The structural outputs of all the short-listed materials are constructed with the help of FEA software package, and the data are comprehensively listed in Table 10.3.

From Figures 10.13 to 10.15, it has been captured that Epoxy Carbon Woven 230 Wet is reacted low at deformation and equivalent stress under static rotational loading and Epoxy Carbon UD-Prepreg is reacted low at shear stress under static rotational loading conditions. Therefore, the comparative analyses have been carried out in between these two materials and thereby Epoxy Carbon Woven 230 Wet is short-listed as best material to resist the natural loading conditions of disc brake.

10.3.7 Numerical Simulation on Brake Pad

Aside from the disc brake, the brake pad is also affected lot from the rotational loads therefore the same frictional stress estimations are extended to

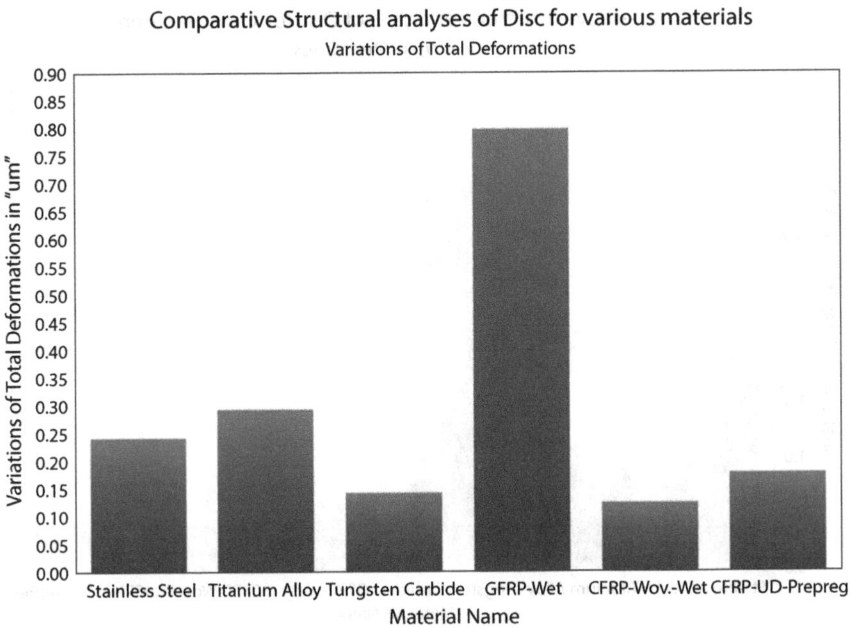

Figure 10.13 Comparative structural analyses of disc for various materials.

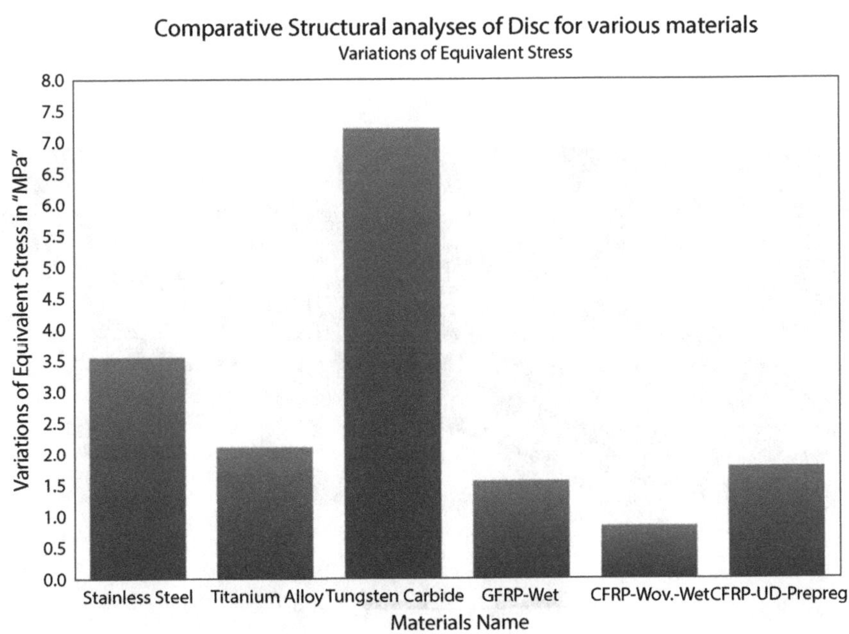

Figure 10.14 Comparative structural analyses of disc for various materials.

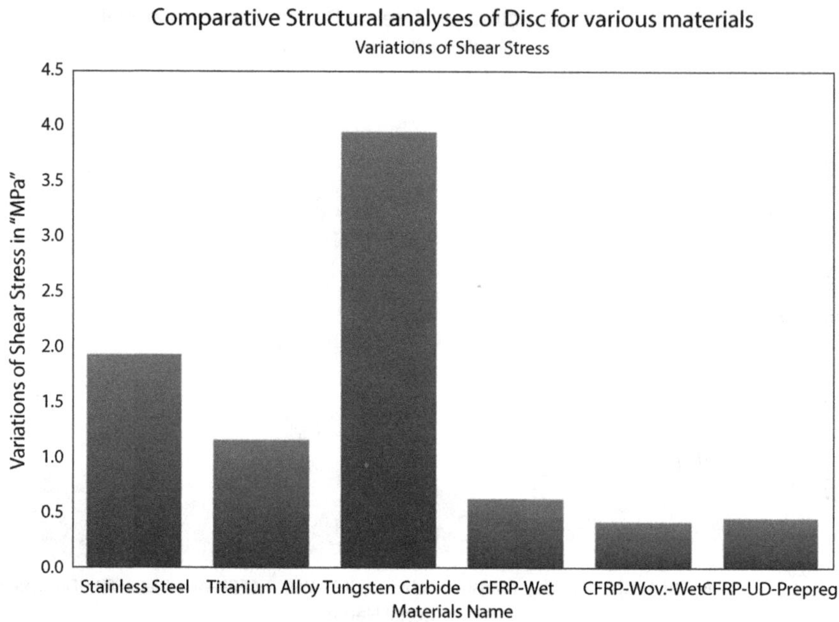

Figure 10.15 Comparative structural analyses of disc for various materials.

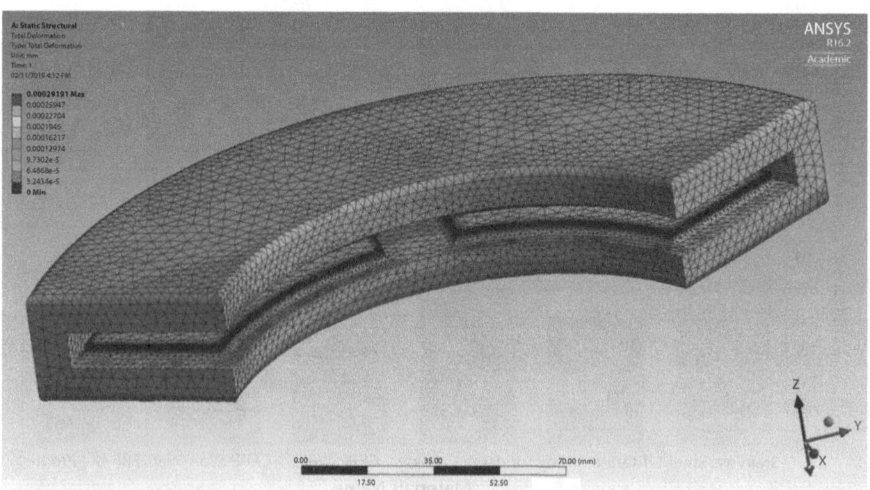

Figure 10.16 Total deformation of titanium alloy.

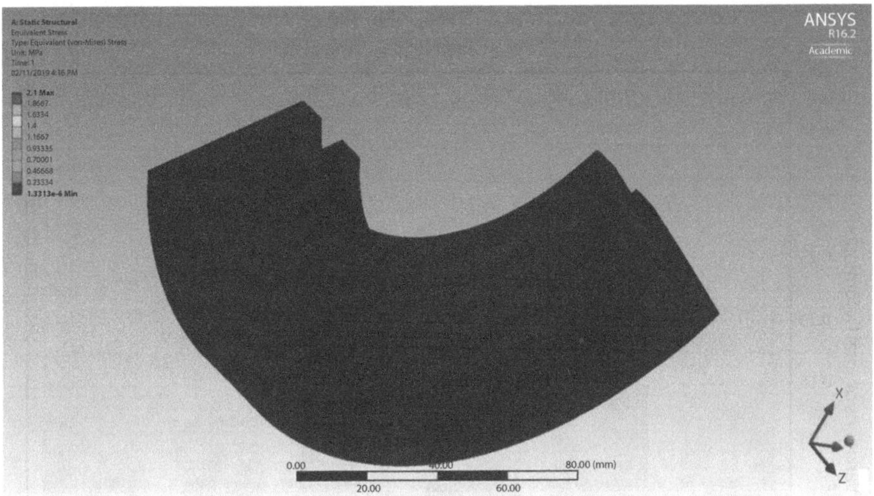

Figure 10.17 Equivalent stress of titanium alloy.

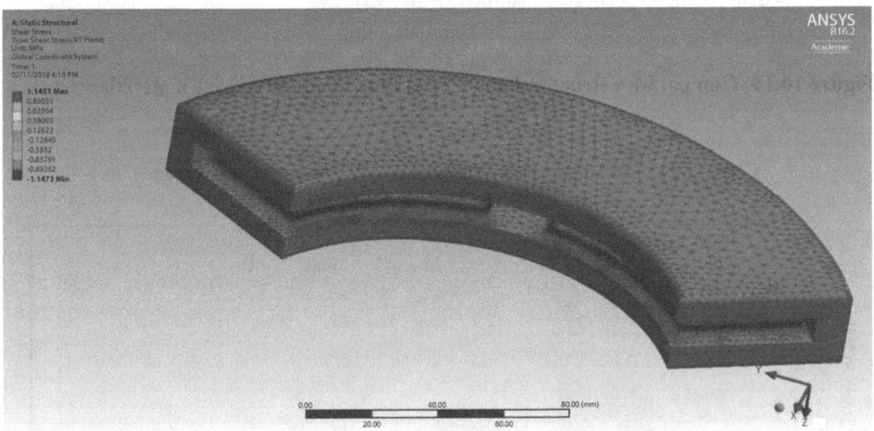

Figure 10.18 Shear stress of titanium alloy.

brake pad in order to understand its structural behavior under rotational force. For the same boundary conditions, the comparative investigations are analyzed on brake for various short-listed materials. In the aerospace filed, titanium alloy implements as a predominant material because of its high thermal resisting capability; thus, the same one is used in thermal analyses on disc brake. The total deformation, equivalent stress, and shear stress distributions of brake pad are revealed in Figures 10.16, 10.17, and 10.18, respectively, in which the material is assigned for this case is titanium alloy.

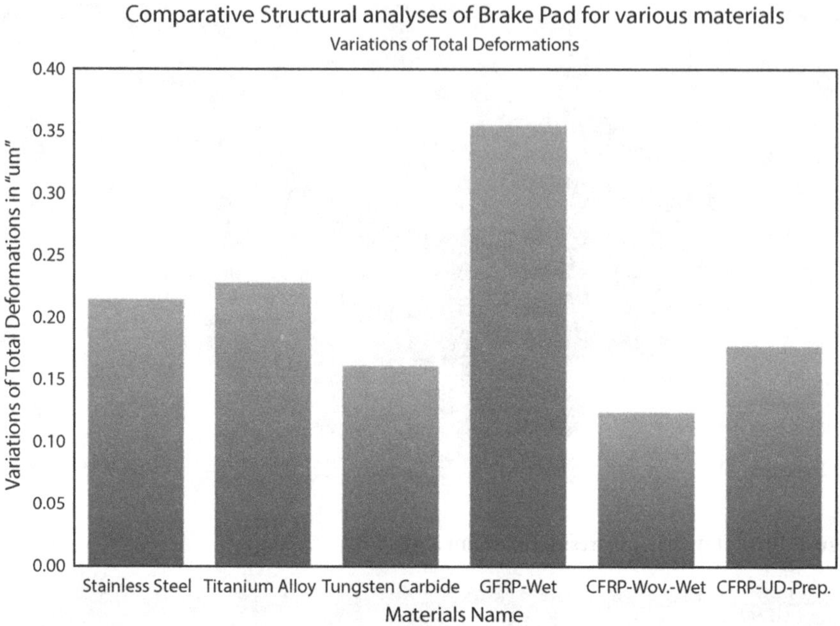

Figure 10.19 Comparative structural analyses of brake pad for various materials.

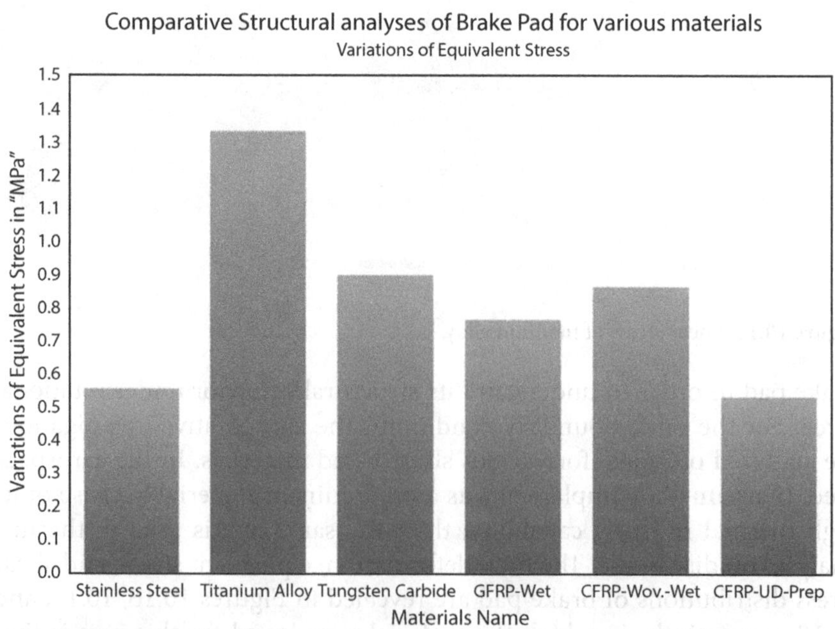

Figure 10.20 Comparative structural analyses of brake pad for various materials.

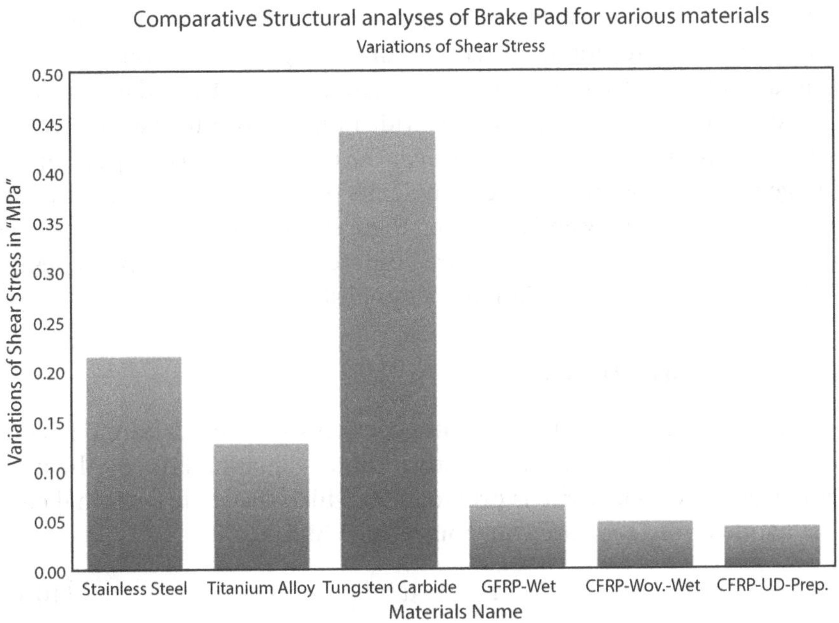

Figure 10.21 Comparative structural analyses of brake pad for various materials.

10.3.8 Comparative Analysis of Brake Pad

The structural outputs of all the short-listed materials are constructed with the help of FEA software package and the data are comprehensively revealed in Figures 10.19 to 10.21. As same as disc brake outcome, the brake pad is also reacted in a similar manner. Thus, carbon fiber with epoxy resin added composites are more preferable to tackle these complicated environments in an efficient manner. Also, the lifespan of the primary components involved in the disc brake and pad may have the chance to increase in a double manner because of these advanced composites.

10.4 Theoretical Investigations

Because of the working principle and platform of computational structural analysis, the outcome may have the chance to get inefficiency in its reliability in a medium probability manner. With inclusion of these characteristics, the comparative structural simulations are carried out on the disc brake, in which the fine-tuned boundary conditions are implemented. Even though the advanced boundary conditions are implemented, the analyses of outcomes are planned to compare with theoretical investigations. In this regard, totally three relevant and good equations are found out and implemented in this work, in which the geometrical data, mechanical properties are supported lot. The wide range of formulae were involved in the estimation shear stresses however the reliability of its output is in the average level only because of the complexity of the real-time problem; thus, the requirement of new and advanced theoretical approaches will emerged anytime. Similarly, in this work, three different theoretical approaches are implemented in the estimation of frictional stress.

10.4.1 Investigation-1

The standard equations such as static friction force and dynamic friction force are implemented in this investigation. In Equation (10.18), the maximum static frictional force is provided, in which the static frictional coefficient and applied load are contributed hugely.

$$F_{max} = \mu * F_n \tag{10.18}$$

where

F_{max} = Frictional force in Newtons
F_n = Applied normal force in Newtons
μ = Static or kinetic frictional coefficient

With the help of Equation (10.18), the static frictional stress equation is formed, which listed in Equation (10.19).

$$\sigma_c = \frac{F_{max}}{\text{Area}} \tag{10.19}$$

Apart from the static frictional stresses, the dynamic frictional stresses are drastically happened in the real-time applications and therefore, in this work, are also implemented, which are listed in Equation (10.20).

The frictional force for dynamic friction can be expressed as follows:

For an object pulled or pushed horizontally, the normal force "N" is simply weight:

$$N = m*g \tag{10.20}$$

where

m = mass in kg
g = gravity in m/s^2

Mass = Density * Volume = 8,000 * 0.000283183267 = 2.265466136 kg

N = m*g = 2.265466136 * 9.81 = 22.42228 N

Contacted area = external diameter – internal diameter – (36 * small diameter)

Area = 0.04154756 – 0.01814584 – (0.00138528)

Area = 0.0232639

Frictional Stress = 22.42228 / 0.0232639 = 963.822919 Pa

10.4.2 Investigation-2

The frictional-based behavioral effects are widely happening in the civil-based structural elements, such as bar, beam, and column. Therefore, in this investigation, the formula is extracted from those elemental-based equations, which is provided in Equation (10.21).

Frictional stress

$$\sigma_f = \frac{\rho * L * c_f}{2} \tag{10.21}$$

where

σ_f = Frictional stress
ρ = Unit weight of the material
L = Length of the object, in which the object is slab
c_f = Average coefficient of friction

In this sample calculation, stainless steel is selected as working material, and thereby, the frictional stress is estimated with the help of Equation (10.21).

$$\sigma_f = \frac{8{,}000 * L * 0.45}{2}$$

Density of stainless steel = 8,000 kg/m^3
Length of the contacted area = external diameter − internal diameter = 230 − 152 = 78 mm

$$\sigma_f = \frac{78453.2002291112 * 0.078 * 0.45}{2}$$

$$\sigma_f = 1376.85367 \text{ N/m}^2$$

10.4.3 Investigation-3

In the third methodology, the formula involved in the shear stress estimation on shaft is implemented because its working nature perfectly matched with disc brake's working nature. Thus, the standard equation is provided in Equation (10.22).

$$\tau = \frac{\frac{2}{3} * \mu_k * F * \left[\frac{R_0^3 - R_i^3}{R_0^2 - R_i^2}\right] * r}{\pi * \left[\frac{D_0^4 - D_i^4}{32}\right]} \tag{10.21}$$

$$\mu_k = 0.75\ \mu_s$$

$$\mu_k = 0.75 * 0.45 = 0.3375$$

$$\tau = \frac{\frac{2}{3} * 0.3375 * 22.42228 * \left[\frac{1520875 - 438976}{13225 - 5776}\right] * 39}{\pi * \left[\frac{2798410000 - 533794816}{32}\right]}$$

$$\tau = \frac{285770.77542721}{222215365}$$

$$\tau = 0.001286008 \text{ N/mm}^2$$

$$\tau = 1286.008 \text{ N/m}^2$$

Table 10.3 provided the comparative results between various theoretical calculation and computational structural analysis, in which the involved theoretical formulae are obtained the results that are settled in the acceptable regions with numerical simulation results. Therefore, the implemented numerical procedures and its meshes are provided the reliable outputs, and thereby, the proposed material is fit for the real-time applications.

10.5 Conclusion

The comparative analyses of shear stress and deformation variations are executed on two-wheeler disc brake with the help of advanced computational tool i.e., ANSYS Workbench 16.2. The standard literature survey is supported a lot in the material selection and its mechanical properties. Totally, five fine-tuned materials such as stainless steel, titanium alloy, tungsten carbide, Epoxy S Glass Wet, and Epoxy Carbon UD 230 GP-Prepreg are used in this analysis, in which Carbon UD 230 GPa Prepreg is most suitable for disc brake. From the comparative analyses, the Carbon UD fiber is short-listed as best one so the same numerical procedure is extended to carbon bidirectional fiber-based composite. Finally, the Carbon UD fiber–based composite is more suitable to withstand in two-wheeler at peak loading conditions. Apart from the results, the test specimen preparation and its loading analysis are very complicated one in this work, and therefore, the discussion on them is mandatory. The dimensions of two-wheeler disc and its brake are extracted from the successful bike model. Also, the conceptual designs of a two-wheeler disc brake are modeled in CATIA. The numerical simulation of this work is executed with the help of FEA tool and composite generation tool. At last, it is clearly understood that, the Carbon UD 230 GPa Prepreg is fit for shear load–based applications because they are good in the provision of high life time, low weight than existing materials, and cost effective.

References

1. Kumar, G.R., Vijayanandh, R., Kamaludeen, M.B., Balasubramanian, S., Jagadeeshwaran, P., Ramesh, M., Comparative Structural Characterization of Fiber Reinforced Composite Rotating Disc: A Validated Investigation. *Tribol. Ind.*, 42, 4, 608–620, 2020.

2. Belhocine, A. and Afzal, A., FEA Analysis of coupled thermo-mechanical response of grey cast iron material used in brake discs. *Rev. Cient.*, 36, 3, 280–296, 2019.
3. Raj Kumar, G., Vijayanandh, R. *et al.*, Comparative Investigations on the Main Elements of Carbon Fiber Based Composites Using Computational Structural Simulations. *J. Phys.: Conf. Ser.*, 1504 012003, 1–11, 2020, https://iopscience.iop.org/article/10.1088/1742-6596/1504/1/012003.
4. Rajkamal, M.D., Dylan Abraham Samson, T., Dhinesh Kumar, P., Chithambaravishnu, S., Structural Analysis of Disc Brake. *Int. J. Mech. Eng. Technol.*, 9, 3, 846–856, http://www.iaeme.com/IJMET/issues.asp?JType=IJMET&VType=9&IType=3, 2018.
5. Naveen Kumar, K., Vijayanandh, R., Bruce Ralphin Rose, J., Swathi, V., Narmatha, R., Venkatesan., K., Research on Structural behavior of Composite Materials on different Cantilever Structures using FSI. *Int. J. Eng. Adv. Technol.*, 8, 6S3, 1075–1086, 2019.
6. Jeyakumar, S. and Karthickmani, P., Structural Analysis of Different Composite Materials in the Application of Automotive Disc Brake. *Int. J. Adv. Res. Manage. Architecture Tech. Eng. (IJARMATE)*, 2, 12, 3–10, 2016.
7. Udhaya Prakash, R., Raj Kumar, G., Vijayanandh, R., Senthil Kumar, M., Ram Ganesh, T., Structural analysis of aircraft fuselage splice joint. *IOP Conf. Ser.: Mater. Sci. Eng. J.*, 149, 1, 012127, 2016.
8. Yevtushenko, A.A., Grzes, P., Adamowicz, A., Numerical Analysis of Thermal Stresses in Disk Brakes and Clutches (A Review). *Numer. Heat Transf. Part A: Appl.*: An International Journal of Computation and Methodology, 67, 2, 170–188, Available from: 10.1080/10407782.2014.923221, 2015.
9. Ramesh, M., Vijayanandh, R., Raj Kumar, G., Mathaiyan, V., Jagadeeshwaran, P., Senthil Kumar, M., Comparative Structural Analysis of Various Composite Materials based Unmanned Aerial Vehicle's Propeller by using Advanced Methodologies. *IOP Conf. Ser.: Mater. Sci. Eng.*, 1017, 012032, 1–10, 2021.
10. Pind, R.R. and Swami, M.C., Analysis and Computational Investigation of Brake Disc for Composite Materials. *Int. J. Eng. Dev. Res.*, 4, 4, 927–946, 2016.
11. Raj Kumar, G., Senthil Kumar, M., Vijayanandh, R., Raja Sekar, K., Mohamed Bak, K., Varun, S., The Mechanical Characterization Of Carbon Fiber Reinforced Epoxy with Carbon Nanotubes. *Int. J. Mech. Prod. Eng. Res. Dev.*, 9, Special Issue 1, 243–255, Jan 2019.
12. Babu, T., Sudharshan, R., Akil, R., Chiranjeev, S., Analysis and Shape Optimization of disc Brake with Alternate Material. *Int. J. Recent Technol. Eng. (IJRTE)*, 8, 1S2, 187–192, 2019.
13. Raj Kumar, G., Vijayanandh, R., Mohammad Bak, K., Shyam Chander, R, Arawinth, R., Experimental Testing On Mechanical Properties Effect of Aluminum Foam. *Int. J. Mech. Prod. Eng. Res. Dev.*, 8, Special Issue 7, 1047–1059, Oct 2018.

14. Ranjith Kumar, G., Thriveni, S., Rajasekhar Reddy, M., Harinath Gowd, G., Design Analysis & Optimization of an Automotive Disc Brake. *Int. J. Adv. Eng. Res. Sci. (IJAERS)*, 1, 3, 24–29, 2014.
15. Raj Kumar, G., Vijayanandh, R., Senthil Kumar, M., Sathish Kumar, S., Experimental Testing and Numerical Simulation on Natural Composite for Aerospace Applications. *AIP Conf. Proc. 1953*, 090045-1–090045-5; https://doi.org/10.1063/1.5032892, 978-0-7354-1648-2.
16. Belhocine, A., Abu Bakar, A.R., Bouchetara, M., Numerical Modeling of Disc Brake System in Frictional Contact. *Tribol. Ind.*, 36, 1, 49–66, 2014.
17. Vijayanandh, R., Venkatesan, K., Ramesh, M., Raj Kumar, G., Senthil Kumar, M., Optimization of Orientation Of Carbon Fiber Reinforced Polymer Based On Structural Analysis. *Int. J. Sci. Technol. Res.*, 8, 11, 3020–3029, November 2019.
18. Mathur, D., Maheshwari, A., Sharma, A., Kumar, C., Nayyar, A., Singh Chundawat, R., Design and Optimization of Disc Brake Rotor. *Int. J. Innov. Res. Sci. Eng. Technol.*, 7, 8, 9157–9164, 2018, Available from:10.15680/IJIRSET.2018.0708042.
19. Vijayanandh, R., Raj Kumar, G., Jagadeeshwaran, P., Mathaiyan, V., Ramesh, M., Jung, D.W., Carbon Nanotubes and Their Polymer Nanocomposites, in: *Nanomaterials and Nanocomposites: Characterization, Processing, and Applications*, Chapter – 9, pp. 139–165, 2021.
20. Challa, G.K.R., Guduru, A.K., Patlori, S., Madhavi, M., Design and Analysis of Carbon Fiber / Epoxy Resin Brake Rotor. *IOSR J. Mech. Civil Eng. (IOSR-JMCE)*, 12, 4, 23–28, Available from: 10.9790/1684-12452328, 2015.
21. Raj Kumar, G., Balasubramaniyam, S., Senthil Kumar, M., Vijayanandh, R., Raj Kumar, R., Varun, S., Crash Analysis on the Automotive Vehicle Bumper. *Int. J. Eng. Adv. Technol.*, 8, 6S3, 1602–1607, 2019.
22. Mohamed Bak, K., Raj Kumar, G., Ramasamy, N., Vijayanandh, R., Experimental and Numerical Studies on The Mechanical Characterization of Epdm/S-Sbr Nano Clay Composites. *IOP Conf. Ser.: Mater. Sci. Eng.*, 912, 1–11, 052016, 2020.
23. Venkatesan, K., Geetha, S., Vijayanandh, R., Raj Kumar, G., Jagadeeshwaran, P., Raj Kumar, R., Advanced structural analysis of various composite materials with carbon nano-tubes for property enhancement. *AIP Conf. Proc.*, 2270, 030005, pp. 030005-1–030005-6, 2020.
24. Mirrudula, P., Kaviya Priya, P., Malavika, M., Raj Kumar, G., Vijayanandh, R., Senthil Kumar, M., Comparative structural analysis of the sandwich composite using advanced numerical simulation. *AIP Conf. Proc.*, 2270, 040005-1–040005-5, 2020.
25. Indira Prasanth, S., Kesavan, K., Kiran, P., Sivaguru, M., Sudharsan, R., Vijayanandh, R., Advanced structural analysis on E-glass fiber reinforced with polymer for enhancing the mechanical properties by optimizing the orientation of fiber. *AIP Conference Proceedings*, vol. 2270, pp. 040006-1–040006-5, 2020.

26. Bhagavathiyappan, S., Balamurugan, M., Rajamanickam, M., Vijayanandh, R., Raj Kumar, G., Senthil Kumar, M., Comparative computational impact analysis of multi-layer composite materials. *AIP Conference Proceedings*, vol. 2270, pp. 040007-1–040007-5, 2020.
27. Venkatesan, K., Ramanathan, K., Vijayanandh, R. *et al.*, Comparative structural analysis of advanced multi-layer composite materials. *Mater. Today: Proc.*, 27, Part 3, 2673–2687, 2020.

11

Studies of Nanomechanical and Wear Testing Methods of Thermal Sprayed Coatings

Zulfequar Ali[1]*, Ashish Kumar Singh[2], Omprakash Thakare[2] and Chitrakant Tiger[3]

[1]Department of Mechanical Engineering, DIETMS, Aurangabad, India
[2]Department of Mechanical Engineering, HITAM, Hyderabad, India
[3]Department of Mechanical Engineering, Shri Rawatpura Sarkar University, Raipur, India
[4]Department of Mechanical Engineering, CEC, Bilaspur, India

Abstract
The paper is aimed at studying the different nanomechanical and wear testing methods for coatings. It includes nanoindentation test and nanoscratch test, which are used for determining the mechanical and wear properties of nanostructured coatings. As the thin films find their applications in aerospace, electronics, and manufacturing industries, it is essential to develop some suitable coatings. High-energy ball milling is often used for producing nanosized powders from the conventional powders which are then sprayed by different thermal spraying techniques. It was found that by controlling the porosity the properties like hardness and wear resistance of the nanostructured coatings may be enhanced. The critical review of some recent developments in this particular field is included in this work, and a new composition is proposed for developing nanocomposite coatings.

***Keywords*:** Thermal barrier coating, HVOF, nanoindentation, nanoimpact, nanoscratch, plasma spraying

**Corresponding author*: zulfee86@gmail.com

Sarfraj Ahmed and Vinayak S. Dakre (eds.) Tribology and Characterization of Surface Coatings, (239–252) © 2022 Scrivener Publishing LLC

11.1 Introduction

Nanocoatings are being used in various applications such as aerospace, electronics, defense, and biomedical industries. Thin coatings find applications in micro- and nanoelectromechanical systems as a wear-resistant material, corrosion-resistant coatings, and bioactive coatings. Nanocoating may increase the service life of aircraft engine components by providing heat shielding and reduce fuel consumption. The surface layers on aircraft engine and gas turbines are called the thermal barrier coating, and they are manufactured by a technique called thermal spraying. A ceramic powder is sprayed onto a surface at a very high temperature 7,000° to 8,000°C, where ceramic particles melt and strike the surface to form a protective layer [1–3]. The ceramic coating experiences high stress due to change in temperature; therefore, it is necessary to add plastic materials to make it more elastic. Plasma spraying is a most widely used thermal spray technique for the deposition of nanostructured coatings. A nanocrystalline powder may be directly sprayed by different thermal spraying methods but mechanical milling of the conventional powders is a most efficient technique to obtain nanosized powders. Nanocrystalline powders obtained by high-energy mechanical milling show better properties, and they are agglomerated prior to the spraying [4–7].

Modifying the surface by depositing the coatings using different methods has become the ground breaking tool to improve the mechanical and wear properties. Nanomechanical characterization of thin films is required to determine the various properties like hardness, impact strength, friction, and wear. When designing a coating for better tribological performance, it is necessary to focus on relationship between mechanical properties of coating and its bonding to the substrate. The mechanical behavior of a coated system depends upon coating-substrate deformation and interfacial adhesion strength. The purpose of this paper is to describe the mechanical and wear testing methods for thermal sprayed nanostructured coatings [8–10]. In recent years, there have been many researches on the development of nanostructured coatings for wide range of applications. This paper also represents a critical review of the research and development of nanostructured coatings by thermal spraying, which provides an overview of fabrication techniques, properties and characterization methods.

11.2 Nanowear Testing Methods

Thin films and coatings are extensively being used to improve the wear resistance of materials in various applications. The instrument consists of an indenter through which load is applied on the sample and friction data is generated. The indenter tip having radius in the range of 1 to 20 μm is generally used and the applied load ranges from 0.5 to 20 mN. The following are the methods of determining the tribological properties of coatings at nanolevel.

11.2.1 Nanoscratch Testing

It is an important testing method for determining the wear behavior and adhesion strength of coatings. Thin films, ceramic coatings, organic coatings, biomaterials, etc., can be tested by using this method, and the effect of experimental variables can be studied. These tests are generally carried out on coatings by using Berkovich indenter system and Nanotest nanomechanical system having different shapes of indenter like spherical and conical [11]. It includes constant load nanoscratch test in which indenter tip moves on a surface with constant load, ramped load nanoscratch test in which after a set leveling distance the load progresses, repetitive unidirectional nanowear test involving multi-pass ramped or constant loading, and progressive three-scan scratch test with linear increasing load. Out of these, the progressive three-scan scratch test is widely used because it may be used to determine the critical load required to initiate deformation which causes removal of materials. The three-scan nanoscratch testing is shown in Figure 11.1 in which the sample slides with respect to indenter.

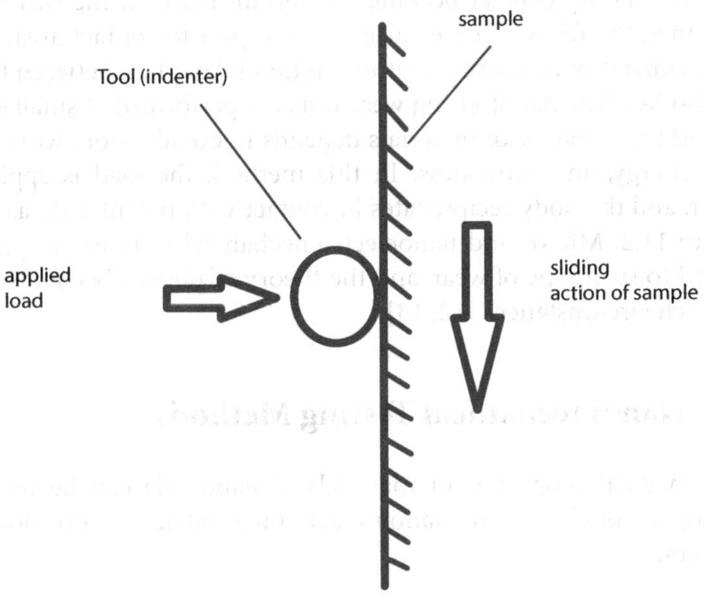

Figure 11.1 Schematic representation of three-scan nanoscratches testing.

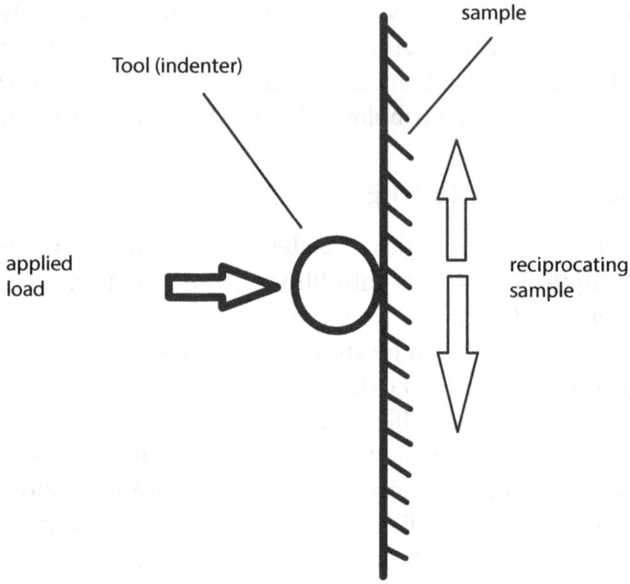

Figure 11.2 Nanofretting wear testing method.

11.2.2 Nanofretting Wear Testing

When the asperity contact becomes important between the two moving bodies, then the friction force varies with respect to contact area. In this process, small displacement vibratory motion takes place between the two sliding bodies. The nanofretting wear tests are performed at small contact scale, and the behavior of materials depends upon adhesion, wear debris, surface energy, and roughness. In this method, the load is applied on indenter, and the body reciprocates in contact with indenter tip as shown in Figure 11.2. Micro- and nanoelectromechanical systems are generally subjected to such type of wear, and the theory of physics becomes invalid under such circumstances [12, 13].

11.3 Nanomechanical Testing Methods

The mechanical properties of materials at nanoscale can be measured by using nanoindentation, nanoimpact, micropillar compression, etc., techniques.

11.3.1 Nanoindentation Technique

Nanoindentation method also known as depth sensing indentation method is based on the theory of contact mechanics. The ratio of maximum load to projected contact area of the indent and the slope of initial curve obtained by the load-displacement graph are used to determine the hardness and elastic modulus. It is an extension of microhardness testing system in which loading unloading curve can be analyzed to find the yield strength, fracture toughness, etc., and it is carried out at a load ranging from 0.1 to 1 mN [14, 15]. It uses projected contact area for the calculation of hardness values, whereas Vickers method uses actual contact area and relation between them as follows:

$$H_{NI} = VHN/0.094$$

where VHN = Vickers hardness number and H_{NI} = nanoindentation hardness.

Generally, the Berkovich indenter with 65.35° projected cone angle is pressed into the materials under given load, and then, the load and displacement is recorded. The hardness is given by the ratio of maximum indentation load to the projected contact area and modulus of elasticity can be determined by using standard formula. Figures 11.3 and 11.4 represent nanoindentation test method and instrumentation system. The international standard for nanoindentation method is described by ISO 14577 (parts 1–4). Even at small penetration, the interaction with substrate may occur, and it has been analyzed that, up to 10% of coating thickness, the substrate does not affect [16].

$$\text{Hardness, } H = P_{max}/A_{proj}$$

$$\text{Elastic modulus, } 1/E_r = (1-\mu^2)/E + (1 - \mu_i^2)/E_i$$

where P_{max} = max applied load, A_{proj} = projected contact area, E = elastic modulus of sample, E_r = $(\sqrt{\pi}.dP/dh.\ 2\sqrt{A})$ = reduced elastic modulus of sample, E_i = elastic modulus of indenter material, μ= poisson's ratio of sample, and μ_i = poisson's ratio of indenter material.

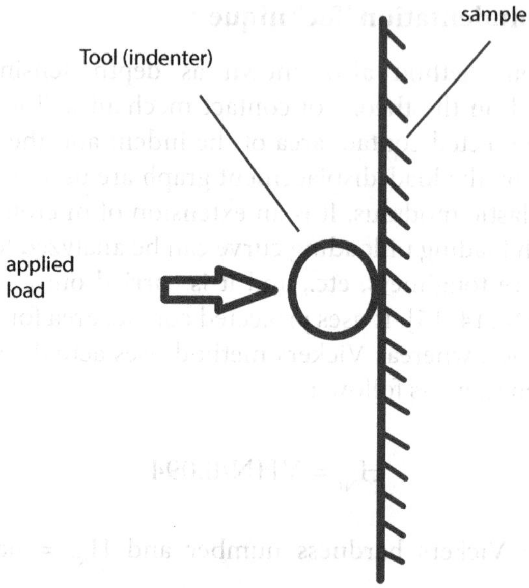

Figure 11.3 Nanoindentation testing method.

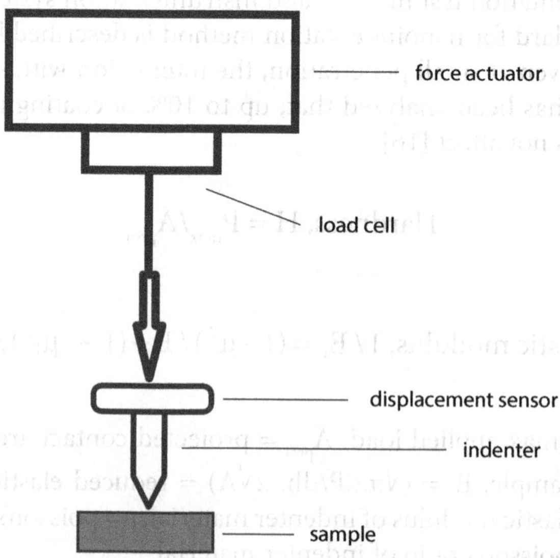

Figure 11.4 Schematic representation of nanoindentation system.

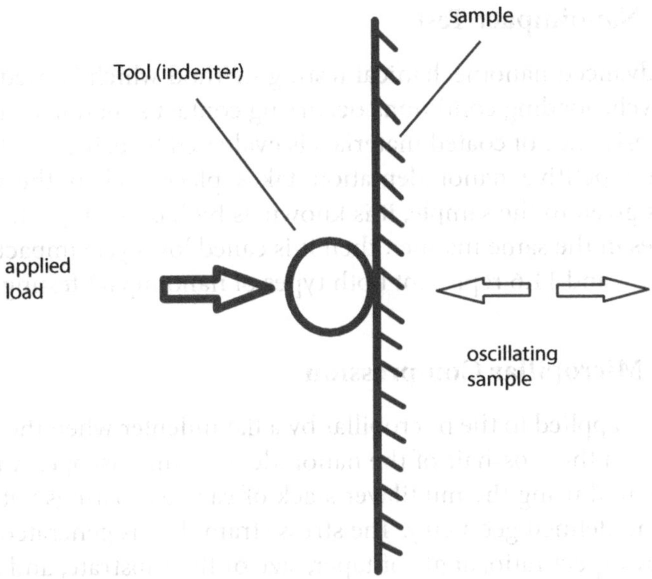

Figure 11.5 High cycle nanoimpact test.

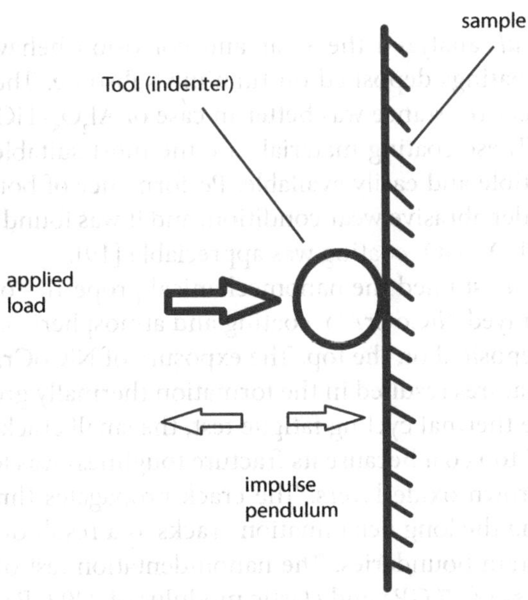

Figure 11.6 Low cycle nanoimpact test.

11.3.2 Nanoimpact Test

It is an advanced nanomechanical testing method which is used to simulate the cyclic loading conditions occurring contact situations. The fatigue fracture resistance of coated materials is evaluated by using this technique in which repetitive nanoindentation takes place. When the repetitive motion is given to the sample, it is known as high cycle impact, and if the tool moves in the same manner, then it is called low cycle impact [17, 18]. Figures 11.5 and 11.6 represent both types of nanoimpact testing.

11.3.3 Micropillar Compression

The force is applied to the micropillar by a flat indenter when the sample is positioned in the crosshair of the nanoindenter's microscope. A micropillar is prepared using the multilayer stack of various coatings with a small volume and defined geometry. The stress strain data is generated which is affected by aspect ratio, angle of taper, size of the substrate, and direction of compression.

11.4 An Overview of Past Research

C. Richard *et al.* analyzed the wear and corrosion behavior of plasma sprayed nanocoatings deposited on titanium substrate. The coefficient of friction and wear resistance was better in case of Al_2O_3-TiO_2 coating than ZrO_2 coating. These coating materials are the most suitable because they are bio compatible and easily available. Performance of both the coatings was similar under abrasive wear condition, and it was found that corrosion resistance of Al_2O_3-TiO_2 coating was appreciable [19].

G. Bolelli *et al.* studied the nanomechanical properties of high velocity oxygen fuel sprayed NiCoCrAlY coating and atmospheric plasma sprayed YSZ coating deposited on the top. The exposure of NiCoCrAlY bond coat to high temperatures resulted in the formation thermally grown oxide layers. During the thermal cycling fatigue test, the small cracks were formed within the YSZ top coat because its fracture toughness was lower than that of thermally grown oxide layers. The crack propagates through the YSZ layer developing the long delamination cracks as a result of inter granular porosity and grain boundaries. The nanoindentation test of YSZ top coat yields a hardness of 17 GPa and elastic modulus of 220 GPa [20].

A. Lanzutti *et al.* worked on different coatings like NiCr, WCCoCr, and NiCr + Cr_2O_3 deposited by atmospheric plasma spray and high velocity oxygen fuel techniques. The reinforcing particles WC led to the highest wear resistance of WCCoCr coating at 300°C and room temperature [21].

I. Garcia *et al.* showed the effect of particle size on wear behavior of Ni/SiC coating at room temperature. At room temperature, the wear resistance of composite coating was decreased with increase in particle size [22].

A. Gupta *et al.* investigated the mechanical and wear properties of ceramic composite coating (Al2O3 + CeO2 + hBN). The mechanisms of material removal were splat cracking and splat detachment under erosive wear conditions. The microstructural changes and generation of new phases were responsible for the improved wear and mechanical properties [23].

Xiaofeng Zhang *et al.* analyzed the mechanical properties and microstructure of high velocity oxygen fuel sprayed Ni-based nano-hBN composite coating. With increase in amount of hBN, the microhardness of the composite coating decreased appreciably. The coefficient of friction increased with increase in temperature, and adhesive strength was found to be lower than the other metal powders [24].

Cecilio Alvares da Cunha *et al.* in his study found that Cr3C2-25(Ni20Cr) nanostructured coating possess excellent corrosion and wear resistance. The particle size of high-energy milled Cr3C2-25(Ni20Cr) powders depends upon milling time and the decrease in particle size by increasing milling time was more pronounced in gaseous nitrogen as compared to liquid hexane. High velocity oxygen fuel technique was used to spray the coating [25].

O. Ali *et al.* developed a mathematical model for high velocity oxygen fuel sprayed WC-12%CO nanocomposite coating, and their microstructure was analyzed using XRD, SEM, and TEM. The heat treatment of hot isostatically pressed coating resulted in increased hardness and modulus which was attributed to the improved bonding between microstructural phases. The wear rate was also reduced due to post heat treatment by hot isostatic pressing and binder phase was strengthened [26].

Lalit Thakur *et al.* studied about the erosive wear performance of HVOF sprayed nanostructured WC-Co-Cr coating deposited on stainless steel specimen. The crack was grown inside the coating and grain detachment led to the eruption of binder. The material was also removed due to brittle cracking of the coating caused by shock loading [27].

X. Sun *et al.* developed a NiCrAlY coating with and without nanoceria by HVOF technique. The adhesive strength and hardness of the above coating was increased due to grain refinement after adding the nanoceria. It was found that 0.2 wt% nanoceria improved the thermal cycling life of the above coating material [28].

X. C. Zhong *et al.* designed two coating materials, namely, CoNiCrAlY-$MoSi_2$ and Ni60-TiB_2 with improved mechanical properties. The prepared nanostructured powders exhibited low porosity, good fracture toughness, moderate wear, and corrosion resistance [29].

A. Aravind and G. R. Kannan investigated the microstructure and mechanical properties of thermal barrier coatings mainly aluminum oxide with titanium oxide and yttria-stabilized zirconia (YSZ). The coatings of 150 microns thickness were applied on aluminum alloy, and it was observed that YSZ coating showed better microstructure and mechanical properties like hardness and yield strength [30].

B. D. Beake *et al.* conducted the nanoindentation and multipass nanowear test on TiN coatings deposited on M42 steel substrate. The coating thickness affects the stress required for the failure of coating. It was found that the critical load depends upon scan direction in nanowear test and coefficient of friction increased with increase in load [31].

C. A. Da Cunha *et al.* showed that nanostructured Cr_3C_2-Ni20Cr coatings possess higher hardness, strength, and wear resistance. High-energy ball milling of the as received conventional Cr_3C_2-Ni20Cr powder was carried out and denser nanostructured coating was obtained by HVOF technique. Microhardness measurement showed that the nanostructured coating had higher hardness than as received powders and crystallite size decreases with increase in milling time [32].

Jenny Hao Hsin Hung *et al.* investigated the wear properties of plasma and high velocity oxygen fuel sprayed TiAl metallic coating and TiAl/Al2O3 composite coating at room temperature. A ball on flat disc tester was used to study the wear behavior of the coating materials. The effect of adding 1% nanoparticle yttria to the composite coating resulted in improved wear performance [33].

Z. G. Ban and L. L. Shaw used high velocity oxygen fuel technique for the synthesis of WC-Co coating and spraying was done at three different ratios of the oxygen to hydrogen flow rate (ROHFR). At ROHFR of 0.5 the porosity was decreased and maximum microhardness was obtained. It was also found that decarburization was increased and porosity was decreased at higher values of ROHFR [34].

11.5 Design of a New Coating

Researchers across the world are developing various types of nanocomposite coatings for tribological applications but still the properties need to be enhanced. Titanium alloys are widely used in biomedical applications due to their significant corrosion resistance and mechanical properties. Wear resistance of Ti-6Al-4V alloy could be improved by depositing the proper coatings on the surface of the alloy. The graphene powders and CeO_2 ceramic powders are mixed together, and then, the high-energy ball milling will be performed to produce nanostructured coatings. The nanocomposite coating of graphene and cerium oxide can be sprayed on a substrate of Ti-6Al-4V using atmospheric plasma spraying technique [35]. Graphene nanopowders could be used in three different compositions mainly 5%, 10%, and 15% by weight. Hardness of the nanostructured coating can be measured by nanoindentation technique and characterization of the above coating could be performed by XRD, SEM, and TEM.

11.6 General Discussions

In order to determine the wear and mechanical properties of thermal sprayed nanostructured coatings, the various methods have been used. Three-scan scratch test with different loading conditions is a most popular method of wear testing at nanoscale, whereas nanowear fretting test defines the behavior of bodies having vibratory motion. Hardness of the nanostructured coating can be measured by nanoindentation method in which indenter's shape, i.e., pyramidal and spherical, also plays an important role. The load is applied on the surface of the coating and nanoindentation hardness is calculated which depends upon the degree upto which indenter penetrates the surface. Thermal spraying parameters like temperature, velocity, and feed stock type also affect the mechanical and wear properties of nanostructured coatings. The proposed nanocomposite coating could represent better wear and corrosion resistance as compared to other surface coatings.

References

1. Beake, B.D., Harris, A.J., Liskiewicz, T.W., Review of recent progress in nanoscratch testing. *Tribology – Mater. Surf. Interfaces*, 7, 2, 87–96, 2013.
2. Malzbender, J. *et al.*, Measuring mechanical properties of coatings: a methodology applied to nano particle filled sol gel coatings on glass. *Mater. Sci. Eng.*, 36, 47–103, 2002.
3. Chao Guo, W. *et al.*, A modified method for hardness determination from nanoindentation experiments with imperfect indenters. *Adv. Mater. Sci. Eng.*, 2016, 1–9, 2016.

4. Benea, L. *et al.*, Fretting and wear behaviors of Ni/nano-WC composite coatings in dry and wet Conditions. *Mater. Des.*, 65, 550–558, 2015.
5. Carr, J.M. *et al.*, Confined crystallization in polymer nanolayered films: A review. *J. Mater. Res.*, 27, 10, 1326–1350, 2012.
6. Chromik, R.R. and Zhang, Y., Nanomechanical testing of third bodies. *Curr. Opin. Solid State Mater. Sci.*, 22, 4, 142–155, 2018.
7. Faisal, N.H. *et al.*, Cyclic Nanoindentation and Nano-Impact Fatigue Mechanisms of functionally graded TiN/TiNi film. *Shape Mem. Superelasticity*, 3, 149–167, 2017.
8. Pandure, P.S. *et al.*, Finite element simulation of nano-indentation of DLC coated HSS substrate. *Proceedia Mater. Sci.*, 6, 1619–1624, 2014.
9. Huang, L.Y. *et al.*, Nano-scratch and fretting wear study of DLC coatings for biomedical application. *Diamond Relat. Mater.*, 10, 1448–1456, 2001.
10. Wang, W. *et al.*, Fabrication and characterization of Ni–ZrO2 composite nano-coatings by pulse electrodeposition. *Scr. Mater.*, 53, 613–618, 2005.
11. Gramajo, J., Gualco, A., Svoboda, H., Study of the welding procedure in nanostructured super-hard Fe- (Cr, Mo, W) - (C, B) hardfacing. *Int. J. Refract. Met. Hard Mater.*, 88, 1–6, 2020.
12. Srinivasa Rao, C. and Eswara Reddy, C., Finite element modeling of nanoindentation to extract load-displacement characteristics of bulk materials and thin films. *Indian J. Pure Appl. Phys.*, 47, 54–59, 2009.
13. Ji, Z. *et al.*, Mechanical and tribological properties of nanocomposites incorporated with two dimensional materials. *Friction*, 8, 5, 813–846, 2020.
14. Zhong, Y., Mechanical Properties Test of Nano Modified Coating Material for Color Antiskid Pavement. *IOP Conf. Ser.: Earth Environ. Sci., 186, 1–8, 2018.*
15. Hammad Nazir, M. *et al.*, Experimental Analysis and Modeling for Reciprocating Wear Behavior of Nanocomposite Coatings. *Wear*, 416, 89–102, 2018.
16. Rueda Ruiz, M. *et al.*, New Instrumentation and analysis methodology for nano-impact testing. *Mater. Des.*, 192, 1–12, 2020.
17. Beake, B.D. and Fox Rabinovich, G.S., Advanced nanomechanical testing for high speed machining of hard-to-cut aerospace alloys. *Int. Heat Treat. Surf. Eng.*, 5, 17–20, 2011.
18. Kennedy, D.M. and Hashmi, M.S.J., Methods of wear testing for advanced surface coatings and bulk materials. *J. Mater. Process. Technol.*, 77, 246–253, 1998.
19. Richard, C. *et al.*, Corrosion and wear behavior of thermally sprayed nano ceramic coatings on commercially pure Titanium and Ti–13Nb–13Zr substrates. *Int. J. Refract. Met. Hard Mater.*, 28, 115–123, 2010.
20. Bolelli, G. *et al.*, Damage progression in thermal barrier coating systems during thermal cycling: A nano-mechanical assessment. *Mater. Des.*, 166, 1–17, 2019.

21. Lanzutti, A. *et al.*, Tribological Behavior of Thermal Spray Coatings, Deposited by HVOF and APS Techniques, and Composite Electrodeposits Ni/SiC at Both Room Temperature and 300 °C. *Tribol. Ind.*, 35, 2, 113–122, 2013.
22. Garsia, J.F. and Celis, J.P., Role of contact frequency on the wear rate of steel in discontinuous sliding contact conditions. *Surf. Coat. Technol.*, 141, 171–178, 2001.
23. Gupta, A. and Kumar, D., Development of Al2O3-based hybrid ceramic matrix composite coating to mitigate the erosive wear of advanced steel. *Proc. Inst. Mech. Eng. Pt. L. J. Mater. Des. Appl.*, 235, 4, 1–11, 2020.
24. Zhang, X. *et al.*, Microstructure Characteristics and Properties of HVOF Sprayed Ni-Based Alloy Nano-h-BN Self-Lubricating Composite Coatings. *Adv. Tribol.*, 2015, 1–7, 2015.
25. Alvares, C. *et al.*, Microstructure and Mechanical Properties of Thermal Sprayed Nanostructured Cr**3**C**2**-Ni20Cr Coatings. *Mater. Res.*, 11, 2, 137–143, 2008.
26. Ali, O. *et al.*, Influence of Post-treatment on the Microstructural and Tribomechanical Properties of Suspension Thermally Sprayed WC–12 wt%Co Nanocomposite Coatings. *Tribol. Lett.*, 65, 33, 1–27, 2017.
27. Thakur, L. and Arora, N., A study on erosive wear behaviour of HVOF sprayed nanostructured WC-CoCr coatings. *J. Mech. Sci. Technol.*, 27, 5, 1–9, 2012.
28. Sun, X. *et al.*, Mechanical Properties and Thermal Shock Resistance of HVOF Sprayed NiCrAlY Coatings Without and With Nano Ceria. *JTTEE*, 21, 5, 818–824, 2012.
29. Zhong, X.C. *et al.*, Nano-composite coatings with improved mechanical properties and corrosion resistance by thermal spraying. *IOP Conf. Ser.: Mater. Sci. Eng.*, 60, 1–10, 2014.
30. Aravind, A. and Kannan, G.R., Experimental studies of mechanical and microstructure properties of plasma sprayed thermal barrier coatings. *Appl. Mech. Mater.*, 2014, 326–332, 2014.
31. Beake, B.D., Shi, B., Sullivan, J.L., Nanoscratch and nanowear testing of TiN coatings on M42 steel. *Tribol.: Mater. Surf. Interfaces*, 5, 4, 141–147, 2011.
32. Alvares Da Cunha, C. *et al.*, Microstructure and mechanical properties of thermal sprayed nanostructured Cr3C2–Ni20Cr coatings. *Mater. Res.*, 11, 2, 1–12, 2008.
33. Hao Hsin Hung, J. *et al.*, Reciprocating wear properties of thermal sprayed titanium aluminide–alumina composite coatings. *Surf. Coat. Technol.*, 202, 5599–5602, 2008.
34. Ban, Z.G. and Shaw, L.L., Characterization of Thermal Sprayed Nanostructured WC-Co Coatings Derived From Nanocrystalline WC-18wt.%Co Powders. *JTTEE*, 12, 112–119, 2001.

35. Li, K. *et al.*, Plasma sprayed cerium oxide coating inhibits H2O2 induced oxidative stress and supports cells viability. *J. Mater. Sci.: Mater. Med.*, 27, 1–10, 2016.

12

Overview of Biological Tissue Properties for Biotribology

L. DhiviyaLakshmi[1]* and V. Muthukumaran[2]

[1]*Biomedical Engineering, Sri Ramakrishna Engineering College, Coimbatore, India*
[2]*Mechanical Engineering, Kumaraguru College of Technology, Coimbatore, India*

Abstract

Biological systems involve in complex interaction while performing even a simple day-to-day activity. There is continuous load absorption, and distribution is happening in our system. This interaction determines the mechanical behavior of the system. Any biological system as a whole is heterogeneous in composition and organization. Hence, the way each component of the system response to a load is quite different and unique from each other. Biotribology is a branch of science that deals with the mechanical behavior such as friction in joints, wear-tear characteristics of hard tissues, and natural phenomenon of lubrication. The purpose of an implant and its performance is highly influenced by the tissue properties.

Hence, when we talk about biotribology, clear understanding of tissue properties is essential to determine the implant material characteristics. Bone tissue and joints are those which are exposed to fracture or failure that need an implant or prosthesis. Here, in bone tissue, loading is highly influenced by the inherent properties such as heterogeneity, porosity, and apparent density of the bone. Other external factors that contribute to the mechanical properties of the tissues are the force, magnitude, direction of load, and time of loading of the tissue.

When the material to work with biological tissue, they should likely to have properties matching the biological tissue. This chapter focuses on the tissue composition, tissue properties, and tissue-material interaction. Metals, ceramics, polymers, and glasses are investigated for decades to be suitable for implantable materials. Especially in the case of bone implants such as bone plates or bone joint replacement implants where the surface property of the implants should match to that of the tissue so that the implant is bio accepted. Most common problem accompanying metal implants are implant fixation, and it is required to ensure that tissue and implant

**Corresponding author*: dhiviyalakshmi.lakshmipathy@srec.ac.in

Sarfraj Ahmed and Vinayak S. Dakre (eds.) Tribology and Characterization of Surface Coatings, (253–280) © 2022 Scrivener Publishing LLC

bonding/interaction is adequate; thus, the material is bioactive. Metal implants, generally, when placed in a biological environment, tend to release metal ions to the surrounding tissue due to friction and wear. The liberated material ions based on its solubility gets accumulated in the local tissue or transported to distant tissue which can lead to adverse effects like implant rejection, periprosthetic osteolysis, and, eventually, implant failure by loosening. Wear resistance and reduction in dissolution of the material will help to increase the life span of the implant and reduce the adverse effect of ion. Tissue properties and tissue response to the implant are detailed in this chapter, highlighting few of the implant materials like titanium, titanium alloy, and stainless steel, with total hip replacement as an example.

Keywords: Tissue composition, mechanical properties, tissue interaction, surface modification

12.1 Introduction

The human body is the organization of complex systems that work in coordination with each other to maintain homeostasis. Homeostasis is defined by the well-being of a person both internally and externally. The system control and regulation are exhibited by the higher-order organs constituting the central nervous system. The peripheral nervous system conducts the signals from the higher-order systems to the effector organs like skeletal muscle, muscles of the heart, GI tract, and blood vessels. The communication is through the conduction of action potential by the nerve cell–to–nerve cell and nerve cell–to–muscle cell signaling through either chemical or electrical conduction interfaces in the cells [1].

The action potential is the signal conducted from the higher center of the brain by the nerve cells and neurons that innervate the effector organs. This signaling is essential for the control and regulation of the function of a tissue, organ, and system [2]. Excitation of the cell is dependent on the cell membrane potential due to the movement of the ions across the membrane. When an action potential is conducted down, a neuron the cell membrane charge is modified by the movement of ions such as Na2+ ions and K+ ions across the membrane. This is the natural physiological behavior of the cell and, in turn, the tissues and the entire system of the body [3].

When it comes to understanding biotribology, this explains the dynamic behavior of the tissue in the presence of foreign or external material such as implants. It emphasizes more on tissue and implant interactions. Since the human body is made of more than 80% fluids. Fluids are everywhere in a tissue environment. The fluid present within a cell is called intracellular fluid and those surrounding the cell are extracellular fluid. Body fluids consist of ions that keep moving in and out of the cell to conduct signals during excitation and to maintain the membrane potential during the resting state.

Biocompatibility and tissue acceptance are important for any material that is introduced in the body's internal environment. Also, the material should be capable enough to function in cooperation with the body tissue to the various kinds of forces and impacts that fall in the location of implantation [4]. The most desired characteristic is that the implant should mimic the original functionality of the region where it is fixed. The interaction of the tissue with the implant may lead to the release of ions from the implanted material due to the activity of the biofluid around the implant. This may even lead to inflammation and loosening of implants. Periprosthetic osteolysis (PPO) is one of the major reasons for implant failure due to the activation of the innate immune system.

Due to its active role in the structural support and load-bearing functions, the tissue that is more commonly exposed to wear-tear mechanism is the musculoskeletal tissues, as they are responsible for providing structural support, integrity, and protection to the body and the internal organs. Whether we involve in an activity or not, our body is continuously subjected to force and tissue loading happens continuously. Tissue loading involves the interaction of the structures which paves the path for wear-tear mechanism.

12.2 Tissue Interaction With Implant

When an implant is introduced into the organism, it is treated as a foreign material by the body, and the process of placement or introduction of the implant involves tear and probing into the healthy tissue. This will initiate the wound healing mechanism at the site of insertion. The introduction of the implant initiates the inflammation and immune response at the host site is the immediate response. The reaction of the body's immune system to the implant depends on the material property of the implant (see Figure 12.1). Anyhow, there is a minimal to severe host response and reaction to the implantable material. Whenever surgical implantation is done, there is tissue damage associated with it.

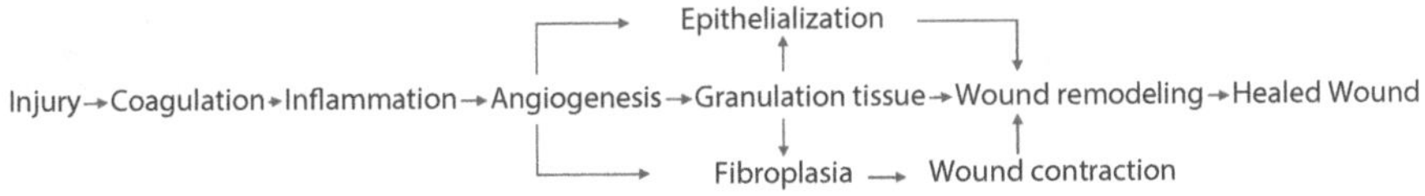

Figure 12.1 Tissue response to wound and the sequence of healing process.

The adjacent tissues to the damaged tissue will start immediate response of inflammation and constriction of capillaries followed by dilation. There is the pooling of formed elements surrounding the capillaries, and due to vasodilation, plasma leakage will occur. This leakage plasma along with the migrating leukocytes and the damaged, dead cells will constitute the exudate. All these reactions of vasodilation, leakage of fluid, and plugging of lymphatics are the sign of immediate response of inflammation—indicated by redness, swelling, heat, and pain. If the tissue injury is extensive due to irritants or infection, then it will lead to tissue destruction by collagenase a proteolytic enzyme. Collagenase is released by the granulocytes which are, in turn, destroyed by the change in the pH (below 5.2) of the surrounding tissue. If no healing process is initiated in 3 to 5 days from injury, then the macrophages arrive to initiate the chronic inflammatory process by forming giant cells. They are phagocytic in nature and remove the bacteria and any infection causing foreign material as shown in the Figure 12.2 [7].

The inflammatory exudate consists of fibrinogen which will convert to active fibrin. Fibrin forms a mesh network surrounding the wound. The migrating fibrinogen utilizes this fibrin framework for new collagen deposition. Capillaries are formed and the fibrin network is removed by fibrinolytic enzymes activated by the endothelial cells. Similarly, at the site of application of an implant due to mechanical loading, friction occurs and materials are liberated and collected as debris which will activate the immune system and, in turn, the macrophages [19].

The implant as such can be biocompatible, but the material derived from the implant due to the engagement of the implant in the bodily loading environment can be highly reactive. The main objective of the development of a biological implant is the identification of material with biocompatibility. The molecular, cellular, and biological characteristics of the materials nature of tissue- implant interaction are important for the development of the implants. Hence, surgical implantation of a biomaterial

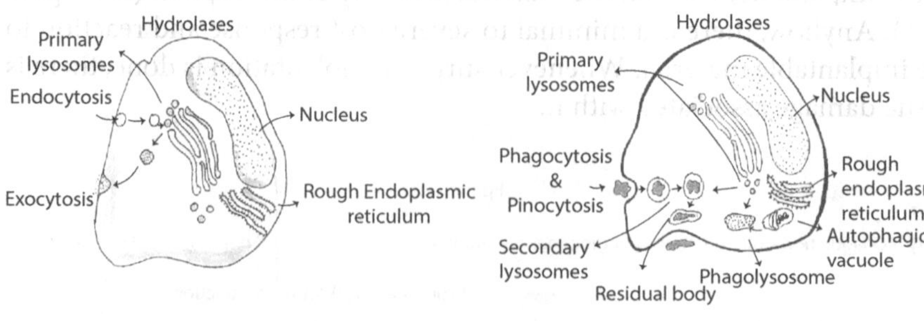

Figure 12.2 Activated macrophage.

always triggers an inflammatory reaction associated with the healing of the damaged tissue. The level of interaction is based on the composition of the implant material, implant surface, type of loading at the site of implant, fatigue mechanism of the implant, and chemical decomposition at the site.

This is an important consideration to make to see how the implant material derived during physical loading reacts to the tissue surrounding the implant, and vice versa. The rate of reaction and acceptance and life span of the implant is decided based on this. Most of the time implant rejection can happen due to the reaction caused by the material debris collected at the site of implantation. Most of the intramuscular implants reported causing severe corrosion and tissue damage in materials such as Cu, Ni, and Co alloys [9]. The rate of dissolution of the metal materials in the tissue is studied by chemical analysis but the rate of dispersion in the soft tissue is faintly identified. One of the attempts to study the elemental distribution of materials in the tissue is done by X-ray scanning analytical microscopy where the energy dispersion of fluorescent X-ray [10]. The amount of metal released from Ti-6Al-4V and Co-Cr alloys are found to be stable [12]. The amount of dissociation of the metal materials into the surrounding tissue gives an idea about the rate of tissue response. It is difficult to detect the low concentration of metal dissociation. Hence, elemental mapping using XSAM is superior in detecting the material dispersion than Electron Probe Microanalysis (EPM).

The composition, function, and mechanical properties of the bone, joints, and cartilages are discussed extensively in the following topics for better correlation of understanding the biotribology of the implants such as total hip joint replacement.

12.2.1 Bone

The bone structures are the primary units for weight-bearing characteristics. Bone is also one of the tissues that are provided with a rich blood supply. Any loading can influence the chemical and mechanical properties of the bone, and it has an excellent self-repair mechanism. Bone is the structure meant for weight bearing as mentioned earlier but to achieve movement joints come to play. Bone density is subjected to change based on the rate of use. Also, the shape of the bone adapts to the mechanical loading acting on it. For example: The dominant limb of a tennis player will be stronger than the opposite limb due to the mechanical demand on the dominant limb. Bone is composed of mostly calcium and phosphate, and it exists in the form of $Ca_{10}(PO_4)_6(OH)_2$ that is the mineral inorganic compound. The organic compound consists of collagen, non-collagen proteins, Osteocalcin, Osteonectin, and Osteoppontin, and bone sialoproteins. The collagen molecules are arranged parallel with each other with a gap or hole zone (40 nm) (Anne Waugh *et al.*). Mineralization begins in the gap zone resulting in mineralized fibril. The most important component governing the self-repair mechanism and that which maintain the equilibrium in wear and tear property of the bone are the bone cells, and the mechanical strength of the bone is determined on the extracellular matrix. The extracellular matrix is nothing but the organic phase made of collagen matrix and the mineral phase composed of hydroxyapatite crystals.

The bone cells are as follows:

Osteoblasts: bone-forming cells,
Osteoclasts: bone-destroying cells, and
Osteocytes: bone-maintaining cells that are inactive osteoblasts trapped in the extracellular matrix.

The structure of the long bone is shown in Figure 12.3.

The bone has relatively high compressive strength but poor tensile strength and is essentially brittle but has a significant degree of elasticity,

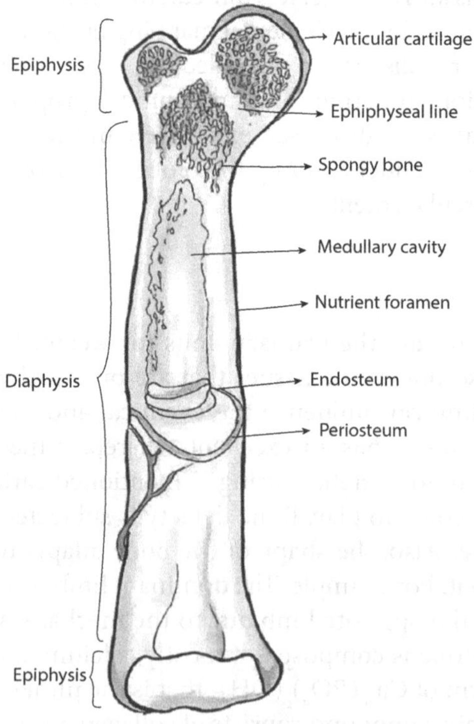

Figure 12.3 Structure of long bone.

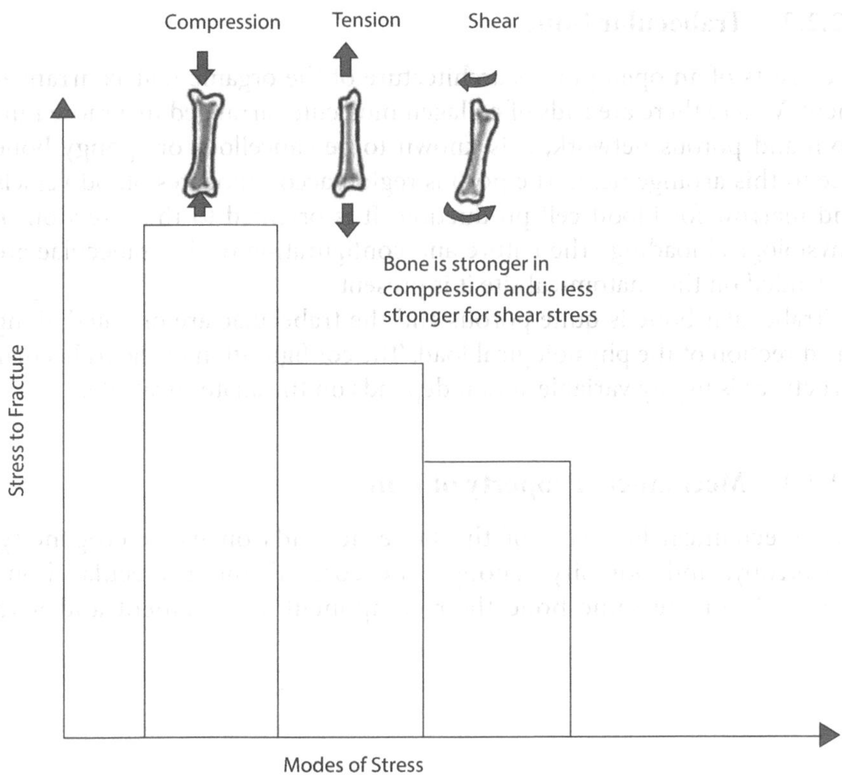

Figure 12.4 Strength of bone.

which is mainly due to collagen. It consists of living cells embedded in the mineralized organic matrix that makes up the osseous tissue. Figure 12.4 shows the relative strength of the bone to various forces. It is that the bone has high strength to withstand compression loading and very low strength to shear stress [4].

12.2.2 Compact Bone

Bones are of two types based on the arrangement of the organic and mineral phases. The hard outer region has compact tissue where the gap between the collagen matrix is small, and it gives the bone a smooth, white, and solid appearance. Due to its compactness, it is also referred to the as dense bone. It is organized in cylindrical shaped elements called osteons composed of concentric lamellae [5].

12.2.3 Trabecular Bone

It consists of an open porous architecture of the organic matrix arrangement. Where there are rods of collagen molecules arranged in a more random and porous network, it is known to be cancellous or spongy bone due to this arrangement. The porous region accommodates blood vessels and marrow for blood cell production. It is oriented to the direction of physiological loading. The nature and configuration of the trabeculae are depended on the anatomical site it is present.

Trabecular bone is quite porous and the trabeculae are oriented along the direction of the physiological load. The configuration of the trabecular structures is highly variable, and it depends on the anatomical site.

12.2.4 Mechanical Property of Bone

The mechanical behavior of the bone depends on its heterogeneity, anisotropy, and porosity. Though the cortical and trabecular bone is present in the same bone their component arrangement and load

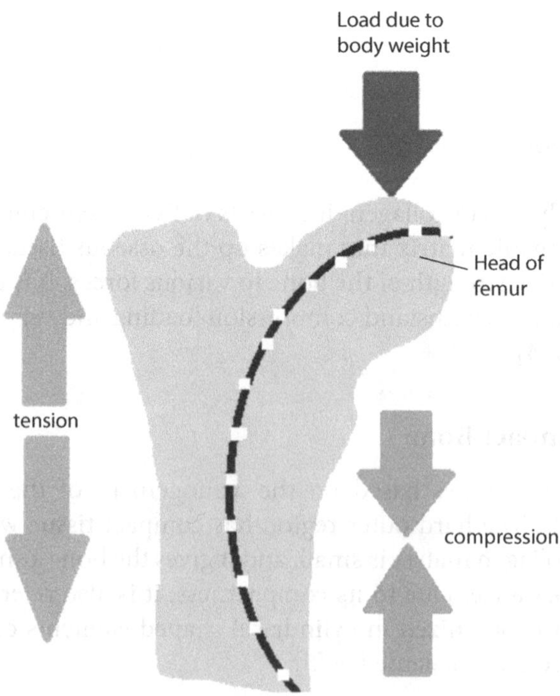

Figure 12.5 Load distribution in bone.

distribution are different from each other. Bone mechanical properties are highly variable according to species, age, anatomical site, liquid content, etc. Each of these properties greatly influences the behavior characteristic of bone.

Figure 12.5 shows the loading of the femur bone due to bodyweight in which the load is distributed differently at different sites of the structure. Here, on the convex side compression force acts with complimenting traction or tension in the concave side of the bone with respect to the given loading direction. Cyclic loading over a longer duration related to occupation will result in adaptation and modification of the structure along the direction of the applied load.

The cortical bone has a high Young's modulus, and it shows higher resistance to torsion and bending and it has low turnover. While the trabecular bone shows higher resistance to compression and lower apparent modulus (apparent density is defined as the ratio of the mass of bone tissue to the bulk volume of specimen, including the volume associated with vascular channels and high-level porosity), and it is more elastic compared to the cortical bone. It shows higher turnover [6].

12.3 Anisotropy

The mechanical property of the material varies according to the direction of loading (Figure 12.6). Cortical bone is anisotropic in nature and is stronger and stiffer in longitudinal loading compared to transverse direction. Also, it is stronger for compression loading than tensile load. The three-point testing done on bone shows the fracture toughness of the bone (Figure 12.7).

The compressive strength of the bovine bone is studied for three anatomical directions as shown in Figure 12.8. The bone is stiffer in the longitudinal direction where the load is evenly distributed along the length of the bone whereas it is tougher in the radial direction. Along the radial direction, there are interstitial lamellae in between the osteon cells which will absorb the load and contribute to the toughness property of the cortical bone. The bone is brittle and resists force in the transverse direction. Hence, the bone during longitudinal loading due to the weight of the body is distributed among the parallel collagen fibril arrangement in the compact bone.

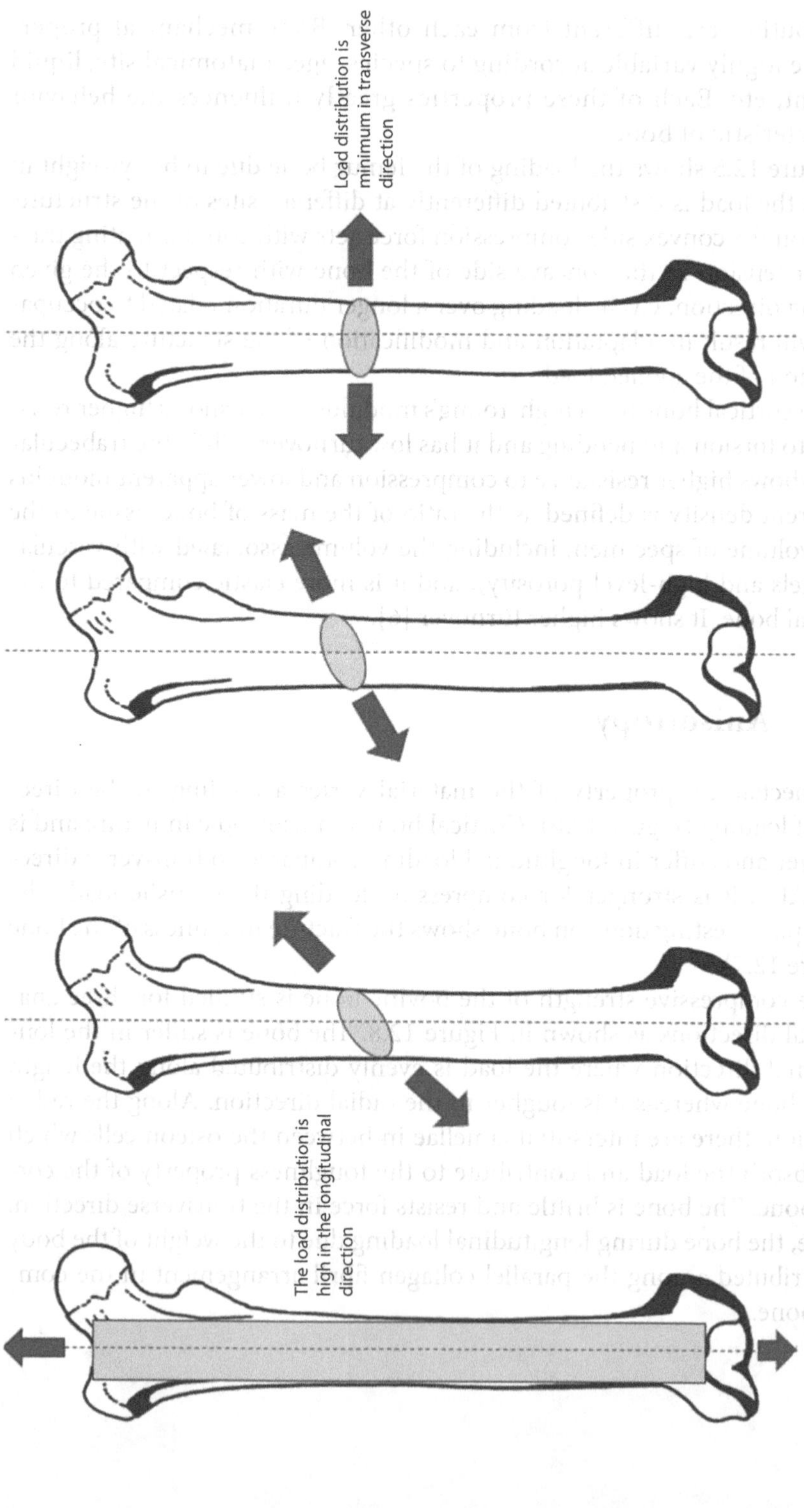

Figure 12.6 Anisotropic behavior of cortical bone.

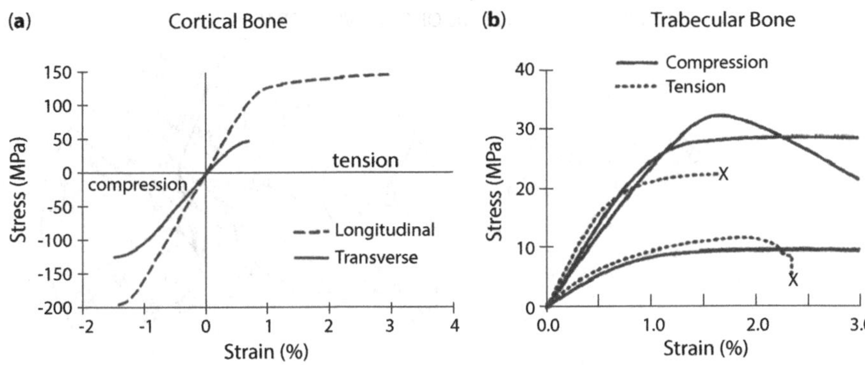

Figure 12.7 Fracture toughness: (a) cortical bone and (b) trabecular bone.

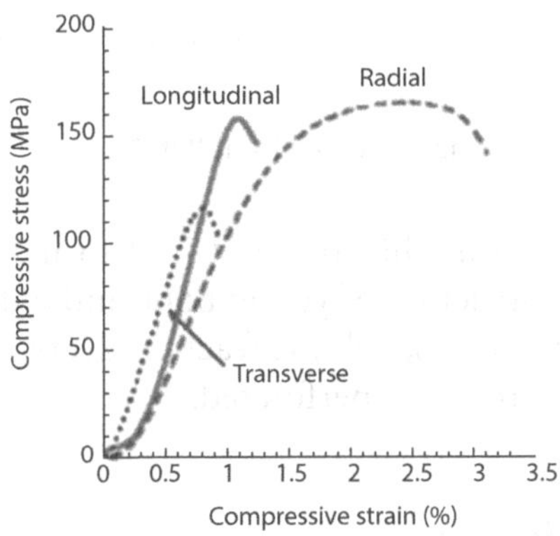

Figure 12.8 Stress-strain response of bovine cortical bone for compression in three directions. Image is modified from Novitskaya *et al.* (2011).

12.4 Heterogeneity

The variation in the microstructure parameters is porosity and percentage of mineralization. The Young's modulus and ultimate strength of the bone decrease to half when the porosity increases from 5% to 30%. Even with a small increase in mineralization, there is a large difference in the strength.

12.5 Aging

The porosity increases, and hence, the modulus and strength decrease and the bone become more brittle. The energy to fracture is less for older bone. Denaturation of collagen happens but the mineralization does not change. As aging happens the bone structure losses its ductility and becomes more

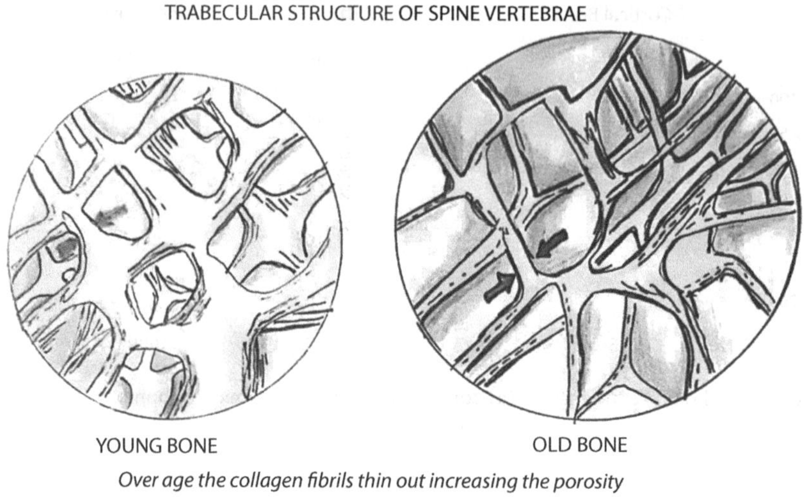

Figure 12.9 Effect of aging (image modified from Ref. 23).

brittle. The collagen fibrils thin out and lead to increased porosity. The change in trabeculae structure of young bone and older bone is shown in Figure 12.9. As age increases, volume fraction decreases and the architecture becomes thinner, rod, and perforated.

12.6 Hysteresis

The loading-unloading and reloading of cortical bone (Figure 12.10a) with tensile load are tested which shows that the loading and reloading curves are similar but there is a residual strain in the bone which leads the modulus near to the perfect damage modulus which is around the secant modulus during unloading. Residual strain is present even between 1 and 2 minutes hold between loading cycles.

Whereas the compressive loading-unloading and reloading of vertebral trabecular bone are shown in Figure 12.10b, where the initial loading causes a large residual strain where the unloading curve is similar to that of the cortical bone during tensile loading. The reloading modulus reduces from a value similar to the intact modulus to a perfect damage modulus. This shows that the trabecular bone can undergo a large deformation before failure, whereas the cortical bones are stiffer and resist deformation. The bone sustains more viscous energy dissipation during compression than in tension and this effect is permanent [14].

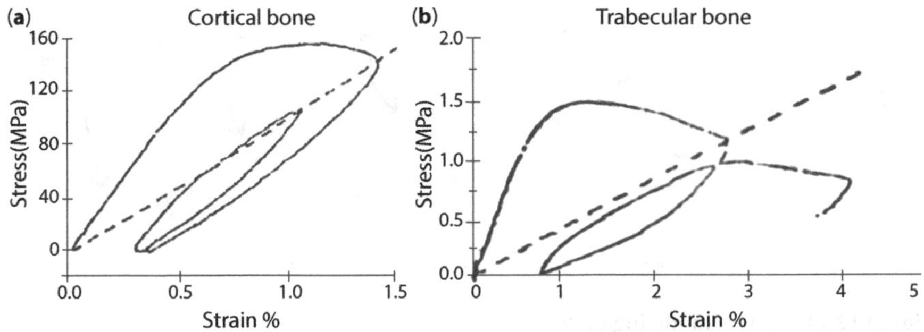

Figure 12.10 Loading-unloading and reloading characteristics of cortical bone during tensile loading (from Ref. 14).

12.7 Bone Remodeling

Since bone is the structure supporting the body with structural integrity and strength. The bones are engaged in continuous loading which, in turn, causes wear and tear due to the exchange of forces during locomotion and various body activities. There is a natural equilibrium between the cells shed and new cells formation which is an intrinsic mechanism known as bone remodeling.

Bone usually adapts and remodels based on the load applied. According to Wolff's Law, bones develop a structure more suitable to resist the force acting and also adapting to both the internal architecture and external confrontment according to the changing external loading conditions. Any change in the stress and strain rate is sensed and adapted by the bone.

The internal architecture of the bone shows the change in density and the trabecular disposition, whereas the cortical region shows a change in the shape of the osteons and hence the external confrontment. When the local strain rate is intensified, then new bones are formed; on the other hand, if the strain rate is lowered, then bone resorption will occur. This is the reason for the reduction in bone density and strength for bedridden patients due to immobilization and lack of activity. When this change in strain is due to a change in the direction of the load, then the microscopic structure gets rearranged in the direction of load. Resorption and deposition are the mechanisms by which remodeling is carried out, and the cells that exhibit these functions are the osteoclast for resorption and osteoblast for deposition of new bone cells (see Figure 12.11).

Thus, the structure of the bone is subjected to change and remodeling based on the nature of load falling at that particular region. Hence, it very

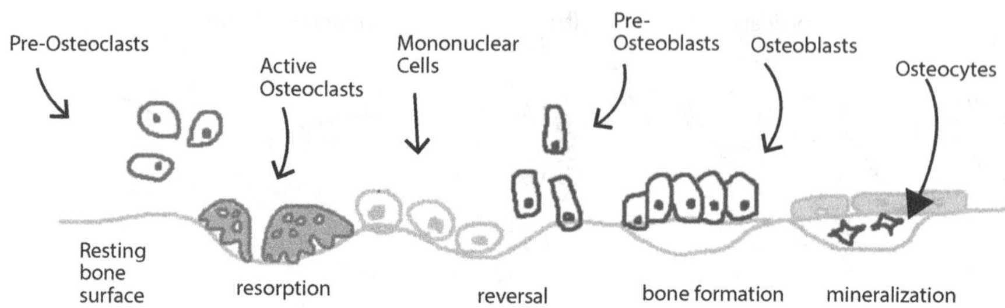

Figure 12.11 Bone remodeling cycle.

much needed to understand the forces and loads applied or acting on the region in which the implant is planned to be applied and henceforth the material selection is to be carried out which can withstand the load and exhibit the functional support it is meant to offer. During some conditions like autoimmune disorders and degenerative disorders, this balance between the tissue resorption and deposition is altered leading to loss of bone tissues.

12.8 Joint

A joint is a gap or contact junction between two bones that are supported with ligaments and tendons. Joints help in the articulation of the bone structures in order to perform an activity. Ligaments and tendons are soft tissues that augment the bone and muscle in order to complete the functional property of the musculoskeletal system. Tendon connects muscles to the bone and the ligament connects bone to bone. They are primarily made up of collagen.

Figure 12.12 shows the parts of a joint. Joints are classified based on structure and function and the number of articulating bones involved. Table 12.1 provides the characteristics of joints based on structure and function.

Structural classification of the joint is done based on the way the bones are connected to each other. Whereas the functional classification is based on the degree of freedom to which movement is achievable between the bones.

Classification based on the number of articulating bones involved are as follows:

Simple Joint: This involves two articulating bones. For example, hip joint.

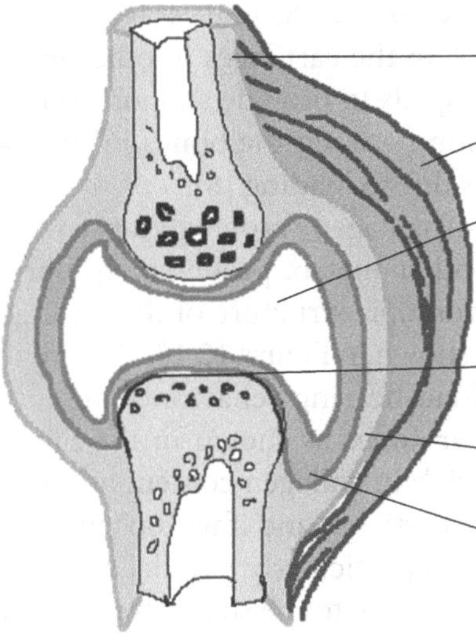

Figure 12.12 Parts of a Joint.

Table 12.1 Classification of structure and function of joints.

Type	Name	Function
Structure based	Fibrous	Fixed
	Cartilaginous	Slightly movable
	Synovial	Freely movable
Function Based	Synarthrosis	Little or no mobility
	Amphiarthrosis	Slight mobility
	Diarthrosis	Variety of movements

Compound Joint: This involves three or more articulating surfaces. For example, radiocarpal joint.
Complex Joint: This involves three or more articulation surfaces and articular disc or meniscus. For example, knee joint.

The synovial joint which is diarthrosis functionally is also known as articulating joints and is the joints that are highly exposed to wear.

The components of the synovial joints are the articular cartilages, joint capsule, joint cavity, and synovial membrane. Articular cartilage is a dense connective tissue that covers the bone at the joint. It is actually hyaline cartilage with 2- to 4-mm thick than the cartilages in other joints. The main function of the articular cartilage is to provide support to the body; most importantly, it acts as a cushion between the bones in the joint and prevents the rubbing of the bones due to loading [8].

The cartilage is made up of collagen, proteoglycans, water, inorganic salts, and small amounts of other matrix proteins, glycoproteins, chondrocytes, and lipids. The microscopic structure of the arrangement of the constituents of the cartilage is shown in Figure 12.13. Each of these components plays a vital role in the load-bearing behavior of the cartilage.

The cartilage lining has four zones as shown in the Figure 12.14, the outermost superficial zone is rich in collagen content and it weighs about 85% of the total collagen of the cartilage, and if we look at the arrangement of the collagen in this zone, it is particularly parallel to the joint surface. This arrangement is to offer resistance to the shear stress. The middle zone occupies around 60% of the space where the collagens are more randomly arranged and it is to dissipate the load. The deep zone contains around 30% of the space and the collagen is more perpendicular compared to the superficial zone and it is ensuring the fixation and ingrowth of the collagen to the interface of bone at the calcified cartilage zone.

The collagen fibrils provide the tensile strength to the joint structure, whereas proteoglycans are hydrophilic and it attracts water, and when compression load falls on the bones, the fluid in the membrane is squeezed out of the permeable membrane in order to provide the cushioning effect. Also, the glycosaminoglycans (GAGs) have a negative charge which will

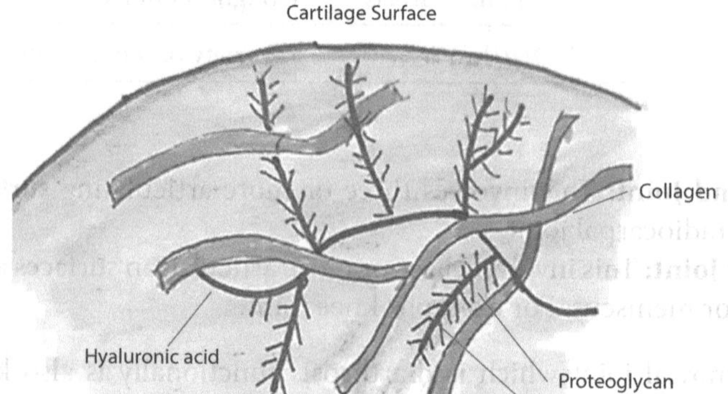

Figure 12.13 Arrangement of constituents of cartilage.

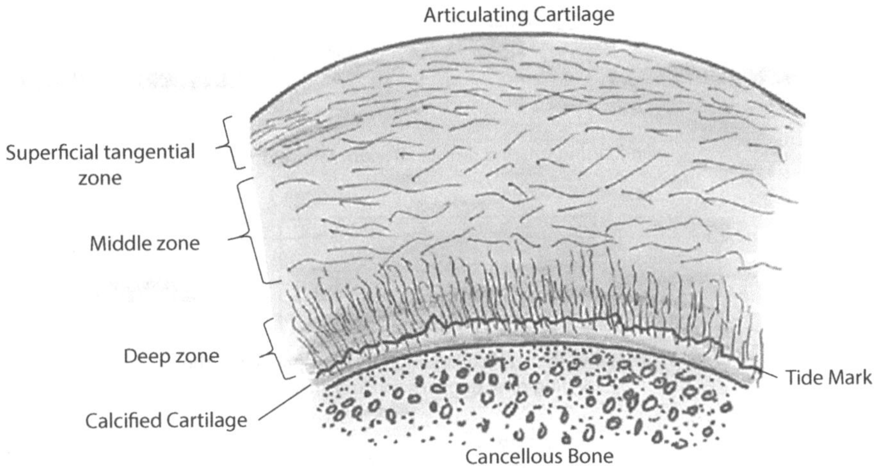

Figure 12.14 The zones of the cartilage layer.

repel each other while compressive loading hence helps in the cushioning. The chondrocytes form the cartilage matrices.

The stress-strain relationship of the cartilage solid phase during tensile loading is shown in Figure 12.15. The arrangement of the collagen shows resistance and exhibits pull-out during initial stress. As there is an increase in stress the collagen fibrils align parallel to one another along the length of the tissue in the direction of application of stress hence the response is linear. On further increase in the stress, the fibrils show a failure by tearing. The arrangement and alignment of collagen fibrils makes cartilage stronger for tensile loading.

Figure 12.16 shows the illustration of the contribution of concentration of water and proteoglycans toward the mechanical load transfer properties. The water forms the part of the solid phase of the extracellular matrices where the proteoglycans have a greater affinity toward water. When a large number of proteoglycans are present, they seal the junction and provide a leak-proof surface that immobilizes the fluid during loading. It is observed that the increase in the water content will increase the load-bearing property of cartilage. It is also that the fluid level at each zone of the cartilage is different. The superficial zone has less fluid level than other zones as they are directly involved in load absorption and distribution which increases their stiffness.

The mechanical behavior of the cartilage is dependent on collagen, water, and proteoglycans. If the amount of these components is modified due to any disease, then it affects the mechanical property of the cartilage.

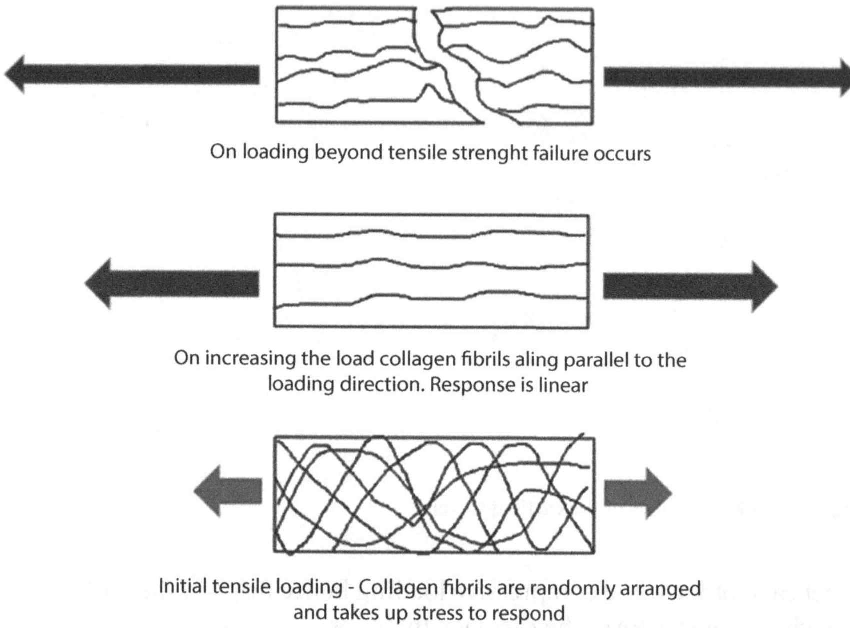

Figure 12.15 Tensile property of solid phase of cartilage.

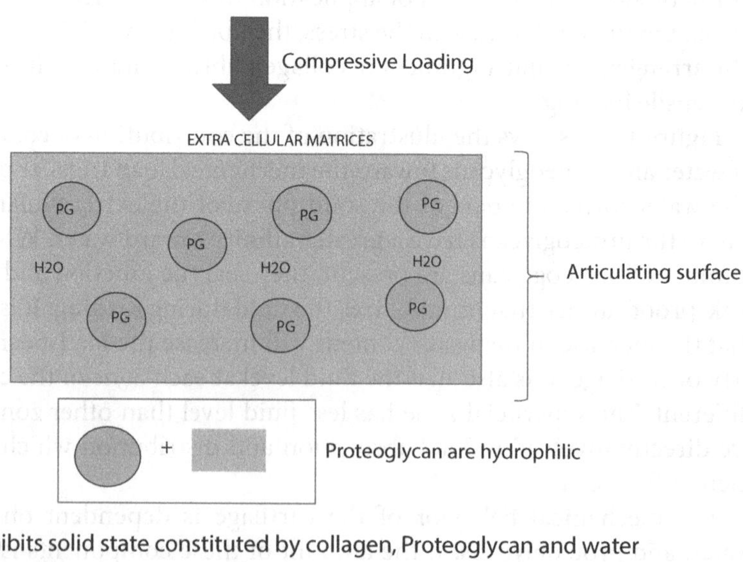

Figure 12.16 Illustration of mechanical property of cartilage vs. concentration.

The flow of fluid out of the tissue creates a drag over the solid surface of the cartilage is responsible for the compressive behavior of the cartilage. The rate of flow of the fluid is dependent on the permeability of the fluid to flow in and out of the tissue, and it follows Darcy's law:

"the rate of volume discharge through a porous solid is related to the pressure gradient applied to the solid and the hydraulic permeability coefficient k".

$$Q = \frac{KA\Delta P}{h}$$

Q is the rate of volume discharge in m^3/s, k is the permeability coefficient in m^4/Ns, A is the area in m^2, ΔP is the pressure gradient in N/m^2, and h is the height of the specimen given in m.

Biphasic theory of cartilage represents the load sharing behavior of the solid and fluid phase driven by permeability. The solid phase is linearly elastic or hyperelastic. When the solid matrix is filled with the interstitial fluid, the cartilage as such is incompressible. It means it resists compression and it is possible only after exudation of the fluid out of the solid matrices. This flow of fluid will cause friction at the interface and this frictional drag force is relatively dependent on the fluid flow velocity.

Under sudden compressive loading like fall, the cartilage experiences a relatively large lateral displacement as it has high Poisson ratio. This expansion is buffered and retarded by the much stiffer subchondral bone, causing high shear stress to act at the cartilage bone interface (see Figure 12.17).

The displacement of cartilage is tested in a confined compression test. The illustration is shown in Figure 12.18 helps us to understand the displacement of the cartilage during constant loading over time [25]. As it is observed at the initial loading, the deformation is rapid as a large amount of the fluid in the cartilage is exuded. When the displacement reaches a constant value, the flow of fluid reduces and approaches zero.

There is an interesting modification that happens in the cartilage when it is loaded. Load absorption and distribution happens within the layers of the cartilage. As shown in Figure 12.19, the first model is the unloaded cartilage (o), when this structure is subjected to compression extrusion of fluid happens immediately and the layers near the surface come closer to each other (a) and as the displacement reaches maximum the fluid flowing out is reduced and is redistributed within the cartilage as shown in the (b, c, d). When the load is maintained in the joint, the new equilibrium is achieved in the cartilage as shown in (e).

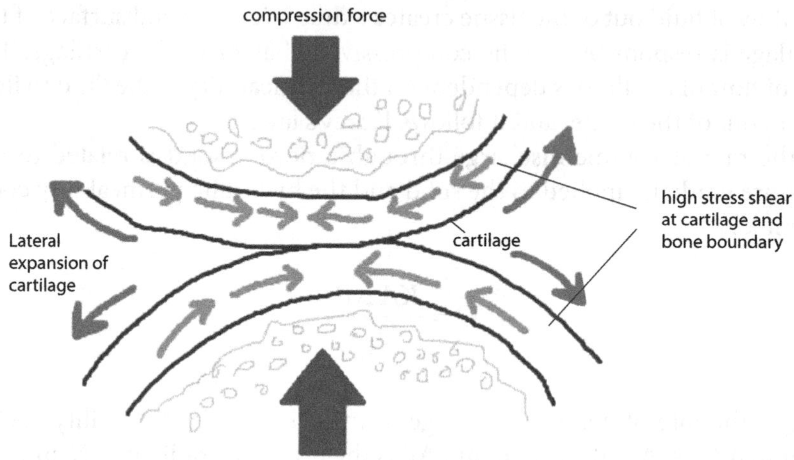

Figure 12.17 Structure of cartilage showing the fluid flow direction during loading (from Ref. 24).

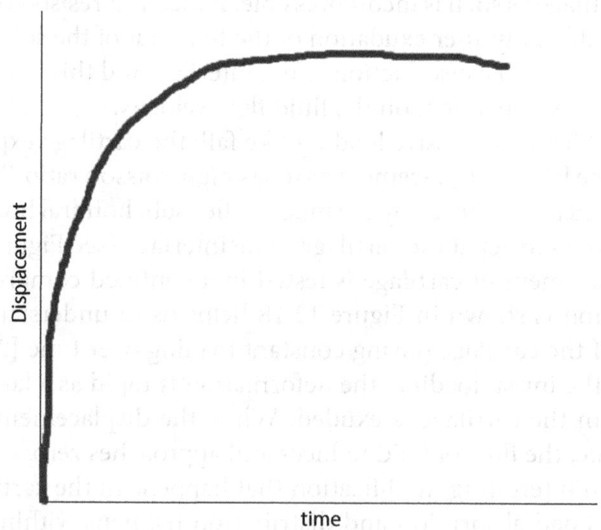

Figure 12.18 Cartilage during compression (image is modified form Ref. 25).

In the stress response curve, we can see maximum peak stress at point (b) in the graph which is due to the fluid exudation. When the fluid redistribution happens, there is stress relaxation happens.

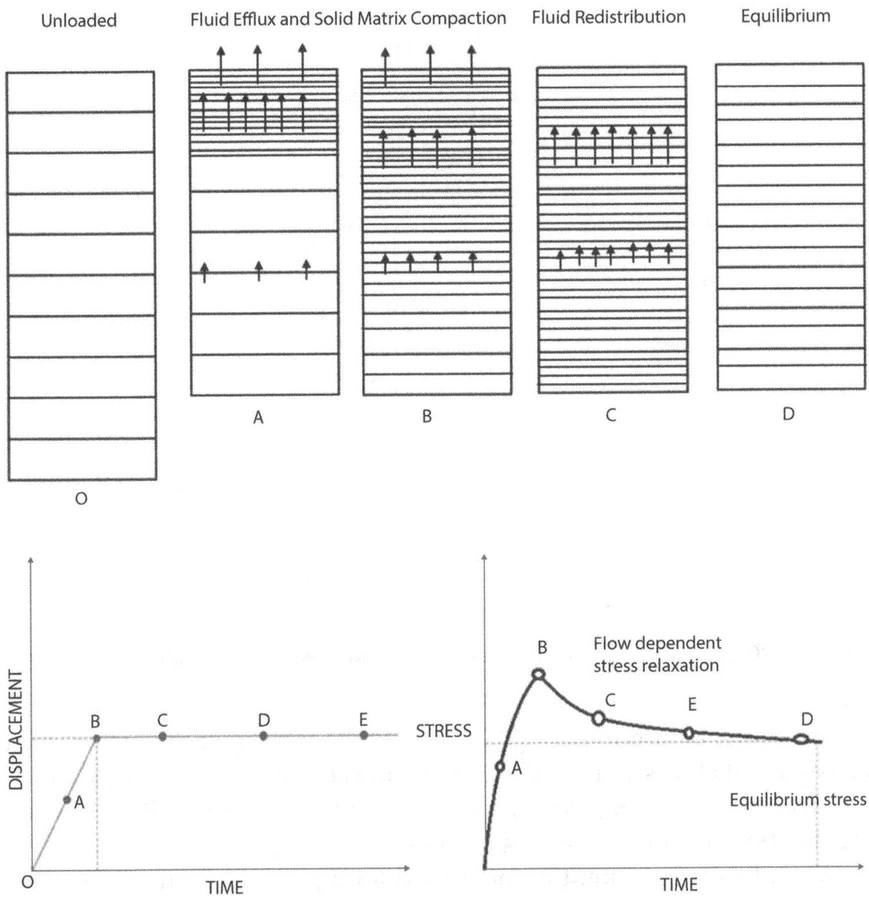

Figure 12.19 Load distribution in cartilage during compression and the stress response curve (image modified from Ref. 24).

12.9 Biomechanics of Cartilage Degeneration

When there are physical activities, it results in joint loads which are transmitted to the chondrocytes through the extracellular matrix (ECM). The flow diagram of the events that happen in the structure and function of cartilage during loading is shown in the Figure 12.20. Normally, when the load is applied, the chondrocyte cellular activities vary in response to the mechano-electrochemical stimuli. For osteoarthritis, there may be intrinsic changes in chondrocytes or the ECM that causes abnormal chondrocyte stimulation and thereof cell activity.

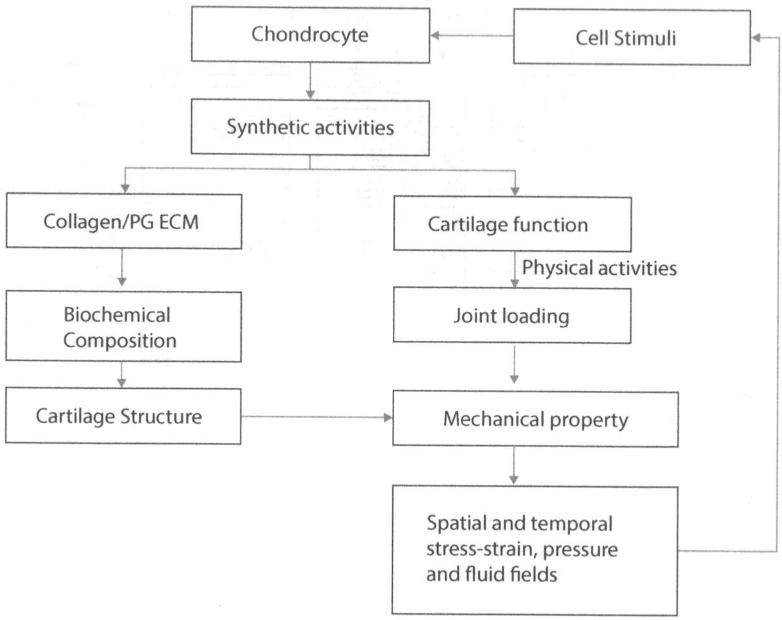

Figure 12.20 Flow diagram of chondrocyte stimulation response to loading.

One of the failure-initiating factors is the loosening of the tile collagen network that allows abnormal PG expansion and tissue swelling. This leads to a decrease in cartilage stiffness and an increase in permeability affecting the functioning of joints during motion.

When this type of joint is anticipated for a joint replacement by using metal-on-metal (MoM) or metal-on-polymer (MoP) kind of implant, the mechanism of loading shall impart compressive loading which can lead to wear and friction of the material. This friction and rubbing of the implant surface may cause the release of materials that constitute the wear debris. The collection of wear debris at the site of implant fixation will have the immune system activation.

12.10 Biotribology of Total Hip Replacement (Metal-on-Metal Articulation)

Due to various reasons for joint-related disease and degeneration, it becomes a requirement to restore the joint function and a decision is taken for total joint replacement [8]. The bearing surface of the implant is subjected to friction, and due to which, there is material wear out. The particles liberated due to friction are released into the surrounding tissue. This can lead to immediate inflammatory reactions which could be one of the major reasons for bone loss. When bone loss is severe and continues, then it will lead to implant loosening. This implant loosening demands surgical revision. One out of five artificial hips is subjected to revision within 15 years [15]. The inflammatory response is the same for any material as its aim is to eliminate the foreign particle [19].

In recent times, joint replacement became common among young and active candidates demands for the durability of the implant [26]. Three classes of materials are being commonly used for hip replacement and they are ceramics, metals, and polymers.

Metals: Cobalt-chrome-molybdenum alloys (CoCrMo).

Ceramics: Aluminium oxide ceramics and zirconia-toughened alumina ceramics.

Polymers: Ultra-high molecular weight polyethylene (PE), both in its conventional and cross-linked (X-PE) forms.

These materials can be used in combination as hard-hard or hard-soft. Due to a greater number of young people getting this implant, more emphasis is on hard-hard type specifically MoM. PPO is one of the common reasons for implant failure and is associated with aseptic loosening [16]. PPO is asymptomatic and the patient may not show any functional difficulty up to the stage to which bone defects occur and implant failure happens [17, 18].

The tissue loading property causes particle production as a result of friction and the body immune system response will eliminate this debris. If the tissue loading exceeds, then the elimination capacity of the body immune system decompensation process begins. Decompensation is chronically activated and disrupts the equilibrium of bone remodeling mechanism by three processes: exaggerated inflammation, disruption of periprosthetic bone formation, and disruption of bone regeneration [20]. When two hard surfaces slide upon each other, wear happens through abrasion, adhesion, and fatigue. The particles liberated due to wear may be bone, cement, metal, metal alloys, corrosion products, etc. (Sinha RK *et al.*). In recent applications, almost all three materials are used in combination; hence, this motivates altered host immune response [21]. When the particles liberated out of the interface surface interact with the surrounding environment by chemical reaction, it is called corrosion. It can also happen due to fretting when the joint functions in different modes [22]. In some situations where the implanted material is like titanium, the metal debris does not dissolve and remain in and around the region of wear [13]. Thus, this local accumulation of the metal particles also contributes for PPO.

PPO is not completely the result of osteoclast acting on the stressed region but it is also the property of the material wear particles that contributes to PPO [11].

Lubrication is the natural mechanism in the joints that prevents friction between the articulating surfaces. The mode of friction varies with the surface roughness, geometry, and material property during motion.

The design consideration for the development of the total hip joint consists of a cup that fits to the acetabular region of the pelvis, and there is a head both in the circular shape with little different diameter. The head diameter is comparatively less than the cup, and it is to accommodate the range of motion and articulation. In Figure 12.21, the radial clearance is observed at the distal edges of the gap between the cup and head to promote movement and avoid friction.

The Stribeck curve (see Figure 12.22) shows different modes of friction in the articulating joint as a function of friction coefficient, fluid film thickness based on viscosity, velocity, and pressure. As we can see in Figure 12.1, the gap is not uniform and it narrows toward the center of the head. The type of lubrication is not the same throughout the lining. The sequence of modes of friction and increasing film thickness is as follows: boundary lubrication, mixed film lubrication, elastohydrodynamic lubrication,

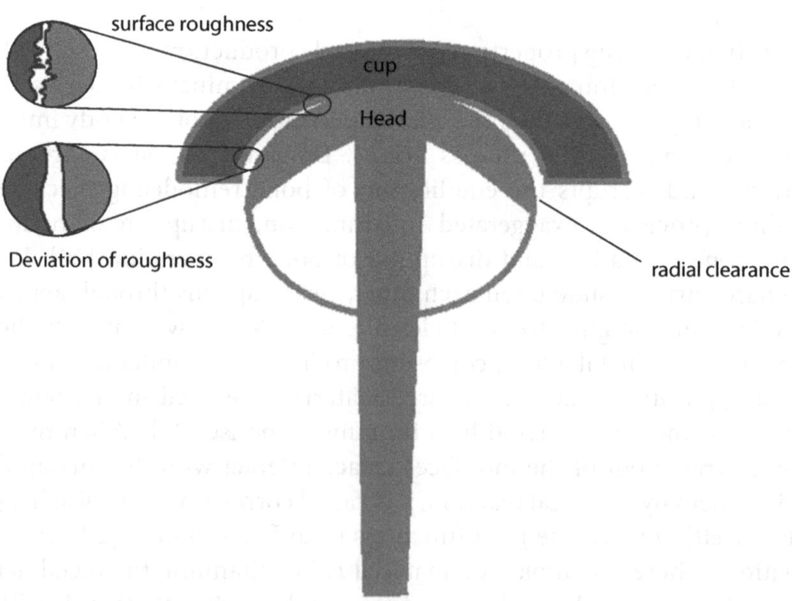

Figure 12.21 Schematic diagram of hip joint design (from Ref. 26).

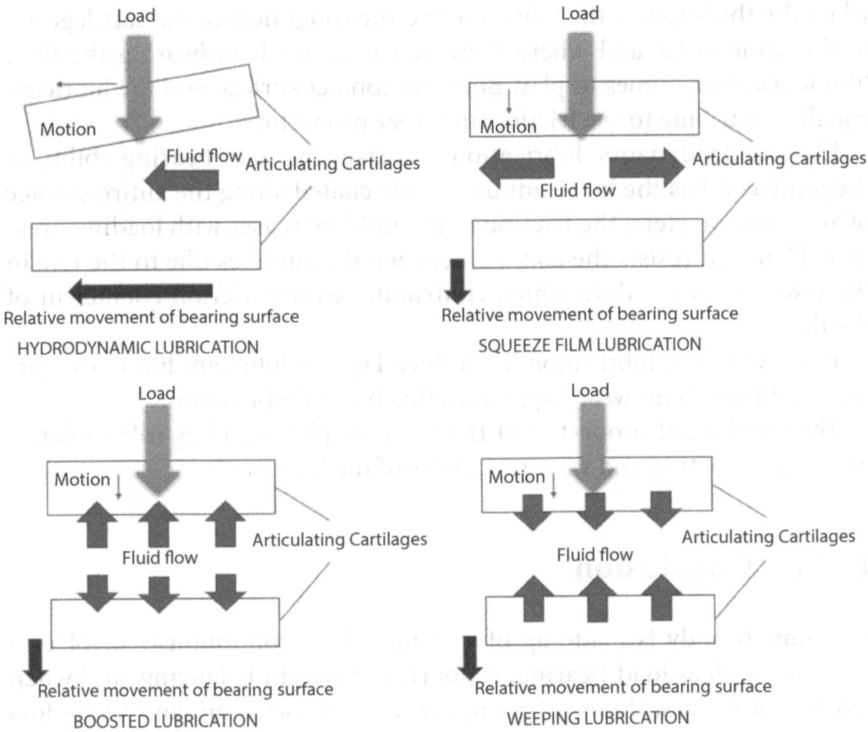

Figure 12.22 Mechanism of lubrication of the articulating joints.

and hydrodynamic lubrication. The coefficient of friction depends on the following:

$$\mu = \frac{\text{Viscosity} * \text{Velocity}}{\text{Stress}}$$

In boundary lubrication, the surface of the cartilage is provided with a lining of lubricating fluid film to avoid direct surface-to-surface contact. It is functional in situations where there are high loads with low velocities. The characteristics of this kind of lubrication depend on the chemical nature of the lubricant. As it is seen in the figure that the fluid thickness is very thin and the coefficient of friction is high for boundary lubrication. It helps in preventing contact between the sliding surfaces [24].

Mixed lubrication is a combination of fluid film and boundary lubrication. Here, the lubricant is partially coating the articulating surfaces. Boundary lubrication occurs where there is more friction in the gap like when the thickness of the lubricant and the roughness of the cartilage are of the same order and where there is a need for load-bearing the fluid film lubrication comes to play. Both the contact surface and the lubricant equally contribute to the friction and wear properties.

Elastohydrodynamic lubrication increases the load-bearing ability of the joint as it has the lubricant uniformly coated along the entire surface of articulation. Here, the friction of the fluid increases with loading pressure. Hence, it resists the contact between the surfaces due to the rise in the viscosity of the fluid which contributes to the friction coefficient of the fluid.

Hydrodynamic lubrication has a thick layer of lubricant lining the surface, and there is no wear happens in this type of lubrication.

The rheological properties of the fluid are playing a key role in determining the friction and wear properties of the joints.

12.11 Conclusion

The human body is made up of systems which continuously involve in activities such as load-bearing, support, and weight balancing, and when movement occurs, there is friction, wear, and corrosion, and tissue loss can happen to lead to the deterioration of the functional capacity of the respective tissue. An attempt to understand biotribology of the body load-bearing characteristics and the mechanical properties of the related structures is needed. Here, the chapter addressed the structure, components, and function of the tissues such as bone, cartilages, and joints involved in load-bearing activities which will give an understanding of the requirements and design consideration of an implant. Tissue response to implant is also important for achieving biocompatibility, improving implant acceptance, etc. The nature of the implantable material and their response to the physio-mechanical loading conditions leads to an altered response of the tissue to the implant, and sometimes, it is confined to the site of application.

References

1. Hall, J.E., *Guyton and Hall, "Text book of Medical Physiology"*, 14th Edition, Elsevier, Pennsylvania, 2020.
2. Waugh, A. and Grant, A., *Ross and Wilson "Anatomy and Physiology in health and illness*, 12th Edition, Churchill Livingstone Elsevier, UK, 2014.

3. Lakshmipathy, D., Human Anatomy and Physiology, in: *Biomedical Engineering and its Applications in Healthcare*, S. Paul (Ed.), Springer, Singapore, 2019, https://doi.org/10.1007/978-981-13-3705-5_1.
4. Knudson, D., *Fundamentals of Biomechanics*, Springer US, New York, 2003.
5. Ghista, D.N. and Roaf, *Orthopedic Mechanics*, Academic Press, Singapore, 2000.
6. Park, J.B. and Lakes, R.S., *Biomaterials - An Introduction*, Second Edition, Plenum Press, New York and London, 1992.
7. Temenoff, J.S. and Mikos, A.G., *Biomaterials - The Intersection of Biology and Materials Science*, Pearson Education, New Jersey, 2009.
8. Williams, D.F., *Medical and Dental Materials: A comprehensive Treatment*, VCH Publishers, Liverpool, UK, 1992.
9. Ratner, B.D., Hoffmann, A.S., Schoen, F.J., Lemmons, J.E., *An introduction to Materials in Medicine*, Academic Press, California, USA, 1996.
10. Uo, M., Watari, F., Yokoyama, A., Matsuno, H., Kawasaki, T., Tissue reaction around metal implants observed by X-ray scanning analytical microscopy. *Biomaterials*, 22, 7, 677–85, 2001.
11. Sukur, E., Akman, Y.E., Ozturkmen, Y., Kucukdurmaz, F., Particle Disease: A Current Review of the Biological Mechanisms in Periprosthetic Osteolysis After Hip Arthroplasty. *Open Orthop. J.*, 10, 241–51, 2016.
12. Wisbey, A., Gregson, P.J., Peter, L.M., Tuke, M., Effect of surface treatment on the dissolution of titanium-based implant materials. *Biomaterials*, 12, 470–3, 1991.
13. Bianco, P.D., Ducheyne, P., Cuckler, J.M., Local accumulation of titanium released from a titanium implant in the absence of wear. *J. Biomed. Mater. Res.*, 31, 227–34, 1996.
14. Nyman, J.S., Leng, H., Dong, X.N., Wang, X., Differences in the mechanical behavior of cortical bone between compression and tension when subjected to progressive loading. *J. Mech. Behav. Biomed. Mater.*, 2, 6, 613–9, 2009.
15. Karrholm, J., Garellick, G., Rogmark, C., Herberts, P., Swedish Hip Arthroplasty Register, Annual Report 2007, Gothenburg, Sweden, 2008, http://www.jru.orthop.gu.se/archive /AnnualReport-2007-eng.pdf
16. Greidanus, N.V., Peterson, R.C., Masri, B.A., Garbuz, D.S., Quality of life outcomes in revision versus primary total knee arthroplasty. *J. Arthroplasty*, 26, 4, 615–20, 2011. http://dx.doi.org/10.1016/j.arth.2010.04.026
17. Pedersen, A.B., Mehnert, F., Odgaard, A., Schrøder, H.M., Existing data sources for clinical epidemiology: The Danish Knee Arthroplasty Register. *Clin. Epidemiol.*, 4, 125–35, 2012, http://dx.doi.org/10.2147/CLEP.S30050
18. Paxton, E.W., Furnes, O., Namba, R.S., Inacio, M.C., Fenstad, A.M., Havelin, L.I., Comparison of the Norwegian knee arthroplasty register and a United States arthroplasty registry. *J. Bone Joint Surg. Am.*, 93, Suppl. 3, 20–30, 2011. http://dx.doi.org/10.2106/JBJS.K.01045.
19. Bitar, D. and Parvizi, J., Biological response to prosthetic debris. *World J. Orthop.*, 6, 2, 172–89, 2015. http://dx.doi.org/10.5312/wjo.v6.i2.172

20. Wataha, J.C., Materials for endosseous dental implants. *J. Oral. Rehabil.*, 23, 2, 79–90, 1996. http://dx.doi.org/10.1111/j.1365-2842.1996.tb01214.x
21. Davies, A.P., Willert, H.G., Campbell, P.A., Learmonth, I.D., Case, C.P., An unusual lymphocytic perivascular infiltration in tissues around contemporary metal-on-metal joint replacements. *J. Bone Joint Surg. Am.*, 87, 1, 18–27, 2005. http://dx.doi.org/10.2106/JBJS.C.00949
22. Jauch, S.Y., Huber, G., Sellenschloh, K. *et al.*, Micromotions at the taper interface between stem and neck adapter of a bimodular hip prosthesis during activities of daily living. *J. Orthop. Res.*, 31, 8, 1165–71, 2013. http://dx.doi.org/10.1002/jor.22354
23. Parkinson, I.H. and Fazzalari, N.L., Characterisation of Trabecular Bone Structure, in: *Skeletal Aging and Osteoporosis. Studies in Mechanobiology, Tissue Engineering and Biomaterials*, M. Silva (Ed.), vol. 5, Springer, Berlin, Heidelberg, 2013. https://doi.org/10.1007/8415_2011_113.
24. Greene, G.W., Banquy, X., Lee, D.W., Lowrey, D.D., Yu, J., Israelachvili, J.N., Adaptive mechanically controlled lubrication mechanism found in articular joints. *Proc. Natl. Acad. Sci. U. S. A.*, 108, 13, 5255–9, 2011.
25. Cutcliffe, H.C. and DeFrate, L.E., Comparison of Cartilage Mechanical Properties Measured During Creep and Recovery. *Sci. Rep.*, 10, 1547, 2020, https://doi.org/10.1038/s41598-020-58220-2.
26. Davim, P.J., Biotribology, ISTE-Wiley, London, 2010.

13

Mechanical Behavior of Single and Multilayer Hardfacing on Ferrous Alloys

B. Venkatesh

Department of Mechanical Engineering, Vardhaman College of Engineering, Hyderabad, India

Abstract

Hardfacing is the most cost-effective technique, and it entails depositing hard, wear-resistant coatings on a worn or new component that is subjected to wear, impact, high temperatures, and oxidation. It protects the base metal against wear, corrosion, and oxidation by acting as a sacrificial layer at the point of impact. Most of the time, it is used to rebuild machinery, but even new equipment can be hardfaced with the purpose of making it more wear-resistant. Hardfacing parts from equipment with many years can mean saving between 25% and 75% of the cost of replacement and extend lifespan of parts. A thick layer of stellites and various types of multilayer electrodes deposited on the base metals (ferrous materials) varying the welding currents by Tungsten Inert Gas welding process. Erosive wear tester and Vickers hardness testing machine are used to measure the rate of erosion and hardness of the hardfaced components and the variation of the erosion and hardness with varying welding current for the substrates. The objective of the work is to determine the erosive wear characteristics, hardness, topology, and microstructural properties of hardfaced ferrous alloys.

Keywords: Hardfacing, multilayer, SEM, XRD, erosion rate, Vickers hardness number, TIG welding process, MS (mild steel)

13.1 Introduction

Initially, fusion surfacing technique [1] is used to improve the component surface properties. Generally, hardfaced layers exhibit better wear, corrosion, and oxidation resistance than the substrate, which is expensive. Due to a strong metallurgical bond with the substrate material, the hardfacing

Email: bvt2275@gmail.com

Sarfraj Ahmed and Vinayak S. Dakre (eds.) Tribology and Characterization of Surface Coatings, (281–306) © 2022 Scrivener Publishing LLC

layers are very resistant to spalling, and different of alloys are used to obtain the greatest performance in a hostile environment.

Hardfacing is economical compared with alternate processing methods like thermal spraying. Cobalt-based hardfacing alloys are of the "stellite" variety [3]. Co-28 Cr-4.3W-1.1C is the nominal composition (wt percent). These alloys are resistant to corrosion, erosion, and sliding wear, and they maintain these capabilities at high temperatures [2]. They also have anti-galling properties. Stellite alloys have exceptional wear resistance due to the high carbide phase distributed in the cobalt alloy matrix, which imparts unique wear properties. These stellites are also utilized on engineering components (valves and spindles) and other worn parts as a weld method.

In test process, impact of particle stream is used to eliminate the material from work piece. This is an uncommon wear mode that affects the longevity of parts exposed to harsh erosive conditions in a variety of industries, particularly aerospace, marine, and mining. The application of a hardfaced stellite layer to the surface of components is an appealing idea for improving performance in an eroding environment.

Liu created a new high-W stellite alloy with a W concentration of up to 32 parts per million [11]. This high W content facilitated the production of W-rich carbide phases in alloy [23], such as (W, Co) 6C and Co4W2C, which are available in huge quantities, have enormous diameters, and are harder than Cr-rich carbides in stellite alloys [4]. This new alloy has a considerable huge amount of W-rich carbides that provide good sliding wear resistance and solid particle erosion. The erosive and corrosive properties for stellite 6 [16] are consistent with an erosive model, where brittle materials are severely destroyed under perpendicular assault while ductile materials are harmed in shallow angles. The massive loss occurred at 60° impingement, indicating that stellite 6 is neither perfectly brittle nor ductile but exhibits a more brittle response when tested [9]. The goal of this research is to see how different processing circumstances affect the pace and processes of erosion on the hardfaced layer of stellite 3, taking into account hardfacing conditions on the erosive stellite layer.

13.2 Experimental Details

13.2.1 Hardfacing Procedure

Hardfacing is done on base metal with a TIG welding setup that operates at 440 V and a current of 70, 80, and 90 amperes as shown in Figures 13.1a–d.

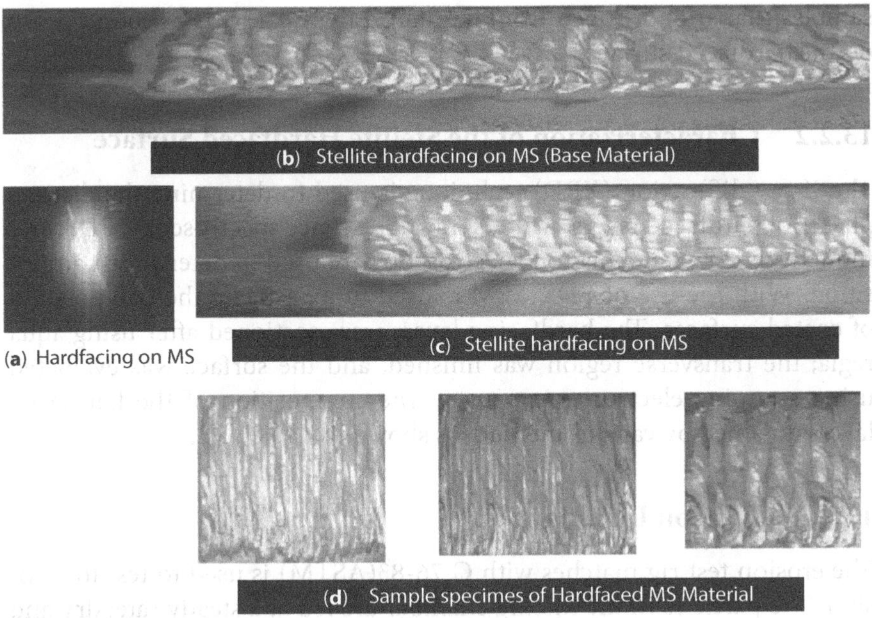

Figure 13.1 (a) Hardfacing on MS. (b) Stellite hardfacing on MS (base material). (c) Stellite hardfacing on MS. (d) Sample specimen of hardfaced MS material.

Samples are ultrasonically cleaned after being slitted to remove dust and corrosion. After the sample was mounted on an appropriate fixture and the argon pressure was set to 10 kg/cm^2, by using a stop watch, hardfacing is done at three various scan speeds. The plate was immediately chilled by soaking it in oil after hardfacing. Table 13.1 shows the current state of hardfacing in detail. The substrate is made of Steel and has dimensions of 100 × 35 × 7 mm. The hardfacing electrode was satellite 3 of 30 × 30 mm

Table 13.1 Conditions of hardfacing.

Parameter	Value
Voltage (V)	440
Current (A)	70, 80, and 90
Speed of scan (mm/s)	2.25, 2.75, and 3.00
Argon pressure (kg/cm^2)	10

samples that are portioned from hardfaced material with a ground finish on the hardfaced surface.

13.2.2 Characterization of the Stellite Hardfaced Surface

The X-ray diffraction (XRD) technique is used to determine the distinct phases of hardfaced layers, while X-ray florescence spectroscopy, EDS, and inductively coupled emission spectroscopy are used to determine composition. With the use of EDS, SEM is used to investigate the composition of coated surfaces. The hardfacing layers were sectioned after using aqua regia; the transverse region was finished, and the surface was evaluated using scanning electron microscopy. The composition of the hardfaced layers obtained by various methods is shown in Table 13.2.

13.2.3 Erosion Investigation

The erosion test rig matches with G 76-83(ASTM) is used to test for erosion. The particles in the mixing chamber are fed at a steady rate; dry and compressed air is allowed to mix with them. By passing the fluidized particles via a converging nozzle, they were accelerated. These accelerated particles will impact the specimen held at a certain angle relative to particles impacting. The compressed air pressure is adjusted to control particle velocity. By altering the sample orientation relative to the hitting particles, a particle hopper and conveyor belt distance that may be adjusted to control the feed rate. Average particle size 150 μm (Knoop hardness, 832.8–880 HV) is utilized in this study, and the test instances are listed in Table 13.3. Using spinning disc approach, the erodent velocity is calculated from particle velocity, calibration curves actually devised by Ruff and Ives [13]. Every erosion test followed the normal test technique. Samples are cleaned ultrasonically and weighed; sample will be kept in test rig for 5 minutes, and later, the weight loss was calculated [24].

Table 13.2 Hardfaced layers composition.

Composition	Co	Cr	W	Fe	Si
XRF (wt %)	37.4	20.5	2.9	36.8	-
EDS (wt %)	35.6	20.07	-	41.08	3.24
ICPOES (wt %)	36.9 ± 0.9	18.79 ± 0.5	3.21 ± 0.8	39.0 ± 1.0	0.72 ± 0.02

Table 13.3 Erosion test conditions.

Parameter	Value
Velocity (m/s)	57, 93
Erodent feed rate	5 g/min
Angle of impingement	30° and 90°
Temperature	Room temperature
Sample size (mm)	30 × 30 × 5
Exposure time	30 minutes

13.2.4 Eroded Sample Characterization

The eroded surface morphology was investigated with assistance of scanning electron microscope (SEM). For this SEM examination, LEO 440 model is employed. After specimen cutting and separation, sectioned surfaces are polished; SEM is used to analyze the areas beneath the degraded surfaces to determine the type of deformation present [24].

13.3 Outcome

13.3.1 Hardfaced Layer Behavior

As shown in Figure 13.2, the image of hardfaced layer pattern is obtained from XRD [7]. The results show the presence of FCC base phase along with MC and $M_{23}C_6$ carbides. Stellite 3 is used as welding rod for hardfacing. The basic compositions are measured by using energy dispersive spectrometer and by optical emission spectroscopy, which are given. In this two methods, composition is valued from hardfacing layer at weld current of 80 amperes and scan speed of 275 mm/s and assumed that obtained composition was not differed. Due to hardfacing, the initial weld material does not have iron; in other words, substantial dilution rises with applied current and downs with speed of scan.

Low-magnification SEM pictures of stellite 3 are represented at various hardfacing conditions, and layer thickness will increase with respective to increase in applied current for process and given speed [7]. Similarly, hardfaced layer thickness reduces vice versa as shown in Figure 13.3.

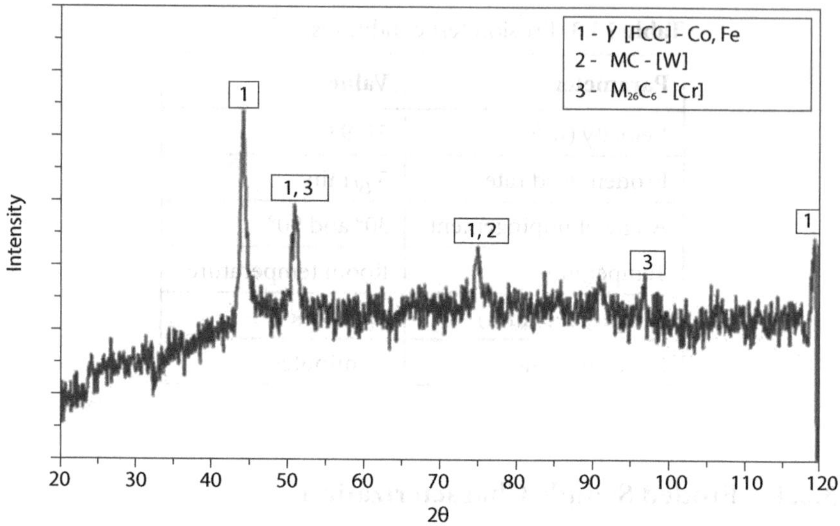

Figure 13.2 The hardfaced surfaces produced an XRD pattern.

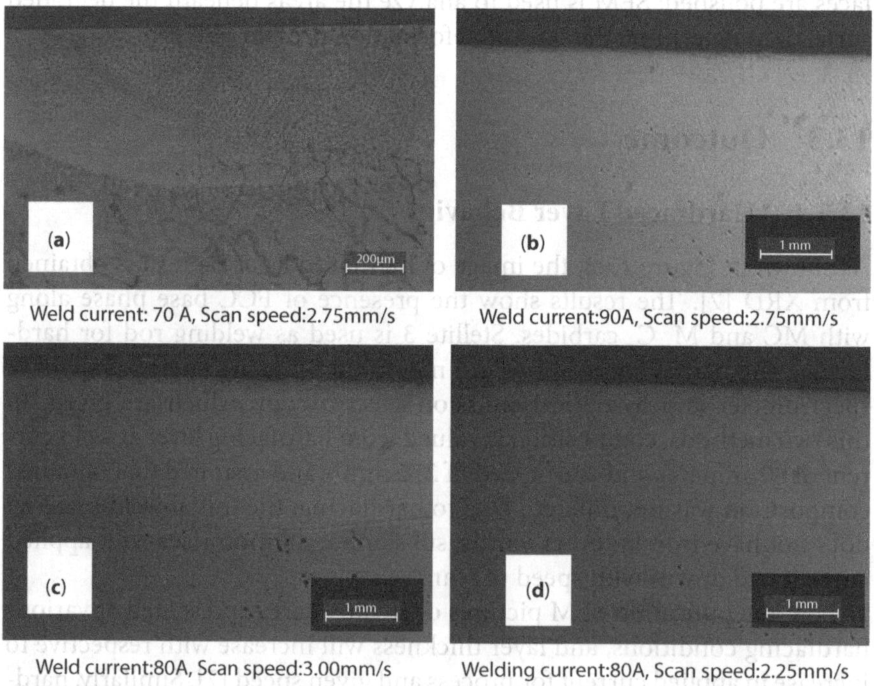

Figure 13.3 Optical image of less magnification of hardfaced layers with various weld currents.

The dendritic microstructure is clearly observed; if hardfacing current increases, then the dendrites size becomes bigger. When the scan speed of the hardfaced layers is increased for an applied current, the hardfaced layer size of dendrites shrinks. With welding current of 80 amperes and a scanning speed of 2.75 mm/s, EDS spectra of dendrites and inter-dendritic area of hardfacing layer. Table 13.4 shows the relative composition. It is obvious from this table that the inter-dendritic zones are enriched in Co and impoverished in Fe. In Table 13.5, the hardness of hardfacing is determined by Vickers hardness tester with a 1-kg load, and the corresponding microstructure for different weld currents is shown in Figure 13.4.

13.3.2 Rate of Erosion

The reciprocal of steady-state rate of erosion is used to calculate resistance of erosion [7]. As shown in bar diagram in Figure 13.5, under various test conditions, erosion resistance [17] is discussed for hardfacing layers, which are displayed. At welding current of 80 amperes, hardfaced alloys

Table 13.4 Hardfaced layer composition of dendritic and inter-dendritic region.

Composition / Regions	**Co**	**Cr**	**W**	**Fe**
Dendritic	48.14	22.3	14.3	15.1
Inter-dendritic	51.8	23.9	12.0	12.3

Table 13.5 Hardfaced layer hardness.

S. no.	**Welding current (amp)**	**Scan speed (mm/s)**	**Hardness (VHN)**	**Hardness (MPa)**
01	70	2.75	650	6922.5
02	80	2.75	440	4686
03	80	2.25	405	4314.43
04	80	3.00	368	3927.48
05	90	2.75	424.4444	4520.33

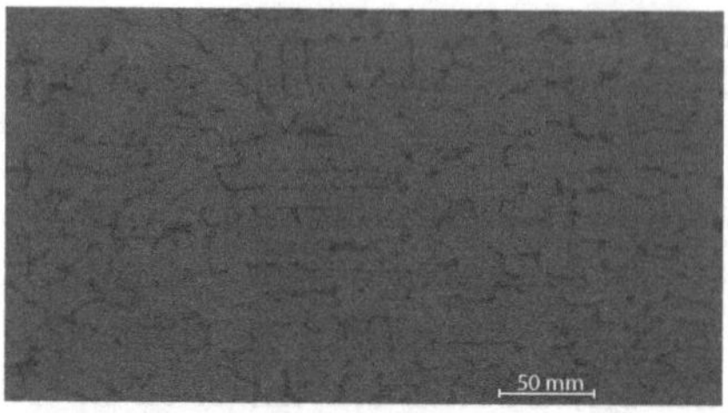

Microstructure-Processing current 70 A

Microstructure-Processing current 80 A

Microstructure-Processing current 90 A

Processing current (A)	Hardness (VHN)
90	390 ± 70
80	450 ± 50
70	650 ± 40

Figure 13.4 Hardness values of hardfaced mild steel specimens.

exhibit better erosion resistance when subjected to more impact velocity (V). Changes in speed of scan have little effect on erosion resistance [10].

At low velocity of impact, the effect of involving condition is insignificant. The responses of erosion depend on rate of erosion on angle of impact and of the coated surfaces, which are shown in Table 13.6. The velocity exponent n is given by

$$E_r = KV^n \tag{13.1}$$

where erosion rate is Er, the impact velocity is V, the velocity exponent is n, and the constant is K. Table 13.7 contains these velocity exponents.

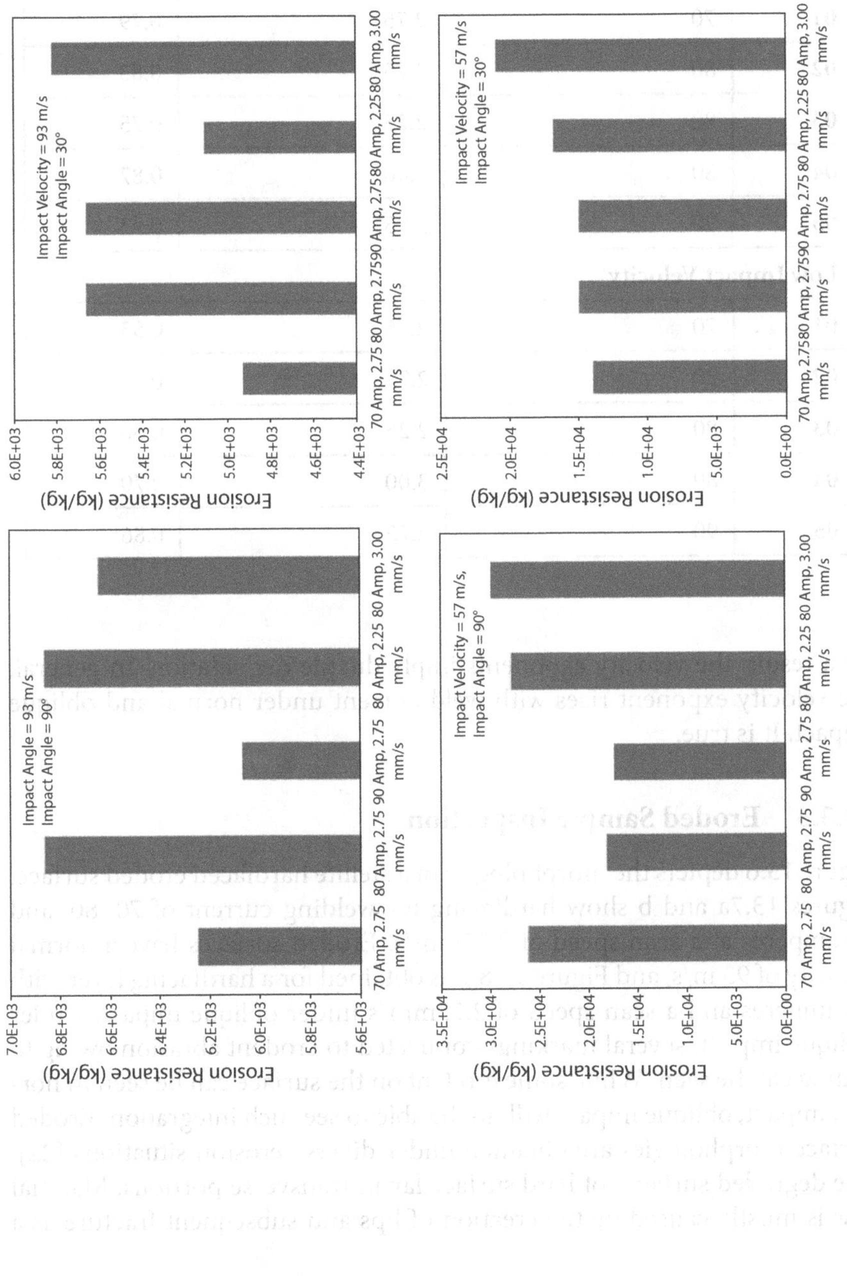

Figure 13.5 The erosion resistance of various layers is depicted as a bar graphic.

Table 13.6 The rate of erosion is affected by the impact angle.

S. no.	Welding Current I (A)	Scan Speed (mm/s)	$E_{90°}/E_{30°}$
High Impact Velocity			
01	70	2.75	0.79
02	80	2.75	0.83
03	80	2.25	0.75
04	80	3.00	0.87
05	90	2.75	0.93
Low Impact Velocity			
01	70	2.75	0.53
02	80	2.75	0.84
03	80	2.25	0.60
04	80	3.00	0.70
05	90	2.75	0.86

As a result, the velocity exponents imply ductile degradation. In general, the velocity exponent rises with weld current under normal and oblique impact. It is true.

13.3.3 Eroded Sample Inspection

Figure 13.6 depicts the morphologies of a stellite hardfaced eroded surface. Figures 13.7a and b show hardfacing for welding current of 70, 80, and 90 amperes at a scan speed of 2.75mm/s. Eroded surfaces have a normal velocity of 93 m/s, and Figure 13.8 was obtained for a hardfacing layer with 80 amperes and a scan speed of 2.75mm/s under oblique impact. Under oblique impact, several markings connected to erodent abrasion owing to sliding can be seen. While some erodent on the surface can be seen in normal impact, oblique impact will not be able to see such integration. Eroded surface morphologies are obtained under diverse erosion situations [22]. The degraded surfaces of hard surface lay in transverse portions. Material loss is mostly caused by the creation of lips and subsequent fracture as a

Table 13.7 How erosion rates are affected by velocity.

S. no.	Welding current I (A)	Scan speed (mm/s)	n	K
Normal Impact				
01	70	2.75	2.92	2.79E−10
02	80	2.75	2.18	7.51E−09
03	80	2.25	2.86	3.33E−10
04	80	3.00	3.00	1.34E−10
05	90	2.75	2.15	9.34E−09
Oblique Impact				
01	70	2.75	2.12	1.33E−08
02	80	2.75	2.21	7.65E−09
03	80	2.25	2.42	3.27E−09
04	80	3.00	2.61	1.22E−09
05	90	2.75	2.00	2.06E−08

result of deteriorated surface morphology. However, in stellite layers, fracturing appears to be along the inter-dendritic zone. The process of factorization always begins in the inter-dendritic region.

13.4 Discussion

13.4.1 Literature Observations

In this chapter, an attempt is made to know the erosion test outcome found in the literature review and to compare with those found in the current inquiry. Because the test conditions, materials, and nature of the test are all different, a qualitative comparison as mentioned below can be performed. Gathering of erosion study on stellite coatings and bulk materials is given in Table 13.9. As per Liu *et al.* [11], the improved erosion resistance of stellite alloy is attributable to formation of huge amounts of large-size W-rich carbides in alloy. As per Romo *et al.* [14], more rate of erosion is achieved at oblique impact, and this is comfortable. Work of Andrew *et al.*

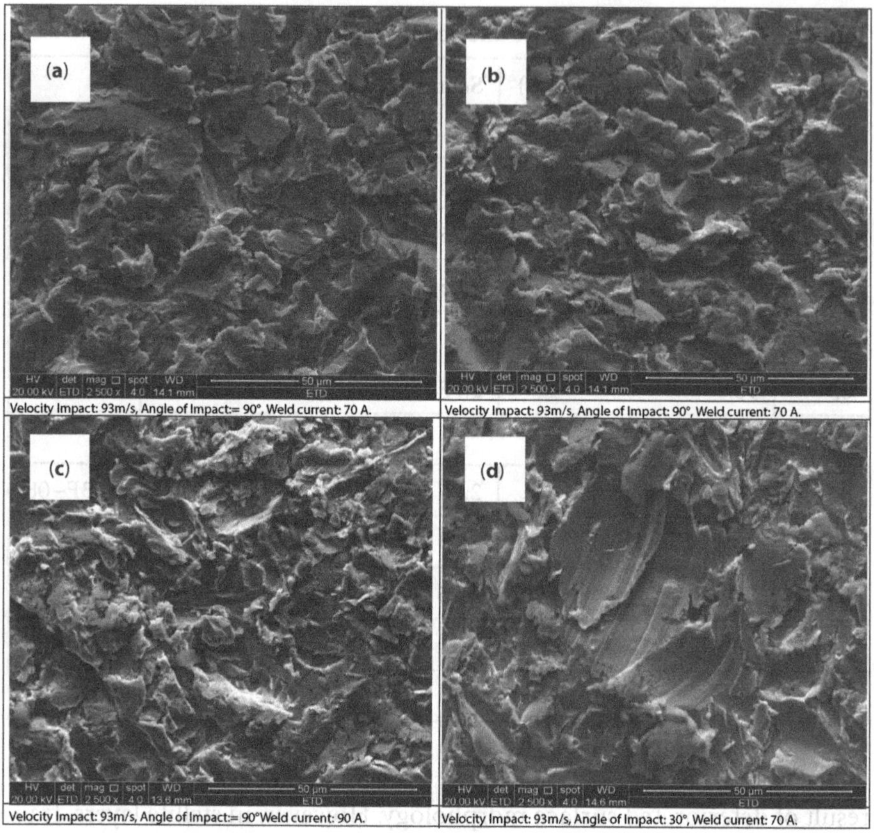

Figure 13.6 Eroded surface morphology.

[12] specified superior erosion resistance of stellite compared to AISI 316 SS, although they found maximum erosion rate under slurry impingement at oblique impact similar to observed. Sidhu and S. Prakash [15] found that corrosion resistance of boiler steel is improved by plasma sprayed stellite 6. As mentioned by Singh *et al.* [17], laser cladding of stellite exhibits enhanced erosion resistance.

13.4.2 Common Erosion Behavior

Table 13.5 shows the erosion rate, which is independent from hardness, and those observations [18, 19] are given by various authors. The material removal mechanism gives small variation from suitable ductile erosion response, and in this, material is removed by lips formation and its subsequent fracture, and in this, fracture happens in inter-dendritic

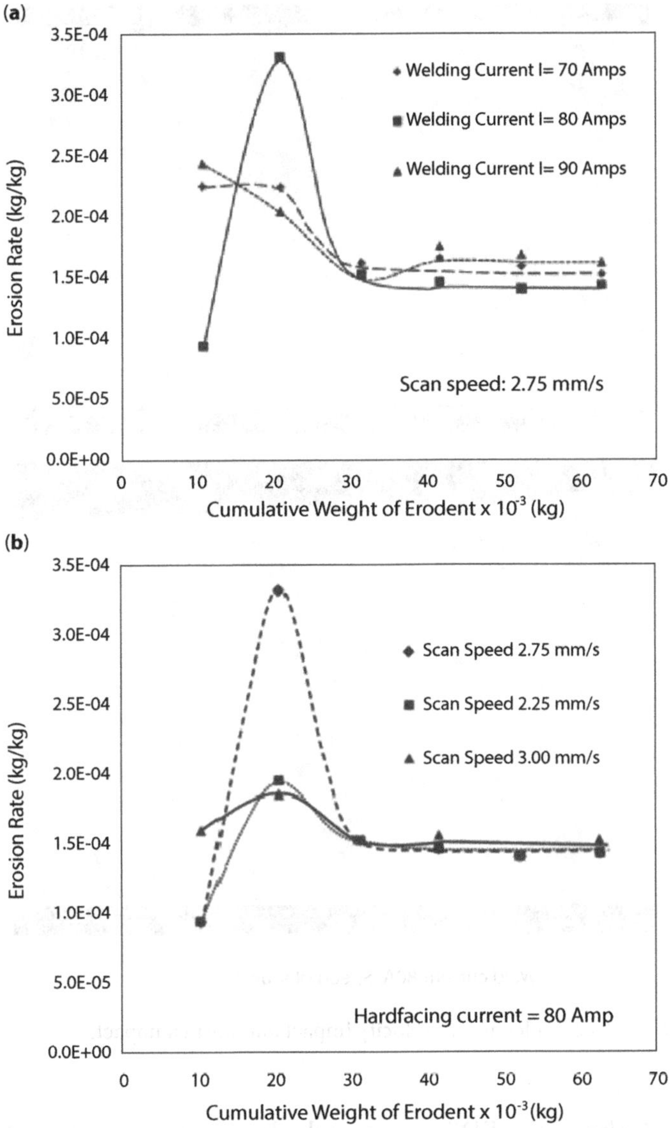

Figure 13.7 (a) Rate of erosion vs. cumulative erodent weight (welding current). (b) Erosion rate vs. cumulative erodent weight (scan speed).

region, which is different by necking observed in other metallics high-magnification image that represents carbide presence in specified regions. The white portions are known as MC carbide, whereas gray portion are $M_{23}C_6$ carbides [20] with dilution; more interaction with the substrate occurs and more amount of chromium is available to form this type of carbide.

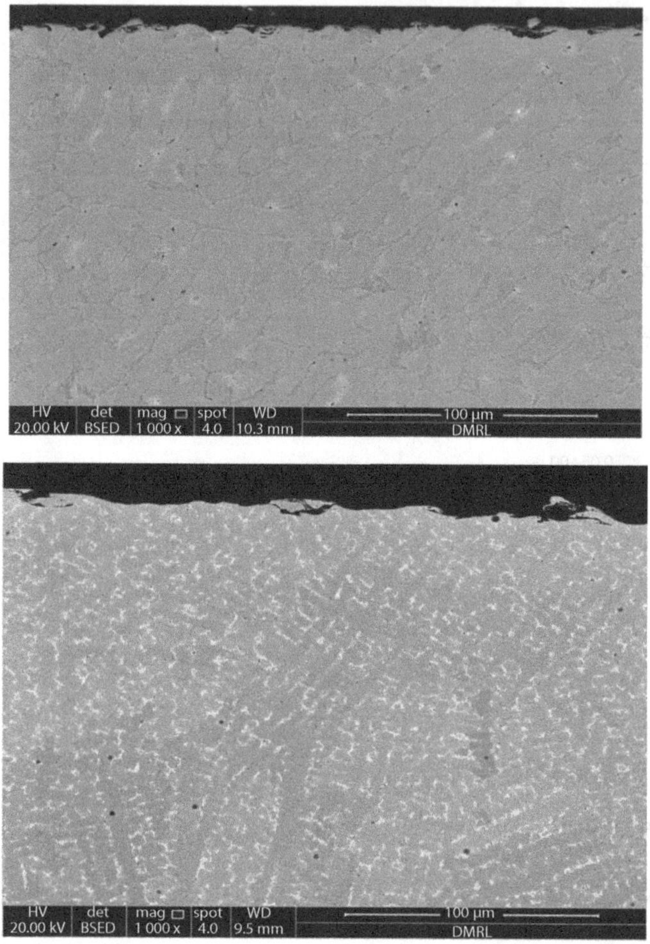

Figure 13.8 Surface eroded under velocity impact and normal impact.

Figure 13.9 shows the EDS pattern, which also confirms more Cr in the inter-dendritic area with respective to dendritic area. Failures may arise because fracture particles or by decohesion of carbide. Lip fracturing is caused by these brittle carbides. But in some alloys, because of $M_{23}C_6$ structure, premature failures may occur but proper heat treatment can avoid that, and it also mentioned that material removal mechanism would not change with changes in hardfacing condition.

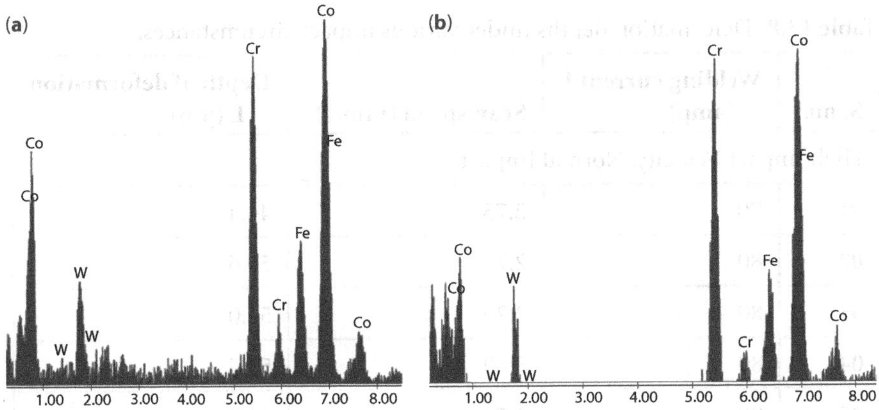

Figure 13.9 EDS spectra from the (a) dendritic and (b) inter-dendritic regions.

13.4.3 Deformation Depth

In this topic, an investigation is made to see if the substrate has any bearing on the erosion rate. The impact deformation depth is estimated in this view. Particle with mass (m) and impact velocity (V) and eroded material hardness (H): particle of mass (m) with an impact velocity (V) impacts the eroded material hardness (H), and volume (U) is obtained [21].

$$HU = \frac{1}{2}(mV^2) = \frac{4}{3}\pi r^3 \rho_p V^2 \tag{13.2}$$

The U = 30°

$$U = \left(\frac{\pi}{4}\right)\left(\frac{W^3}{\tan\theta}\right) = \left(\frac{\pi W^3}{12}\right) \tag{13.3}$$

W is the diameter of the crater. The crater diameter is derived by substituting the following formula for U:

$$W = 2\rho_p^{\frac{1}{3}} r V^{\frac{2}{3}} / H^{\frac{1}{3}} \tag{13.4}$$

Carter diameter is proportional to depth of deformation (L) [24]; L is written as

Table 13.8 Deformation depths under various impact circumstances.

S. no.	Welding current I (amp)	Scan speed (mm/s)	Depth of deformation L (μm)
High Impact Velocity, Normal Impact			
01	70	2.75	44.4
02	80	2.75	50.6
03	80	2.25	52.0
04	80	3.00	53.7
05	90	2.75	51.2
Low Impact Velocity, Normal Impact			
01	70	2.75	32.1
02	80	2.75	36.5
03	80	2.25	37.5
04	80	3.00	38.7
05	90	2.75	37.0
High Impact Velocity, Oblique Impact			
01	70	2.75	28.0
02	80	2.75	31.9
03	80	2.25	32.3
04	80	3.00	32.8
05	90	2.75	33.8
Low Impact Velocity, Oblique Impact			
01	70	2.75	20.2
02	80	2.75	23.0
03	80	2.25	23.3
04	80	3.00	23.6
05	90	2.75	24.4

$$L = \beta W = 2\beta \rho_p^{\frac{1}{3}} r V^{\frac{2}{3}} / H^{\frac{1}{3}} \tag{13.5}$$

where r and p are constants.

Using Equation (13.5) and assuming $\beta = 1.5$, $\rho_p = 2600$ kg/m^3, and r = 10^{-4} m, for various coatings [5] and test settings, the L has been estimated. The deformation depth of coatings is listed in Table 13.8, and based on a review, the deformation depth is estimated to be between 20 and 53 μm. Deformation depth is contained within the hardfaced layer due to its thickness of more than 1 mm, and erosion is reported as intrinsic erosion rate.

13.4.4 Efficiency of Erosion

Sundararajan *et al.* suggested a new measure, efficiency of erosion, to describe the type and erosion mechanism [24]. This statistic indicates the efficiency with which the volume displaced by the impacting erodent particle is really removed. This efficiency parameter [25] can be obtained from

$$\eta = \frac{2EH}{\rho V^2} \tag{13.6}$$

If erosion occurs by lip formation and fracture, then erosion efficiency has range of 0 to 1. If spalling and removal of large chunks of material creates erosion by interlinking cracks, then, in case of brittle material, erosion efficiency is even more than 100%.

The erosion efficiency values of the coatings developed in this study were computed using Equation (13.6) and reported in Table 13.9. Efficiency of erosion is much low in few cases for being consistent with ductile erosion response. During microcutting, it is observed that erosion efficiency is high. This is owing to the fact that lip fractures occur along the inter-dendritic region rather than as a result of necking. When compared to normal impact, oblique impact erosion efficiency is higher, and it decreases as weld current increases. At the greatest scan speed, it is low, and erosion efficiency is low at lower impact velocity, indicating a more ductile response. In this aspect, erosion efficiency agrees with the fact that erosion rate angular dependency is considerably more ductile at lower velocity of impact.

This is consistent with the fact that fracturing lips at low impact velocity is more challenging. This circumstance could be caused by a lower strain rate for deformation at a lower impact velocity.

Table 13.9 Tested layers' erosion efficiencies.

S. no.	Welding current I (Amp)	Scan speed (mm/s)	Erosion efficiency (%)
Normal Impact, High Velocity			
01	70	2.75	2.96
02	80	2.75	1.83
03	80	2.25	1.68
04	80	3.00	1.58
05	90	2.75	0.47
Normal Impact, Low Velocity			
01	70	2.75	1.88
02	80	2.75	1.84
03	80	2.25	1.10
04	80	3.00	0.94
05	90	2.75	0.45
Oblique Impact, High Velocity			
01	70	2.75	15.04
02	80	2.75	8.86
03	80	2.25	9.05
04	80	3.00	7.23
05	90	2.75	2.03
Oblique Impact, Low Velocity			
01	70	2.75	14.14
02	80	2.75	8.93
03	80	2.25	7.34
04	80	3.00	5.35
05	90	2.75	2.05

13.4.4.1 *Abrasive Wear*

Coating wear rates [12] due to abrasion are measured using a rubber wheel (DSRW) abrasion test setup in accordance with ASTM standards at varied applied loads (G 65); as abrasives, dry sand is employed [7]. The abrasives are in irregular shape, with an estimated size of 200 ± 50 μm. The specimens are placed in a sample holder and firmly pressed against the rubber wheel rim using a lever arrangement after a known normal force is applied. The abrasion test follows a well-defined procedure. The sample was cleaned and weighed precisely before being abraded in the rig for 200 revolutions, equating to about 125 g of abrasive particles. After that, the samples are cleaned and weighed again to determine how much weight has been lost. The incremental wear rate of the coatings owing to abrasive is calculated as the ratios of weight loss incurred by the specimens to the mass of abrasives are responsible for loss of mass. The same test was repeated until constant rate of wear due to abrasion is attained.

13.4.4.2 *Sliding Wear*

The sliding wear test was carried out using a ball-on-flat disc setup. The ceramic ball (SiN), diameter of 6 mm was slide against the test material in the form of a 40-mm-diameter disc under dry atmospheric condition [6]. The wear rate of the disc was estimated at the end of the test using a non-contact optical profilometer. Ultrasonically cleaned and weighed disc was subjected to sliding for a predetermined number of cycles. The temperature during the wear testing was between 30°C and 36°C. Humidity was noted to vary between 55% and 70%.

13.4.4.3 *Erosive Wear*

The erosion test was carried out in a facility with compressor and drying machine, and a test rig that met G 76-83(ASTM) and erosive test conditions are shown in Figure 13.10. The particles are allowed to mix with the dried and compressed air. When these fluidized particles are accelerated, they travel through a converging nozzle [24]. The accelerated particles contact the specimen at a specific angle in relation to the impacting particles. Particle velocity is regulated by adjusting compressed air pressure. The angle between the striking particles and the sample is changed by the impact angle. The feeder is conveyor system with a decreased particle feeder, and rate of feed is controlled with regulating the distance in between conveyor belt and hopper. The erodent in this study was angular silicon

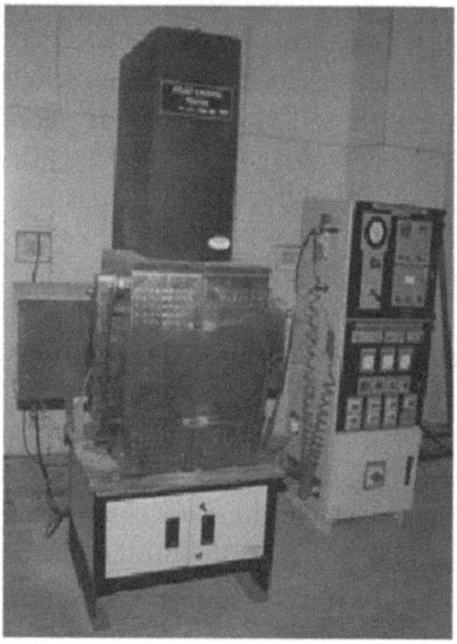

Erosion test conditions

Parameters	Values
Test Temperature	Room Temp
Impact Angle (°)	30°, 90°
Impact Velocity (m/s)	61.3, 88.2
Erodent	SiO_2
Erodent size (μm)	150 50

Figure 13.10 Erosion tester.

Topography of Substrates

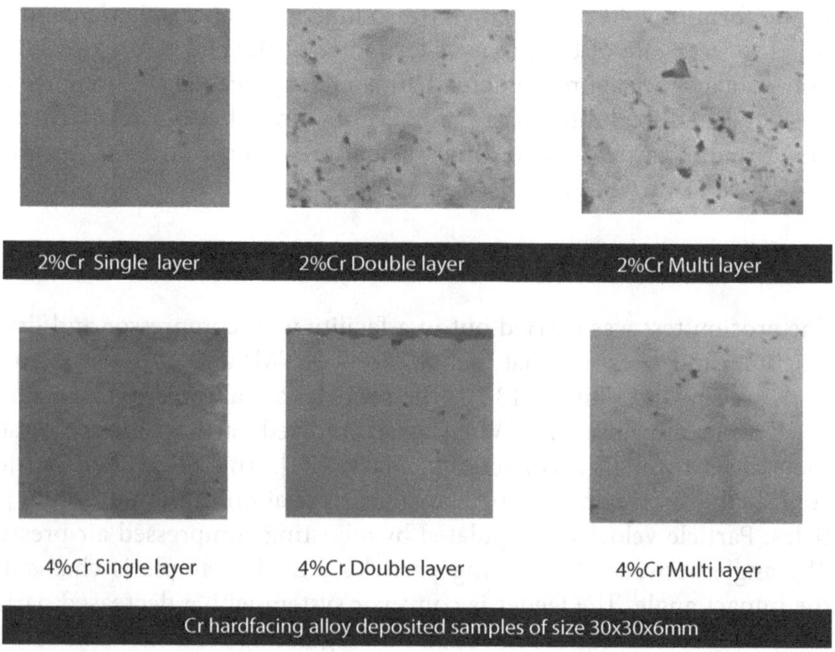

Figure 13.11 Topography of substrates.

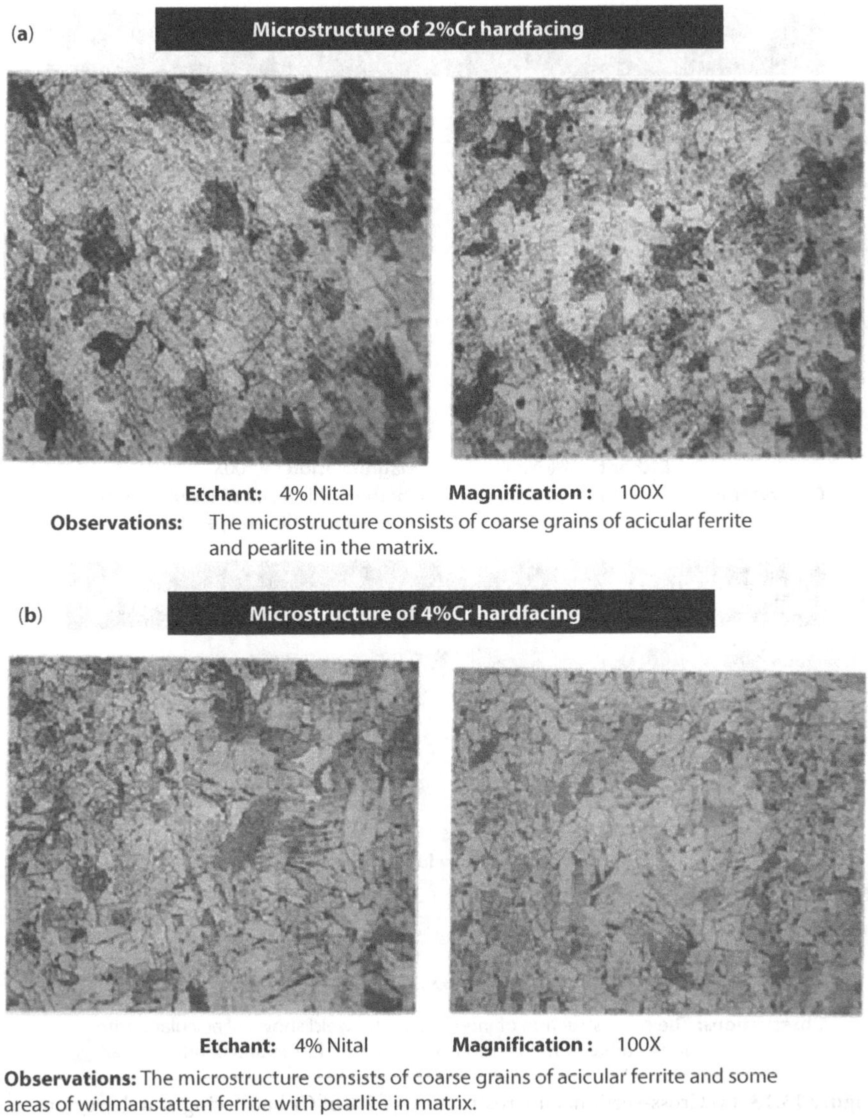

Figure 13.12 (a) Microstructure of 2% of Cr hardfacing on MS specimen. (b) Microstructure of 4% of Cr hardfacing on MS specimen.

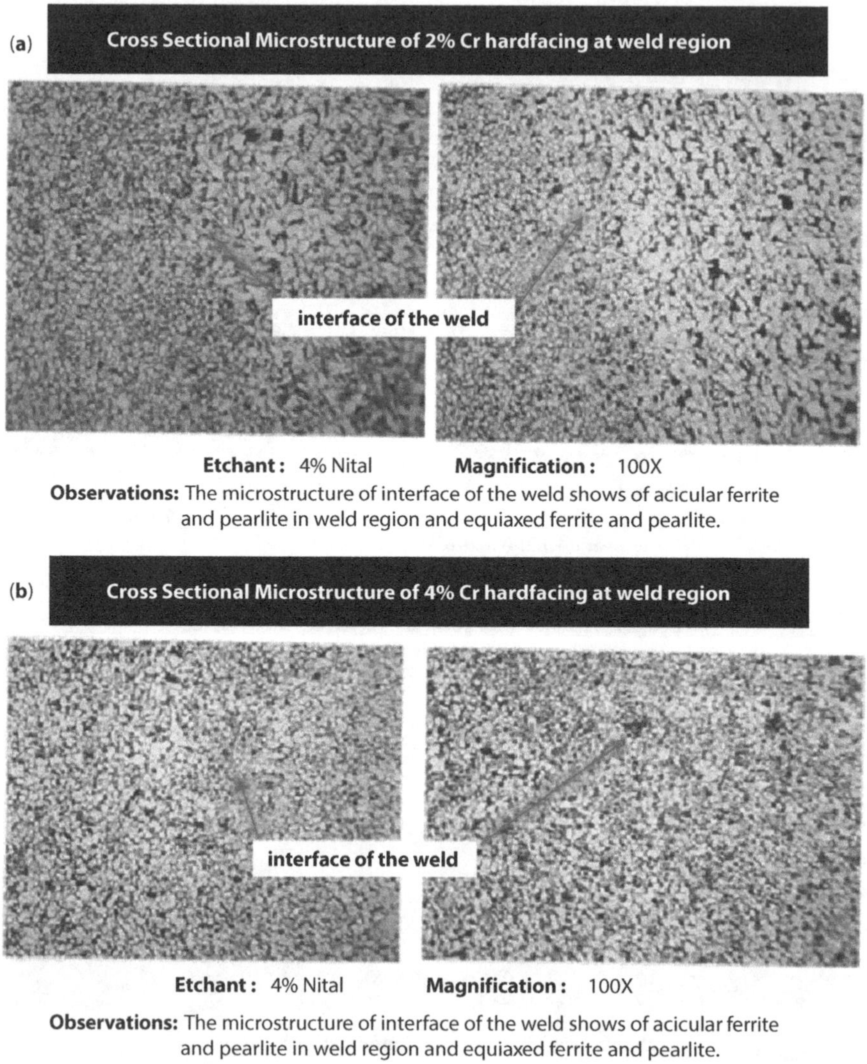

Figure 13.13 (a) Cross-sectional microstructure of 2% of Cr hardfacing at weld region. (b) Cross-sectional microstructure of 4% of Cr hardfacing at weld region.

and ρ = 2,600 kg/m3, and hardness (832.8–880 HV). Table 13.1 shows test conditions. Rotation velocity calibration curves are used to determine erodent velocity. Erosion test samples are ultrasonically cleaned and weighed to a precision of 0.01 mg. After that, the specimen was eroded in a test rig for 4 minutes with 10-g erodent before being ultrasonically cleaned and the weight loss calculated. The sample-to-abrasive weight loss ratio was used to calculate the rate of erosion. This process was continued until the rate of erosion remained constant [24].

13.4.4.4 Microstructures for Fe-Cr Hardfaced Layers

The chromium hardfacing alloy deposited samples of size 30 × 30 × 6 mm by TIG welding technique; the substrate topography is shown in Figure 13.11, and the microstructural examination on mild steel samples is performed, which are coated by 2% and 4% of Cr hardfacing deposition as shown in Figures 13.12a and b, and also the cross-sectional microstructure of 2% and 4% of Cr hardfacing at weld region is shown in Figures 13.13a and b.

13.5 Conclusions

Hardfacing is an adaptable process to enhance the life of worn-out component. In these advanced days, it is a best process to reduce the cost and replacements, and it also reduces downtime because parts last longer and few shut downs are required to replace them. Hardness values are improved; when composition of chromium hardfacing varies, oxidation properties are drastically changed with addition of chromium layer, as it is the element having oxidation resistance property. High processing current gives thick coating, and the microstructure becomes rough. Coarser microstructure gives less hardness. Multilayer depositions improve characteristics by increasing the percentage of components, but they limit the formation of Fe3C carbides by increasing the fraction of Si. The eroded surface morphology and the areas below the eroded surface are examined through SEM. Stellite coatings improve erosion resistance through hardfacing. On mild steel, a hardfaced cobalt-based coating provides superior erosion resistance. In welding conditions, hardfaced characteristics will differ, but the rate of erosion will not varied much. Material elimination occurs in this hardfaced layer via the creation of lips, which will eventually shatter. In inter-dendritic zones, fracture begins occurs, and erosion efficiency is somewhat higher than perfect microcutting. Due to confined impact of particles within hardfaced layer, the depth of deformation extends.

References

1. Oliveira, A.S.C.M. D', Pareds, R.S.C., Santos, R.L.C., Pulsed current plasma transferred arc hardfacing. *J. Mater. Process. Technol.*, 171, 167–174, 2006.

2. Catsich, C., Badisch, E., Roy, M., Heath, G.R., Franek, F., Wear behaviour of hardfaced Fe-Cr-C alloy and austenitic steel under 2-body and 3-body conditions at elevated temperature. *Wear*, 267, 1856–1864, 2009.
3. Schwarz, W. and Warlimont, H., Effects of silicon on the wear behaviour of cobalt-based alloys at elevated temperature. *Mater. Sci. Eng. A*, 226–228, 1098–1101, 1997.
4. Aoh, J.N. and Chen, J.C., Constitutive relationship and kinetics model of DRX during thermal deformation of Stellite 6B alloy. *Wear*, 250, 611–620, 2001.
5. Shin, J.C., Doh, J.M., Yoon, J.K., Lee, D.Y., Kim, J.S., Influence of coating thickness and temperature on mechanical properties of steel deposited with Co-based alloy hardfacing coating. *Surf. Coat. Technol.*, 166, 117–135, 2003.
6. Ashrafizadeh, H. and Ashrafizadeh, F., Damage of different tungsten carbides under impact-sliding motion. *Wear*, 276–277, 75–84, 2012.
7. Venkatesh, B., Malvi, B., Roy, M., Sarkar, P., Effect of welding conditions on erosive wear of hard-faced Co-based alloy layer. *J. Eng. Tribol.*, 232, 1–13, 2018.
8. Martiny, P. and Juss, K., Deposition of Electrically Conductive Coatings on Castable Polyurethane Elastomers by the Flame Spraying Process, in: *Proceedings of the NACE International Corrosion Conference and Expo*, New Orleans, LA, USA, 2008, Paper # 08667.
9. Ninham, A., The effect of mechanical properties on erosion. *Wear*, 121, 3, 307–324, 1988.
10. Hattori, S. and Mikami, N., Experimental comparison of cavitation erosion rates of different steels used in hydraulic turbines. *Wear*, 267, 1–9, 2009.
11. Liu, R., Yao, J.H., Zhang, Q.L., Yao, M.X., Collier, R., Sliding wear and solid-particle erosion resistance of a novel high-tungsten Stellite alloy, Collier. *Wear*, 322–32341–32350, 2015.
12. Andrew, N., Giourntas, L., Galloway, A.M., Pearson, A., Items where subject is Mechanical engineering and machinery. *Wear*, 320, 143–151, 2014.
13. Ruff, A.W. and Ives, L.K., Measurement of solid particle velocity in erosive wear. *Wear*, 35, 1, 195–199, 1975.
14. Romo, S.A., Santa, J.F., Giraldo, J.E., Toro, A., Cavitation and high-velocity slurry erosion resistance of welded Stellite 6 alloy. *Tribol. Int.*, 47, 16–24, 2012.
15. Sindhu, B.S. and Prakash, S., Surface Engineering Analysis of D-Gun Sprayed Cermet Coating in Aggressive Environment. *Wear*, 260, 1035–1044, 2006.
16. Malayoglu, U. and Neville, A., Comparing the performance of HIPed and Cast Stellite 6 alloy in liquid–solid slurries. *Wear*, 255, 181–194, 2003.
17. Singh, R., Kumar, D., Mishra, S.K., Tiwari, S.K., Resent Research Status on Laser Cladding as Erosion Resistance Technique - An Overview. *Surf. Coat. Technol.*, 251, 87–97, 2014.
18. Levin, B.F., Dupont, J.N., Marder, A.R., Weld overlay coatings for erosion control. *Wear*, 181–183, 810–820, 1995.

19. Sundararajan, G. and Roy, M., Solid particle erosion behaviour of metallic materials at room and elevated temperatures. *Tribol. Int.*, 30, 339–359, 1997.
20. Finnie, I., Effect of welding conditions on erosive wear of hard-faced Co-based alloy layer. *Wear*, 3, 87–106, 1960.
21. Da, L., Ferreira, S., Graf, K., Scheid, A., Microstructure and Properties of Nickel-based C276 Alloy Coatings by PTA on AISI 316L and API 5L X70 Steel Substrates. *Mater. Res.*, 18, 1, 1439–1460, 2015.
22. Roy, M., Micro-and nano-mechanical properties of diamond film with various surface morphologies. *J. Appl. Phys.*, 39, R101–R124, 2006.
23. Roy, M., Dynamic hardness of detonation sprayed WC-Co coatings. *J. Therm. Spray Technol.*, 11, 393–399, 2002.
24. Sundararajan, G., Roy, M., Venkatraman, B., Erosion efficiency-a new parameter to characterize the dominant erosion micromechanism. *Wear*, 140, 369–381, 1990.
25. Sundararajan, G., Roy, M., Venkataraman, B., Erosion efficiency-a new parameter to characterize the dominant erosion micromechanism. *Wear*, 140, 2, 369–381, 1990. https://www.sciencedirect.com/science/journal/00431648/140/2.

Index

σ-bond, 117

Abrasion, 74, 83
  coefficient, 77
  law of, 76
  three-body abrasion, 76, 90
  two-body abrasion, 75, 90
  wear test, 90
Action potential, 254
Adhesion, *see* wear, 72, 83, 93
  abrasive erosion, 80
  laws of, 74
Adsorbed, 139, 142–144, 146, 147, 151–152, 154, 157–158
Adsorbent, 139, 141, 143, 146–148, 157
Adsorption, 140, 143
Advancement, 140
Affect, 140, 142
Algorithm, 144, 156–158
Analysis, 141, 143
Analytical, 140
Angular, 148
Anisotropy, 260, 261
Approach, 139, 142
Approximation, 141, 143
Archard, 73, 83
Articular cartilage, 258, 267, 268
Articulate joints, 161
Artificial blood plasma, 167
Aspect, 139, 147
Asperity, 72, 90
Assembly, 144, 152, 154–157
Assessment, 140, 146–147
Assumption, 143
Atomic force microscopy, 56–57
Atomic layer deposition, 25, 26
Attitude angle, 139, 145, 147, 151, 157
Axial, 141, 157

Bearing, 139, 141, 143, 145, 147, 149, 151, 153, 155, 157–158
Behavior, 139–140, 143, 147, 157
Bifurcate, 139
Bioactive, 254
Bioactive coating, 240
Bioactive glasses, 171
Biocompatibility, 255, 256, 278
Biotribology, 253, 254, 257
Bond strength, 33, 37
Bonded solid lubricants (BSL), 34
Bone remodeling, 265, 275
Bone resorption, 265
Boron nitride nanotube, 171
Boundary, 140, 142, 144, 147
Boundary lubrication, 124

Calculated, 141, 143, 145
Cancellous bone, 260, 269
Capacity, 139, 141, 145, 147, 150–157
Carrying, 145, 147, 157
Cathodic arc deposition, 16
CFRP, 212
Characteristics, 139, 141, 145, 147
Chemical deposition, 3
Chemical passivation, 128
Chemical vapor deposition, 18, 19, 190

atmospheric pressure CVD, 19, 20
laser CVD, 23, 24
low-pressure CVD, 20, 21
metal organic CVD, 21–23
plasma enhanced CVD, 23
Circumferential, 144, 148, 157
Clearance, 139, 141–142, 147, 157
Coating deposition, 2, 3
Coating deposition techniques,
bonding, 34, 41
burnishing, 34
particle embodiment, 34
thin-film coating, 34
Coatings, 139–140, 142, 144, 146, 148, 150, 152, 156, 158
Cold spray method, 189
Collagen fibril, 261, 264, 268, 269
Compact bone, 259, 261
Composite coatings, 33
MoS_2-TiO_2, 36, 39
MoS_2-TiO_2-ZrO_2, 33, 36, 40, 45
Pure MoS_2, 41
Compressive loading, 264, 269, 271, 274
Condition, 141, 144, 147, 157
Contact stress, 126
Corrosion, *see* wear, 79, 80
Counter bodies/surface, 126
Cracks, 78
Crazing, 84
Crystallite size, 36
CVD techniques, 122

Dangling bond, 125, 128
Deformation, 82, 91
Depicted, 144, 147
Design, 140–141, 158
Detrimental, 146
Developed, 139–140
Diamond coatings,
MCD, 119, 127
NCD, 119, 127
UNCD, 119, 127
Diamond-like carbon, 117–118, 121
Diamond like carbon coating, 162
Dimensionless, 142, 144, 147–148, 157
Dip coating, 28, 29, 192
Direction, 141–142, 145
Disc brake, 212
Discretize, 139
Distribution, 139, 141, 144, 148
Dynamic, 143, 147

Eccentricity, 147–149, 151–155, 157
Effect, 143, 145–147
Effectiveness, 139
EHL, 158
Elastic modulus, 101
Elastohydrodynamic, 140
Electrodeposition coating, 188
Electron probe microanalysis, 62
Electroplating, 27
Endossceous implants, 159
Endurance life, 33, 45
Energy, 139, 146–147, 152–153, 156–157
Erosion, *see* wear, 80
abrasive erosion, 80
mechanisms, 81
ratio, 82
wear test, 92
Evaporation, 6
electron-beam, 7, 8
flash, 7
laser, 8
thermal, 6

Fatigue, *see* wear, 78, 83
empirical relation, 79
rolling elements, 78
sliding, 79
Film, 139, 143, 150–153, 155–158
Finite, 139, 144
Finite element analysis, 215
Flow, 139–142, 144–147, 154–155, 157
Force, 145–147, 152, 157
Fracture, 253, 259, 261, 263

Fretting, 82
Frictional, 139, 141, 145–147, 153, 156–157
Frictional stress, 232-234

Geometry, 143
GFRP, 212
Glow discharge, 166
Governing, 142, 144
Governing equations used in FEA, 218–220
Graphitization, 124–125
Grid convergence test, 221
Gumbel, 142, 147

Hardness, 74–76, 82, 89
H-DLC coatings, 125–126
Hexagonal phase boron nitride, 164
Hip joint, 257, 266, 276
Humidity, 124, 131
Hybridization, 116–117, 123
Hydrodynamic, 139–145, 147–148, 150–152, 156–157
Hydrogen peroxide, 167
Hydrogen terminated bonds, 125
Hydrophobic, 185
Hypersensitive reaction, 161

Impact energy, 84
Impact wear, 83
Impact wear test, 93
Implant loosening, 274
Incompressible, 141
Influence, 139, 143
Interlayers, 124

Journal, 139, 141, 143, 145, 147

KFRP, 212

Layers, 139, 141, 158
Load, 139–141, 144–145, 147, 150–151, 156–157

Low friction, 126
Lubricant, 139–144, 146, 149–150, 154–155, 157–158
Lubrication, 139, 141, 143, 149, 158
  boosted, 277
  hydrodynamic, 276–278
  squeeze film, 277
  weeping, 277

Mean free path, 8, 9
Measurement, 140–141, 158
Mechanical properties, 213
  elastic modulus, 117
  hardness, 117, 128
Metal doped DLC, 129
Metal implants, 253, 254
Metal oxynitrides, 162
Metal-on-metal (MoM), 274, 275
Microhardness, 243, 247, 248
Microstructure, 51, 55, 61, 64, 131
Model, 140, 143–144, 158
Molecular beam epitaxy, 17
Multilayer, 129

Nano, 139
Nanocoatings, 183
Nanocomposite coatings, 248
Nanocomposites, 196
Nanofretting, 242
Nanoindentation, 239, 242, 243
Nanoscratch test, 239, 241
Newtonian, 141, 144
Numerically, 141, 144

Obtained, 143–144, 146
Operating, 140–141
Operating conditions,
  load, 40
  sliding speed, 40
Osteoblast, 258, 265, 266
Osteoclast, 258, 265, 266, 276
Osteogenic growth factor, 161
Osteons, 257, 259, 261, 265

Parameters, 139, 144, 147, 154, 157
Performance, 139, 141, 143–145, 147, 149, 151, 153, 155, 157
Periprosthetic osteolysis (PPO), 255, 275, 276
Permeability, 271, 274
Physical deposition, 2, 3
Physical vapor deposition, 190
Plasma detonation technology, 188
Polyether ether ketone, 163
Polymers, 253, 275
Porosity, 253, 260, 261, 263, 264
Potentiodynamic polarization, 169
Pre-treatment processes,
  abrasive blasting, 36
  anodizing, 36
  etching, 36
  phosphating, 36, 37
  salt-bath nitriding, 36
  sand-blasting, 36
  shot peening, 36
Proposed, 140, 142–143, 145, 148, 149, 151–152
Proteoglycan, 268–270
PVD techniques, 121

Raman spectroscopy, 49, 51, 65–66
Rate, 140, 145–147
Regime, 139–141, 143–144, 146–149, 151–153, 155–156
Reinforcement material, 36
Residual stress, 55, 100, 124
Reynolds, 139–144, 158

Scanning electron microscopy, 51, 57
Scherrer equation, 53
Secondary ion mass spectroscopy, 64–65
SEM/EDAX analysis, 45
Shear, 140, 145, 157
Shear stress, 259, 268, 271
Short, 141–145, 147–148, 150–152, 155–156
Simulate, 143–144, 147
Small joint implants, 160
Soft contact lenses, 160
Sol-gel technique, 189
Solution dispersion, 191
Spalling, 79
Spin coating, 28
Spray coating, 192
Sputtering, 9, 10
Steady-state, 141, 144, 147
Surface, 139–140, 142, 144, 146, 148, 150, 152, 154, 156, 158

Tensile loading, 264, 265, 269, 270
Tetrahedral a-C coatings, 126
Texture, 54
TFL, 139–141, 143, 145–147, 149, 151, 153–155, 157
Thermal effect, 41
Thermal spray method, 191
Thermal spraying, 239, 240, 249
Thermal wear, 84
  coefficient, 84
Thickness, 139–148, 149–158
Thin film, 2, 139–140, 143, 146–148, 158
  characteristics, 4
  demand, 29
Tissue response, 254, 255, 257, 278
Torque, 139, 145–146, 152–153, 156–157
Transformation toughening, 204
Transmission electron microscopy, 51, 60
Tribochemical, 125
Tribo-corrosion, 162
Tribological performance, 34, 36
Tribological properties,
  COF, 41, 45

wear rate, 41, 45
Tribometer, 85, 86

UHMWPE, 162
Ultra-sonication, 37
Under, 139–141, 143–147, 149, 151, 153, 155, 157
Unit cell program, 52
Utilized, 139, 141–143, 147

Vacuum arc deposition, 189
Vapor deposition,
CVD, 50
PVD, 50, 52, 55
Variation, 140, 142–144, 148–150, 152, 155–156
Viscosity, 139–141, 143–144, 147, 157–158

Wear,
abrasion, 74, 83
adhesion, adhesive, 72, 76, 78, 80, 81, 83
coefficient, 74
corrosion, or corrosive wear, 79
curve, 85
debris, 73
definition, 72
erosion, 80
factor, 85
fatigue, or surface fatigue, 78, 83
fretting, 82
impact, 83
mechanisms, 72, 90
running-in, 85
steady-state, 85
test, *see* wear test
testing methods, 84
thermal, or diffusive, 84
Wear mechanism, 200
Wear rate, 85
Wear resistance, 124
Wear test, 85
abrasion, 90
erosion, 92
impact, 93
rolling, 88
rotating, 87
scratch, 89
sliding, reciprocating, 87
sliding, unidirection, 86
Wear tracks, 58–59, 61
Williamson-Hall relation, 53

X-ray diffraction, 49, 51–52, 55
X-ray photoelectron spectroscopy, 52, 62
XRD, 247, 249

Printed in the USA/Agawam, MA
September 16, 2026

910661.044